P9-EDW-139

Current Issues in Biomechanics

Mark D. Grabiner, PhD
Biomedical Engineering & Applied Therapeutics
Cleveland Clinic Foundation
Editor

Human Kinetics Publishers

Library of Congress Cataloging-in-Publication Data

Current issues in biomechanics / Mark D. Grabiner, editor.

p. cm.
Includes index.
ISBN 0-87322-387-X
1. Human mechanics. I. Grabiner, Mark D.
QP303.C87 1992
612.7'6--dc20 92-1588
CIP
AC

ISBN: 0-87322-387-X

Developmental Editor: Larret Galasyn-Wright
Assistant Editors: Laura Bofinger, Moyra Knight, and Julie Swadener
Copyeditor: Jane Bowers
Proofreader: Dawn Barker
Indexer: Barbara Cohen
Production Director: Ernie Noa
Typesetter and Text Layout: Yvonne Winsor
Text Design: Keith Blomberg
Cover Design: Jack Davis
Printer: Braun-Brumfield

Printed in the United States of America

10 9 8 7 6 5 4 3 2 1

Human Kinetics Publishers
Box 5076, Champaign, IL 61825-5076
1-800-747-4457

Canada Office:
Human Kinetics Publishers
P.O. Box 2503, Windsor, ON N8Y 4S2
1-800-465-7301 (in Canada only)

Europe Office:
Human Kinetics Publishers (Europe) Ltd.
P.O. Box IW14
Leeds LS16 6TR
England
0532-781708

Australia Office:
Human Kinetics Publishers
P.O. Box 80
Kingswood 5062
South Australia
374-0433

Contents

Contributors

Steven D. Bain
Musculo-Skeletal Research Laboratory
Department of Orthopaedics
State University of New York, Stoney Brook
Stoney Brook, New York

Roger M. Enoka
Departments of Exercise & Sport Sciences and Physiology
University of Arizona
Tucson, Arizona

Andrew J. Fuglevand
Departments of Exercise & Sport Sciences and Physiology
University of Arizona
Tucson, Arizona

Mark D. Grabiner
Department of Biomedical Engineering & Applied Therapeutics
The Cleveland Clinic Foundation
Cleveland, Ohio

Robert J. Gregor
Department of Physiological Sciences
University of California at Los Angeles
Los Angeles, California

Ted S. Gross
Musculo-Skeletal Research Laboratory
Department of Orthopaedics
State University of New York, Stoney Brook
Stoney Brook, New York

David Hawkins
Department of Physical Education
University of California at Davis
Davis, California

Stuart M. McGill
Occupational Biomechanics Laboratories
Department of Kinesiology
University of Waterloo
Waterloo, Ontario, Canada

Robert W. Norman
Occupational Biomechanics Laboratories
Department of Kinesiology
University of Waterloo
Waterloo, Ontario, Canada

Mary M. Rodgers
Institute for Rehabilitation Research and Medicine
Wright State University of Medicine
Dayton, Ohio

Michael D. Sussman
Shriners Hospital for Crippled Children
Portland, Oregon

Christopher L. Vaughan
Department of Orthopaedic Surgery
University of Virginia
Charlottesville, Virginia

Keith R. Williams
Department of Physical Education
University of California at Davis
Davis, California

Preface

Various sport biomechanics texts emphasize the application of engineering principles and technology to sport. This application typically includes efforts to enhance skill execution by analyzing the skill, improving the athlete, or improving the sport implement used.

In planning this book, the contributors agreed to depart from that approach. We wanted to contribute to the literature a unique volume that emphasizes the fundamentally interdisciplinary nature of biomechanics and its interaction with sport without restricting the content to sport.

Our purpose is to highlight what we feel to be an inevitable conclusion: We're all in this together. That is, the future for those in exercise and sport science is entwined with the scientific and clinical interests and goals of each other as well as of those in other life and physical sciences and medicine. Given that perspective, the traditional biomechanical methods of experimental design, data collection and analysis, and selection of an appropriate model may require augmentation to fully integrate the interdisciplinary philosophy.

Each chapter of the book presents a thought-provoking treatment of a topic relevant to sports biomechanists. The contents of the chapters demonstrate the metamorphosis occurring in the study of human motion. In a sense, the contributors have tried to foretell the future. By discussing the advances as well as some murky areas in their specializations, they suggest new avenues for research that may address conclusively questions that have thus far gone unanswered. The contributors enjoyed having license to speculate without being shackled to reporting and interpreting established findings. These instances of speculation are driven by scientific experience and intuition and research in progress.

The book's contributors represent a spectrum of biomechanics endeavors, ranging from studies of cellular and tissue mechanics to whole body analysis, from bioelectrical analysis to mathematical modeling and computer simulation. Many of the authors have a sporting element in their history, and all have integrated biological and mechanical components in their diverse study of the human musculoskeletal and neuromotor systems.

We hope this book will illuminate a variety of nontraditional research paths for sport biomechanists, both aspiring and established.

Credits

Figure 2.1 is from *Dynamics of Human Gait* by C.L. Vaughan, B.L. Davis, and J. O'Connor, 1992, Champaign, IL: Human Kinetics Publishers, Inc. Copyright 1992 by Human Kinetics Publishers, Inc. Adapted by permission.

Table 2.2 and Figure 2.4 are from "Plantar Pressure Distribution Measurement During Barefoot Walking: Normal Values and Predictive Equations," by M.M. Rodgers, 1985, Unpublished doctoral dissertation, Penn State. Copyright 1985 by Mary McIntyre Rodgers. Reprinted by permission.

Figure 2.5 is from M.M. Rodgers and P.R. Cavanagh, "Pressure Distribution in Morton's Foot Structure," *Medicine and Science in Sport and Exercise*, **21**(1), pp. 23-28, 1989, © by The American College of Sports Medicine. Adapted by permission.

Figure 2.6 is adapted with permission from *Journal of Biomechanics*, **13**, P.R. Cavanagh and M.A. Lafortune, "Ground Reaction Forces in Distance Running," Copyright 1980, Pergamon Press plc.

Figure 3.1 is from "A Hypothesis for the Role of the Spine in Human Locomotion: A Challenge to Current Thinking," by S. Gracovetsky, 1985, *Journal of Biomedical Engineering*, **7**, pp. 205-216. Copyright 1985 by Butterworth-Heinemann Ltd. Adapted by permission.

Figures 3.2 and 3.3 are from "Rectus Femoris Transfer to Improve Knee Function of Children With Cerebral Palsy," by J.R. Gage, J. Perry, R.R. Hicks, S. Koop, and J.R. Werntz, 1987, *Developmental Medicine and Child Neurology*, **29**, pp. 159-166. Copyright 1987 by the American Academy of Cerebral Palsy and Developmental Medicine. Adapted by permission.

Figure 3.4 is from "Computer Simulation of Human Motion in Sports Biomechanics," by C.L. Vaughan, *Exercise and Sport Science Reviews* (Vol. 12) (pp. 373-411) by R.L. Terjung (Ed.), 1984, New York: Macmillan. Adapted by permission of Macmillan Publishing Company.

Figure 3.5 is adapted from *Journal of Biomechanics*, **14**, R.D. Crowninshield and R.A. Brand, "A Physiologically Based Criterion of Muscle Force Prediction in Locomotion," Copyright 1981, with kind permission from Pergamon Press Ltd., Headington Hill Hall, Oxford, OX3 0BW, United Kingdom.

Figures 3.6 and 3.7 are adapted from *Journal of Biomechanics*, **25**, F. Sepulveda, D.M. Wells, and C.L. Vaughan, "A Neural Network Representation of Electromyography and Joint Dynamics in Human Gait," Copyright 1992, with kind permission from Pergamon Press Ltd., Headington Hill Hall, Oxford, OX3 0BW, United Kingdom.

Figures 4.1, 4.2, 4.3, 4.4, 4.8, and 4.13 are reprinted with permission from "Loads on the Lumbar Spine and Associated Tissues," by S.M. McGill, In *Biomechanics of the Spine* by V. Goel and J. Weinstein (Eds.), 1990, Boca Raton, FL: CRC Press. Copyright CRC Press, Inc. Boca Raton, FL.

Figures 4.6 and 4.7 are reprinted with permission from *Journal of Biomechanics*, 1992, Cholewicki and S.M. McGill, "A Myoelectrically Based Dynamic 3-D Model to Predict Loads on Lumbar Spine Tissues During Lateral Bending," Copyright 1992, Pergamon Press plc.

Figure 4.9 is from *Journal of Orthopaedic Research*, "Electromyographic Activity of the Abdominal and Low Back Musculature During the Generation of Isometric and Dynamic Axial Trunk Torque: Duplications for Lumbar Mechanics," S.M. McGill, **9**, 1991, pp. 91-103. Reprinted by permission of Raven Press, Ltd.

Figures 4.11 and 4.12 and Tables 4.2, 4.3, and 4.4 are from "Kinetic Potential of the Lumbar Trunk Musculature About Three Orthogonal Orthopaedic Axes in Extreme Postures," By S.M. McGill, 1991, *SPINE*, **16**(7), pp. 809-815. Reprinted by permission of J.B. Lippincott Company.

Figure 4.14 is from "Reassessment of the Role of Intraabdominal Pressure in Spinal Compression," by S.M. McGill and R.W. Norman, 1987, *Ergonomics*, **30**(11), pp. 1565-1588. Reprinted by permission of Taylor & Francis Ltd.

Figure 4.15 is from "The Effect of an Abdominal Belt on Trunk Muscle Activity and Intraabdominal Pressure During Squat Lifts," by S.M. McGill, R.W. Norman, and M.T. Sharratt, 1990, *Ergonomics*, **33**(2), pp. 147-160. Adapted by permission of Taylor & Francis Ltd.

Figure 4.16 is from "The Potential of Lumbodorsal Fascia to Generate Back Extension Moments During Squat Lifts," by S.M. McGill and R.W. Norman, 1988, *Journal of Biomechanical Engineering*, **10**, pp. 312-318. Reprinted by permission of the publishers, Butterworth-Heinemann Ltd. ©.

Figure 5.1 is from "The Multicomposite Structure of Tendon," by J. Kastelic, A. Galeski, and E. Baer, 1978, *Connective Tissue Research*, Ines Mandl (Ed.), **6**(1), pp. 11-23. Copyright 1978 by Gordon and Breach Science Publishers, Inc. Reprinted by permission from Gordon and Breach Science Publishers, Inc.

Figure 5.2 is from *Living Anatomy* (2nd ed.) (p. 157) by J.E. Donnelly, 1990, Champaign, IL: Leisure Press. Copyright 1990 by Joseph E. Donnelly. Adapted by permission.

Figures 5.3 and 5.4 are from F.R. Noyes, C.S. Keller, E.S. Grood, and D.L. Butler, "Advances in the Understanding of Knee Ligament Injury, Repair, and Rehabilitation," *Medicine and Science in Sports and Exercise*, **16**(5), pp. 427-433, 1984, © by the American College of Sports Medicine. Reprinted by permission.

Figure 5.5 is from "Biomechanical Analysis of Human Ligament Grafts Used in Knee Ligament Repairs and Reconstruction," by F.R. Noyes, D.L. Butler, E.S. Grood, R.F. Zernicke, and M.S. Hefzy, 1984, *Journal of Bone and Joint Surgery*, **66A**(3), pp. 344-352. Copyright 1984 by The Journal of Bone and Joint Surgery, Inc. Reprinted by permission.

Figure 5.8 is from "The Strength and Failure Characterstics of Rat Medial Collateral Ligaments," by R.D. Crowninshield and M.H. Pope, *Journal of Trauma*, **16**(2), pp. 99-105, © by Williams & Wilkins Co., 1976. Adapted by permission.

Figure 5.9 is from "The Time and History-Dependent Viscoelastic Properties of the Canine Medial Collateral Ligament," by S.L.Y. Woo, M.A. Gomez, and W.H. Akeson, 1981, *Journal of Biomechanical Engineering*, **193**, pp. 293-298, published by The American Society of Mechanical Engineers. Copyright 1981 by The American Society of Mechanical Engineers. Reprinted by permission.

Figure 5.10 is from *Journal of Orthopaedic Research*, "Tensile Properties of the Medial Collateral Ligament as a Function of Age," S.L.Y. Woo, C.A. Orlando, M.A. Gomez, C.B. Frank, and W.H. Akeson, **4**(2), 1986, pp. 133-141. Copyright 1986 by Raven Press. Reprinted by permission.

Figure 5.11 is reprinted with permission from *Biorheology*, **9**, S.L.Y. Woo, M.A. Gomez, Y-K. Woo, and W.H. Akeson, "Mechanical Properties of Tendons and Ligaments," Copyright 1982, Pergamon Press plc.

Figure 6.1 is from "Bone Cell Biology: The Regulation of Development, Structure, and Function in the Skeleton," by S.C. Marks and S.N. Popoff, 1988, *American Journal of Anatomy*, **183**, pp. 1-44. Copyright © 1988 by Alan R. Liss. Adapted by permission of Wiley-Liss, a division of John Wiley and Sons, Inc.

Table 6.1 is from "The Cellular Basis of Bone Remodeling: The Quantum Concept Reexamined in Light of Recent Advances in the Cell Biology of Bone," by A.M. Parfitt, 1984, *Calcified Tissue International*, **36**, pp. S37-S45. Copyright 1984 by Springer-Verlag. Adapted by permission.

Figure 7.1a is from D.W. Fawcett, *Textbook of Histology*, 11th Ed. © D.W. Fawcett. Reprinted by permission.

Figures 7.1b and 7.4 are from *Muscles, Reflexes, and Locomotion* by T.A. McMahon, 1984, Princeton, NJ: Princeton University Press. Copyright 1984 by Tom McMahon. Reprinted by permission.

Figure 7.2 is from *Neuromechanical Basis of Kinesiology* by R.M. Enoka, 1988, Champaign, IL: Human Kinetics Publishers. Copyright 1988 by Roger M. Enoka. Reprinted by permission.

Figure 7.3 is from *The Biology of Physical Activity* by D.W. Edington and V.R. Edgerton, 1976, Boston: Houghton Mifflin. Copyright 1986 by D.W. Edington and V.R. Edgerton. Reprinted by permission.

Figure 7.5 is from *Energetic Aspects of Muscle Contraction* by R.C. Woledge, N.A. Curtin, and E. Homsher, 1985, London, England: Academic Press. Copyright 1985 by Academic Press. Reprinted by permission.

Figure 7.7 is from *Muscles Alive: Their Function Revealed by Electromyography* by J.V. Basmajian and C.J. DeLuca, 1985, Baltimore: Williams & Wilkins Co. © 1985, the Williams & Wilkins Co., Baltimore. Reprinted by permission.

Figure 7.8 is reprinted with permission from *Journal of Biomechanics*, **16**, M.H. Sherif, R.J. Gregor, L.M. Liu, R.R. Roy, and C.L. Hager, "Correlation of Myoelectric Activity and Muscle Force During Selected Cat Treadmill Locomotion," Copyright 1983, Pergamon Press plc.

Figure 7.9 is reprinted with permission from "Muscle and Tendon: Properties Models Scaling and Application to Biomechanics and Motor Control," by F.E. Zajac, 1989, *Biomedical Engineering*, **17**, p. 377. Copyright 1989 by CRC Press, Inc. Boca Raton, FL.

Figure 7.10 is from "Comparison of Stiffened Soleus and MG Muscles in Cats," by B. Walmsley and U. Proske, 1981, *Journal of Neurophysiology*, **46**, pp. 535-577. Copyright 1989 by Journal of Neurophysiology. Reprinted by permission.

Figure 7.12 is from J. Tjoe. Unpublished figure. Reprinted by permision of J. Tjoe.

Figure 7.13 is reprinted with permission from *Journal of Biomechanics*, **21**, R.J. Gregor, R.R. Roy, W.C. Whiting, J.A. Hodgson, and V.R. Edgerton, "Mechanical Output of Cat Soleus During Treadmill Locomotion In Vivo vs. In Situ Characteristics," Copyright 1988, Pergamon Press plc.

Figures 8.2 and 8.3b are from A.J. Fuglevand, Unpublished doctoral thesis, University of Waterloo, Ontario, Canada, 1989. Reprinted by permission of A.J. Fuglevand.

Figure 8.3a is from "Maximum Bilateral Contractions are Modified by Neurally Mediated Interlimb Effects," by J.D. Howard and R.M. Enoka, 1991, *Journal of Applied Physiology*, **70**, pp. 306-316. Reprinted by permission of the American Psychological Society.

Figure 9.1 is from "The Biomechanics of Anterior Cruciate Ligament Rehabilitation and Reconstruction" by S.W. Arms, M.H. Pope, R.J. Johnson, R.A. Fischer, I. Arvidsson, and E. Eriksson, 1984, *American Journal of Sports Medicine*, **12**(1), pp. 8-18. Adapted by permission.

Figure 9.3 is from "Strain Within the Anterior Cruciate Ligament During Hamstring and Quadriceps Activity" by P. Renstrom, S.W. Arms, T.S. Stanwyck, R.J. Johnson, and M.H. Pope, 1988, *American Journal of Sports Medicine*,

14(1), pp. 83-87. Copyright 1988 by American Orthopaedic Society for Sports Medicine. Adapted by permission.

Figure 9.4 is from "Hamstrings Activation in the Presence of Anterior Cruciate Ligament Deficiency and Reconstruction," by M.D. Grabiner and G.G. Weiker, in press, *Clinical Biomechanics*, in press. Reprinted by permission of the publishers, Butterworth-Heinemann ©.

Part I

Mechanics of Human Motion

This first part of *Current Issues in Biomechanics* has four chapters that share anatomical and analytical connections. The anatomical connections are the various biomechanical roles of the lower extremities, whereas the analytical connections are the methods of modeling the human body to answer research questions. Chapters 1 through 3 analyze various aspects of human locomotion, and chapter 4 focuses on the low back and the mechanics of lifting. Like the others, chapter 4 highlights the integration of anatomical and mechanical modeling and the application of engineering principles and technology, but it addresses industrial rather than sport biomechanics.

In the 1970s, sport biomechanics experienced a growth in interest and productivity that was unrivaled in the exercise and sport sciences. New instrumentation allowed researchers to undertake more studies, and more complex ones, than ever before. Film analysis became a standard technique for deriving kinematic, and then kinetic, descriptions of sport performance, and in combination with force plates allowed more in-depth descriptions of motor skills.

Of all the motor skills biomechanists have analyzed, locomotion (walking and running) has received the most attention. In industrial biomechanics, the relationship between lifting tasks and costly low back injuries is important from commercial, social, and medical standpoints. But despite long research histories and voluminous publications, much remains to be discovered in both of these areas of interest.

Chapter 1

Biomechanics of Distance Running

Keith R. Williams

This chapter by Keith Williams on the biomechanics of competitive distance running seems a perfect point of departure for the entire text, because the topic is one studied by most exercise and sport science students. In presenting an overview of the current state of affairs, the chapter also notes what still requires attention. You will get a peek into the future to the integration of existing and evolving biomechanical techniques with physiological and tissue-level mechanics and other cutting-edge methodologies. The biomechanical techniques may help runners improve their levels of participation, whereas the physiological and tissue-level mechanics will affect the balance between training intensity and potential for injury.

The focus for an examination of distance running mechanics depends on the type of runner and the intent of the analysis. For the competitive runner, performance enhancement is the most important objective, and efforts toward improving times might be enhanced either directly, by making changes in running style that improve running performance, or indirectly, by reducing susceptibility to injury and allowing the runner to perform and train to maximal capabilities. For someone who runs primarily for health, the objectives might be different; improving performance is not as important as reducing susceptibility to injury so that a consistent running program can be maintained. The intent of a research study aimed at this population of runners thus might have a different focus than one for competitive runners.

A more basic objective for the study of distance running biomechanics might be to gain insight into how and why human movement patterns are optimized. Running is a complex, vigorous, highly repetitive movement well suited for optimization. Although general movement sequences are similar in most distance runners, specific implementation (e.g., specific muscle firing patterns, joint kinematic ranges, magnitudes of forces produced) often varies widely. Understanding how an individual develops and maintains a finely tuned movement pattern will aid our general understanding of neuromusculoskeletal optimization.

If one were devising a model of distance running, the first approximation would be derived from those features common to all runners. All runners with normal neuromusculoskeletal function exhibit lower extremity segmental movements and muscle activation patterns that follow a general pattern like those shown in Figure 1.1. Although there are variations in these patterns, the general characteristics are similar. If running conditions are altered, the manner in which running mechanics change is similar for most runners. For example, when running speed increases, the maximal angle to which the knee flexes during swing increases; when running downhill it decreases, and when going uphill it increases (Milliron & Cavanagh, 1990). The structural and functional characteristics of the human body define a range of acceptable movements for ''efficient'' running, and the first level of a model would aim to describe how general anatomical features define limits to this range. Using a large and diverse group of subjects would make the results of such a model more generally applicable, though it might also be desirable to examine general characteristics of more specialized subgroups, such as elite runners or runners with physical handicaps.

How similar or different should we expect running patterns to be between individuals? Are there specific patterns of movement associated with successful running that everyone should try to emulate, or is running style highly personal? Consider the data presented in Table 1.1 showing values for a variety of biomechanical variables for seven elite female middle distance runners during running at 5.36 $m \cdot s^{-1}$. These runners are of similar ability and have competed successfully nationally and internationally for years. For a given variable there are substantial differences between the runners and often considerable deviations from group means (these seven runners were taken from a group of 54 runners). Though a given individual will show the general characteristics of these segmental movement patterns, specific characteristics of individual running patterns are highly repeatable and differ from running styles of other individuals.

The second level in developing a model of distance running might incorporate factors specific to individual runners and be aimed at understanding how structural and functional capabilities cause deviations from ''average'' running mechanics. This approach would still involve analyzing groups of subjects but would incorporate measures that distinguish between individuals. It would also be useful to analyze individuals in detail, using a case study approach. Although such an analysis would be narrow in scope, it would be specific and applicable to the individual. Slight differences in skeletal structure, flexibility, strength, and neural control would all contribute to subtle differences in running mechanics, and the

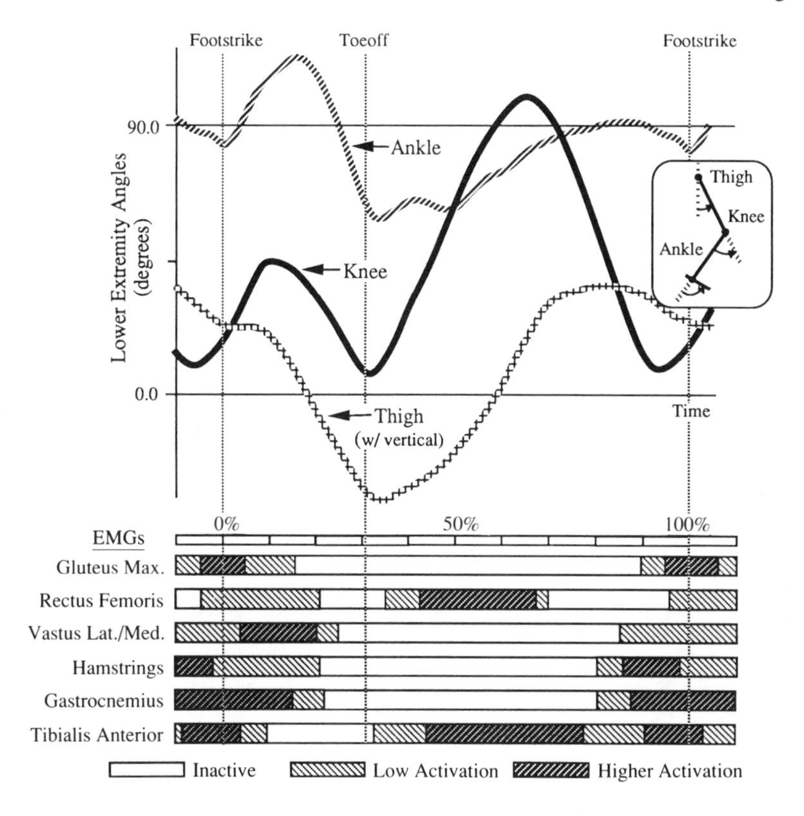

Figure 1.1 Thigh (with the vertical), knee, and ankle angles versus time for a single individual running on a treadmill at 4.48 $m \cdot s^{-1}$, and general patterns for electromyographic activity of lower extremity muscles. Periods of EMG activity are approximate.

examination of individual characteristics could lead to an understanding of how such factors interact with running mechanics.

Optimization of Running Mechanics

We might assume that runners, through experience and other influences, try, consciously or subconsciously, to optimize their running mechanics. Many factors influence the adoption of specific aspects of running style. Some depend on functional body characteristics: flexibility, strength, neuromuscular function, body structure. Others may be the result of years of training and racing, influenced by training methods, coaching, self-experimentation, and adaptations to injuries. Preconceptions of what constitutes ''correct'' mechanics might also influence them. The influences of these various interconnecting factors make understanding the optimization of distance running mechanics a very complex matter.

Table 1.1 Biomechanical Variables for Seven Elite Female Middle Distance Runners Compared to a Group of Elite Female Middle Distance Runners

	Subject no.							Elite group	
	81	37	73	42	89	46	27	mean	*SD*
Height (cm)	168.9	171.2	172.1	172.7	174.4	173.4	165.1	165.4	(6.5)
Mass (kg)	59.1	54.0	60.9	59.0	62.3	58.2	48.5	51.1	(7.1)
Leg length (cm)	87.8	91.5	92.4	90.4	88.6	90.7	85.9	85.8	(4.8)
Support time (ms)	168.0	191.7	179.5	192.5	167.0	178.5	175.0	179.9	(16.8)
Maximum thigh flexion with vertical (°)	38.4	45.9	46.9	53.6	53.9	43.5	50.8	48.2	(5.2)
Maximum thigh extension with vertical (°)	–39.1	–30.9	–39.9	–39.5	–42.9	–29.2	–25.8	–34.0	(4.6)
Maximum knee flexion swing (°)	109.1	137.1	132.3	136.0	139.5	120.5	130.9	127.9	(8.9)
Maximum knee flexion support (°)	44.9	47.8	51.4	64.9	48.5	38.5	32.2	45.5	(6.9)
Maximum pronation angle (°)	–4.3	–12.4	–7.0	–0.1	–11.3	–7.5	–9.7	–7.7	(3.8)
Stride length (% LL)[a]	387.2	384.7	406.1	417.5	403.2	374.0	377.2	395.6	(23.1)
Vertical oscillation (% LL)	8.0	9.1	9.4	9.0	9.0	8.9	9.0	7.6	(1.4)
Strike index (%)	55.3	13.0	36.8	50.7	52.0	23.0	74.0	42.2	(21.2)
First vertical force peak (BW)[b]	2.72	2.56	2.77	2.28	2.69	2.14	3.29	2.57	(0.44)
Second vertical force peak (BW)	2.98	3.22	3.24	2.84	2.08	3.01	3.83	3.19	(0.28)
Change in vertical velocity ($m \cdot s^{-1}$)	1.48	1.45	1.78	1.50	1.47	1.30	1.79	1.51	(0.21)
Peak braking force (BW)	–0.99	–0.73	–0.70	–0.94	–0.57	–0.56	–1.28	–0.82	(0.17)
Peak propulsive force (BW)	0.54	0.51	0.54	0.53	0.64	0.52	0.56	0.57	(0.07)

Note. Seven elite middle distance runners were compared as individuals to a group of elite runners. For mean values, $N = 54$ except for maximum pronation angle, where $N = 28$.

[a]LL = leg length.

[b]BW = body weight.

Optimization implies that some variable or system is being fine-tuned to a status that best facilitates some objective. It is somewhat unclear, however, which criteria are involved in the running optimization process and how the criteria might be different between individuals. Several optimization criteria could be important in refining distance running style: metabolic economy, minimizing tissue stress, reducing muscle fatigue, maximizing muscle efficiency. For competitive athletes improving performance is the ultimate objective, but optimizing performance involves the interaction of many criteria. In some cases optimization related to one variable might be counterproductive with regard to another. For example, developing the capabilities of the cardiovascular system to improve performance might best be done using a training regimen that involves adopting a specific movement pattern. However, if used too frequently, this movement pattern might result in an injury that would be counterproductive to improving performance. As a result the runner may have to adopt running mechanics that are a compromise of several factors. The fitness runner may have optimization criteria completely different from those of the competitive runner, but with similarly complex interactions. The arduous task for the biomechanist, physiologist, psychologist, physician, coach, and athlete is to understand how all contributing factors interact and to evaluate whether there is a better method of training, moving the limbs, exerting muscular forces, or mentally preparing for racing and training that might further enhance a given athlete's abilities.

The biomechanist's task is to develop the research base that improves the understanding of whether there are general rules that can be applied to runners as a group and of how individuals optimize their own specific running styles. This chapter presents a view of how biomechanists should address important questions about distance running biomechanics. The approach taken is not only to examine the results and implications of various investigations, but also to look at the process involved in evaluating distance running biomechanics. To proceed beyond descriptive or superficial levels of study, one must carefully consider the design and methodology involved in a distance running study. Once a study is done, one must realize the limitations in interpreting results so that findings are not generalized beyond an appropriate scope.

Research Base in Distance Running Biomechanics

Considerable research is available on distance running biomechanics, and recent publications cover the current state of knowledge well (Cavanagh, 1990; Frederick, 1986; Nigg, 1986; Vaughan, 1985; Williams, 1985a). Although distance running biomechanics research is a relatively mature area of study based on the number and diversity of scientific works, many questions remain unanswered. Much of the work to date has been descriptive, and many studies investigating conceptual relationships have raised as many questions as have been answered. Obtaining descriptive data of running mechanics is a routine activity because of technological advances in kinematic and kinetic data collection, but biomechanists must resist the temptation to make description the primary objective of studies in distance running mechanics.

A major goal for research in running biomechanics should be identifying causative mechanisms that affect running performance. A scientific basis is needed to explain the mechanism of various injuries, to understand why one running style might use less energy than another, or to provide insight as to whether a runner should or should not change running style when fatigued. Currently, many relationships between biomechanical factors and injury are anecdotal or inferred, with little scientific evidence of causal relationships (Frederick, 1986; McKenzie, Clement, & Taunton, 1985; Nigg, 1987; Valiant, 1990). The lack of precise knowledge of mechanisms does not mean that current levels of understanding are not useful. Knowing that running on hard surfaces or in poor shoes may increase risk of injury for a runner is of great practical interest to the runner and coach, but science demands that we attempt to understand the underlying mechanisms responsible for greater injury susceptibility. We need to know why. Such understanding will pave the way for practical benefits and will provide basic understanding of human body functioning that descriptive analyses will never provide.

Experimental Design and Methodology

The intent in this section is not to review in detail the varied methodologies used in distance running research but to highlight aspects of data collection and analysis that are particularly relevant. The major areas of data acquisition and analysis for biomechanical studies of running include kinematic analysis; ground reaction force and pressure measurement; electromyographic assessment; energy, work, and power analysis; modeling of internal muscle moments and joint forces; and evaluation of structural and functional capabilities such as anthropometric measures, flexibility, and strength. References for methodology can be found in Dainty and Norman (1987), as well as in many of the other articles referenced in this chapter.

Kinematic Data Analysis

Cinematographic, Video, and Optoelectronic Analyses. Analysis of kinematics has been the most frequently used approach in distance running studies. Whereas cinematography has historically been the predominant method, advances in video (Motion Analysis, Peak Performance, Vicon) and optoelectronic (Selspot, Watsmart) technology in the past decade have made high-speed automated analysis a more viable and sometimes preferred methodology. Several differences in the methods used to collect and analyze cinematographic, video, or optoelectronic data warrant discussion.

Particularly relevant is whether data of similar accuracy can be obtained using each system. Whereas cinematographic and optoelectronic techniques typically have very high spatial resolution, the method of processing the video images can limit the accuracy of video analyses. If manual digitizing of video images is used, resolution depends on the number of pixels on the video screen, although smoothing of the data and detection of multiple pixels by the cursor may be able to overcome some of the problems associated with the relatively poor resolution found on most video screens. Automated video systems typically activate a number of pixels, and resolution to a fraction of a pixel is achieved through an averaging process. Although definitive studies have not been reported comparing video, cinematographic, and optoelectronic techniques, there are indications that all three methods can produce similar results.

Speed of data acquisition is another relevant question. Using standard 60-Hz video cameras could result in poor temporal resolution, particulary if the analysis system does not allow access to each interlaced field of video data separately (resulting in an effective speed of 30 Hz). If the cameras are not shuttered, some blurring of the image may occur with faster running movements. Figure 1.2 shows data for the horizontal and vertical velocity of the heel during a distance running cycle obtained at 200 Hz using a high-speed video camera. A vertical line indicates footstrike, and the enlarged portion of the curve near footstrike shows the specific data points collected at this speed. The darker marks simulate a data collection speed of 67 Hz. At this slower rate, footstrike might be detected at the same point as with the 200-Hz sample rate, or it might be detected up to two frames later, depending on the absolute timing of the video acquisition at the slower speed. For this example, the velocity of the heel in the horizontal direction at footstrike would change from 218 $m \cdot s^{-1}$ to either 177 $m \cdot s^{-1}$ or 136 $m \cdot s^{-1}$, a 38% range. Velocity and acceleration values would be the most affected by sampling frequency, especially when one is determining maximum and minimum values or values at specific instants. Although peak displacement angles are not markedly affected by sampling rate (because there is a dwell time near the end of the range of movement of a joint), getting values for a joint angle at a specific instant in time, such as footstrike or toeoff, can result in substantial differences. If the study design indicates that the expected changes in kinematic measures between conditions may be small, slower sampling rates may be inappropriate and care should be taken. Whereas 60 Hz can be appropriate for some measures, 100 Hz has been more commonly used in cinematographic studies, and 200 Hz is recommended for automated analyses in which digitizing time is not a problem.

The technique used to obtain and analyze the video or optoelectronic images is a potential source of problems, depending on how coordinates for points on the body are obtained. In either video or cine techniques that use a manual digitizing method, the operator typically must perceive the centers of various joints between body segments and digitize them. As perspective changes for different regions of the body at different times during a running cycle, the perception of the joint center can be adjusted for the changing view. In contrast, automated video and optoelectronic techniques that track the movement of highly reflective markers on the body provide measures of movement kinematics for the body surface rather than for joint centers. If the joint center is along a line

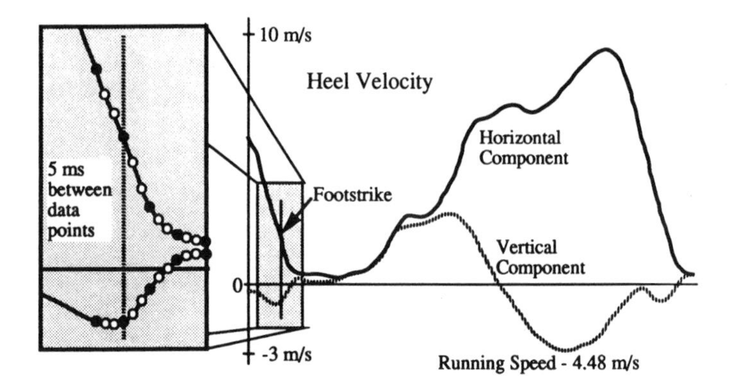

Figure 1.2 Heel-velocity components for a running cycle for one subject. The insert shows specific data points and illustrates how values at footstrike could be affected by sampling rate.

projected from the camera center to the marker, a good estimation of the joint center position can be obtained. However, when perspective misaligns the joint center and the vector from the camera to the marker, or when the subject rotates a marker out of the plane of forward progression, the automatically produced marker coordinates may poorly reflect the joint center position—potentially affecting any analysis of data that assumes movement of joint centers. How much error this might introduce has not been detailed. One can overcome some of these limitations by not assuming that surface markers provide information relevant to joint centers and by providing alternative means of obtaining such information. Typically either knowledge of the relationship of the surface markers to the joint centers is needed so that a correction to coordinates can be made, or three noncollinear points are used to describe a segment's orientation in three-dimensional (3-D) space. Both techniques markedly increase the complexity of the analysis. One way to minimize the perspective problems introduced by having fixed camera positions and a moving runner is to use treadmill rather than overground running. The horizontal displacements of the runner's segments are minimized and perspective errors are reduced.

Data Smoothing. In kinematic data acquisition methods, a certain amount of noise is introduced that can potentially confound true measurement values. A variety of filtering methods have been suggested to reduce the influence of noise and to isolate more clearly actual displacement, velocity, and acceleration patterns. Although displacement values are not appreciably affected by the noise, a measure such as maximal velocity of the knee during flexion can be markedly affected by the amount of smoothing applied, and acceleration data derived from displacements are changed even more. Several filtering techniques exist (Wood, 1982); digital filters and spline functions are most commonly used. Low-pass digital filtering frequencies for running kinematic data involving cinematography are typically in the range of 2 to 10 Hz (for a 100-Hz sampling rate). Because the exact contribution of noise to the unfiltered data is unknown and the degree of smoothing necessary depends on the rate of data acquisition and the characteristics of the movement pattern, a quantitative method should be used to determine appropriate filtering levels, and smoothing of each body location should be determined separately (Hinrichs, 1982; Jackson, 1979). Hinrichs reported appropriate cutoff frequencies for digital filtering of running data varying from 2 to 8 Hz for different body locations.

Three-Dimensional Analysis. In recent years the direct linear transformation (DLT) method (see Dainty & Norman, 1987, for references) has become the routine method for 3-D analysis, and many commercially available kinematic systems include software for such analyses. There is still a high cost in time and equipment to perform 3-D analyses, and little research is available to demonstrate the specific deficiencies in two-dimensional analyses or the advantages of 3-D studies. For primarily planar movements, a 2-D analysis may provide information highly correlated to that obtained using three dimensions, and the increased effort of a 3-D analysis may not be warranted. Obviously, a number of measures cannot be obtained from planar analyses, and some investigations necessitate 3-D data. The exact methods employed in the 3-D analysis should be carefully considered, and if coordinates for points on the body surface are determined, rather than for joint centers, the influence this will have on the analysis should be examined.

Studies are needed that demonstrate the differences betweeen 2-D and 3-D analyses of distance running mechanics relevant to economy, injury, or understanding of the basic movement patterns in running. For example, 3-D analyses of rearfoot motion have been performed on movement occurring at the subtalar joint during running (Engsberg & Andrews, 1987; Soutas-Little, Beavis, Verstraete, & Markus, 1987). These studies have provided information about nonfrontal plane movements that is not available from a traditional rearview analysis. However, it has not been convincingly demonstrated that the more limited set of data available from a planar analysis is misleading or incorrect when evaluating performance characteristics of individual running style, assessing differences in motion control offered by shoes, or identifying factors related to injury. It may be that important information is being overlooked in 2-D analyses, but such deficiencies have not been demonstrated, and it would be valuable to know whether the extra time and effort of a 3-D analysis are warranted.

Variability Within and Between Subjects. An important benefit of automated kinematic analysis systems is the ability to record and analyze a relatively large number of cycles of movement quickly and easily. Few studies of running biomechanics have analyzed more than one or two cycles of running for sagittal plane analysis due to the enormous amount of time involved in manual digitizing, and the question of cycle-to-cycle variability is relatively unexplored (Bates, Osternig, Mason, & James, 1979; Bates, Osternig, Sawhill, & James, 1983; Milliron & Cavanagh, 1990). The faster analysis capabilities of automated video and optoelectronic systems allow data averaged over a larger number of cycles to be obtained and provide a better opportunity to investigate the variability present in locomotor mechanics. Figure 1.3 shows thigh (relative to the vertical), knee, and ankle angles for a runner on a treadmill over seven cycles. Whereas the movement patterns for each joint appear very repeatable, examination of the magnitudes of maximum and minimum values reveals potentially important variability between cycles. Figure 1.4 shows the same knee angle curve with an expanded vertical scale. Values vary considerably for maximum knee flexion angle during support and swing, and for maximal knee extension prior to footstrike and following toeoff. Analyzing only one or two cycles could result in the use of atypical data for this subject, depending on which cycles were chosen. Differences from cycle to cycle are as large as 5°, which is much larger than might be expected for differences between conditions in many studies. There also are differences in cycle-to-cycle variability for one individual compared with another, as pointed out by Milliron and Cavanagh. Obtaining data for more than one or two cycles of running for an individual could provide insight not otherwise available concerning neuromuscular control of cyclic movement patterns. Care must be taken to refine methodology and technique so that variability introduced by the collection and analysis system is minimized.

The variability between subjects is typically greater than the variability between cycles for a given subject (Bates, Osternig, Mason, & James, 1979; Bates, Osternig, Sawhill, & James, 1983; Milliron & Cavanagh, 1990). The data in Table 1.1 highlight the variability between subjects. It is of interest to know whether data collected during a single session of running are representative of the running mechanics for a given athlete, either relative to day-to-day variability or changes over a period of time. J. Jones (1989) found no significant differences for a variety of ground reaction force and kinematic measures between 2 days of testing

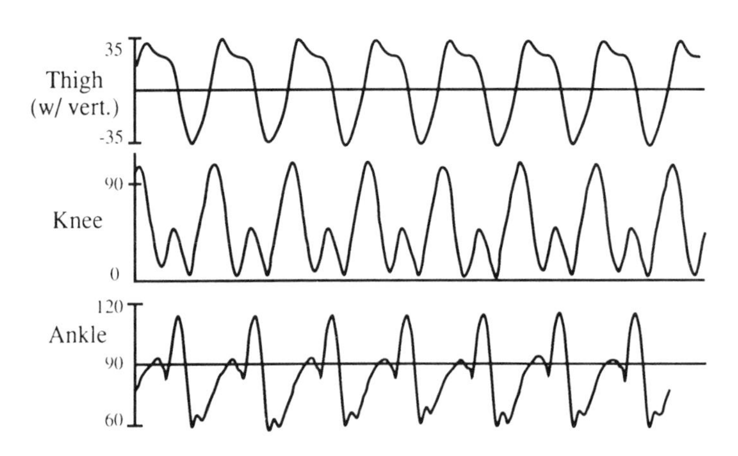

Figure 1.3 Thigh (with the vertical), knee, and ankle angles versus time for seven cycles of running on a treadmill for a single subject showing generally repeatable patterns from cycle to cycle.

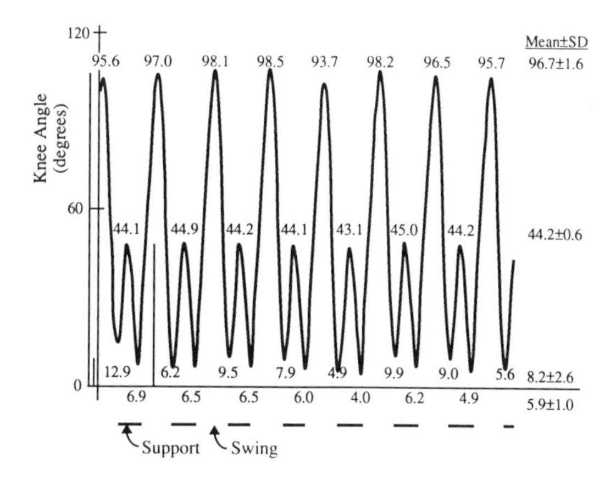

Figure 1.4 Knee angle versus time from Figure 1.3 with an expanded vertical scale showing variability in several specific knee angle measures between successive footstrikes.

for a group of 10 subjects. Data for individuals can show day-to-day differences, but with a sufficiently large sample, data from a given test session would appear to be representative of group results. For studies aimed at assessing changes in mechanics for individual runners, the day-to-day variability is very important. Movement patterns for a given runner are fairly consistent over a long period of time, with some variation found between measurement periods. Several factors may change the exact values obtained for a set of variables at different times. Some variation results from factors such as state of training, injury status, or voluntary changes in mechanics, and some may be the result of imprecise repeatability in the methodology and technology used. Some features of running mechanics appear to be stable over time.

Measures of Loads Applied to the Body

Force platform measurements have been in common use for many years; both piezoelectric (Kistler) and strain-gauge (AMTI) technologies have been used to provide ground reaction forces and center of pressure data during running (Cavanagh & Lafortune, 1980; D.I. Miller, 1990). Although early systems were sometimes subject to unacceptable vibration noise, current commercially available platforms have a stiffness and resonant frequency suitable for most running analyses. The resonant frequency of the platform should be considerably higher than the highest frequency contained in the ground reaction force patterns, and the resonant frequency may be different in the vertical and shear directions. In some situations unwanted oscillatory noise may be evident in the data, due to inadequate stiffness. This can be more of a problem with the larger versions of commercially available platforms and is most likely to show up in mediolateral forces.

During the past decade breakthroughs in pressure-measuring technology have allowed for a more localized mapping of forces under the feet during running. Pressure systems can provide more specific information relative to foot loading that may be used to investigate the relationships between load and injury. A variety of transducers use piezoelectric, piezoceramic, light diffraction, and capacitance technology. Three basic categories of systems are currently in use. One involves pressure measurement from a rectangular area of dimensions slightly longer than the foot (Novel Electronics, Inc.; Biokinetics, Inc.). These mats are useful for measuring pressures under the bare foot or in certain types of footwear studies but cannot measure pressures under the foot inside shoes. Another category uses a pressure-sensitive insole that can fit inside the shoe and provides discrete measures of pressure between the foot and insole in the shoe (Novel Electronics, Inc.). The resolution of these systems is typically not as fine as that found with the pressure mats. Both the mat and insole systems require high-performance computers, and the amount of information that can be collected may be affected by memory storage and sampling rate limitations. Sampling rate is typically low (30 to 100 Hz) compared with that used with force-measuring systems, and this limits the temporal precision of the data. The third category of pressure systems uses individual piezoelectric or piezoceramic transducers attached under the foot (h.a.l.m. elektronik GmbH, Frankfurt, Germany). Rather than providing a full pressure distribution map under the foot, transducers are placed at specific locations, such as under the heel, the metatarsal heads, or the hallux, and provide information about pressures in those areas. Although transducers do not provide a pressure data set for the entire plantar surface of the foot, they do provide pressure information about specific locations of interest under the foot. Also, with individual transducers fast sampling rates are possible and memory storage is less of a problem.

Selection of Measurement Variables

Many running studies have used a set of variables specific to temporal or spatial characteristics of running. Examples include stride length, time to maximal vertical force, angle of the ankle at footstrike, energy transfer between joints of the lower extremity, and maximal electromyogram (EMG) during knee extension. Having a well-defined purpose for a study helps in selecting appropriate variables, but often it is not clear which variables should be selected for a given type of biomechanical analysis. Having a large number of variables increases the likelihood of finding significant relationships based on chance alone, whereas selecting only a few increases the risk of not measuring important aspects of running.

Some studies have used mechanical power as a measure of running mechanics. Mechanical power might be considered analogous to a global measure of metabolic energy expenditure, $\dot{V}O_2$. An individual biomechanical measure, such as knee angle range of motion, contributes to mechanical power in a manner similar to the way the metabolism of a single muscle contributes to whole body $\dot{V}O_2$ (Williams, 1985b). Although measures of power based on changes in mechanical energy (Winter, 1978) have been used often, a number of questions about calculation methods remain unanswered (Vaughan, 1985; Williams & Cavanagh, 1983). Factors confounding analyses include questions about the best method to account for transfer of energy between segments, the differences between positive and negative work, and the influence of the stretch-shorten cycle. An alternative method for assessing power involves using linked-segment analysis to calculate measures of joint force power and muscle moment power (Wells, 1988; Winter, 1983). This method can identify the energy-absorbing and energy-generating functions of muscle groups, but it is experimentally more complex.

Immediate Adaptation by Runners and Its Effect on Research Findings

The capacity of the human body to adapt to changes in stress levels can confound investigations into the relationships between biomechanical variables and injury. Table 1.2 shows several ground reaction force variables for racing and training shoes from a group of elite runners. Although the use of racing shoes might be expected to result in higher forces than the use of training shoes because racing shoes typically have poorer cushioning properties, there are no marked differences for the group as a whole. The only significant difference was for strike index, the percentage of shoe length measured from the heel to where the first center of pressure point within the shoe appears. It may be that runners slightly alter their kinematics when wearing racing shoes to compensate for the differences in footwear and to maintain applied forces at safe levels.

Adaptation is suggested in other studies as well. Investigators have reported little or no differences in the impact peak of the vertical ground reaction force during running in comparisons of shoes of varying midsole hardness (T.E. Clarke, Frederick, & Cooper, 1983; Nigg, Bahlsen, Luethi, & Stokes, 1987). These findings suggest that the runners may have adapted to the varying properties of the shoes. J. Jones (1989) found no significant differences in ground reaction force measures or tibial deceleration at footstrike for a group of runners tested initially in new running shoes and retested in the same shoes after 250 mi of running. Shock absorption properties of the shoes had degraded significantly in both the rearfoot (5.7%) and the forefoot (33.6%), as evaluated using a drop impact tester. This suggests that the subjects may have altered their running mechanics in response to changes in the shoe properties to maintain acceptable levels of force applied to the body, or that external forces applied to the body were not appreciably affected by shoe characteristics. Surprisingly, there were only minimal changes with wear in a variety of kinematic measures. Perhaps only subtle changes are necessary in kinematic data to effect marked changes in kinetic responses.

Table 1.2 Ground Reaction Force Variables for Racing Shoes Compared to Training Shoes for Elite Runners at Speeds of 5.36 and 5.96 m · s^{-1}

	Support time (ms)	1st vertical force peak (BW)[c]	2nd vertical force peak (BW)	Change in vertical velocity (m/s)	AP[a] braking peak (BW)	AP propulsive peak (BW)	ML[b] medial peak (BW)	ML lateral peak (BW)	Strike index (%)
Racing shoes	163.9	2.71	3.20	1.50	–0.82	0.58	0.27	–0.27	49.5*
(*SD*)	15.0	0.45	0.27	0.19	0.15	0.08	0.11	0.11	23.5
Training shoes	165.0	2.75	3.22	1.50	–0.83	0.57	0.32	–0.32	45.5
(*SD*)	16.7	0.50	0.27	0.21	0.18	0.08	0.09	0.09	19.9

Note. N = 23.

[a]AP = anteroposterior.

[b]ML = mediolateral.

[c]BW = body weight.

*Significantly different from training shoes ($p < 0.05$).

Stacoff, Denoth, Kaelin, and Stuessi (1988) studied the influence of shoe cushioning properties and effective lever arm distance at initial footstrike on forces and moments acting on the subtalar joint. Because material properties of the shoes affected measures related to pronation more than they affected measures related to external and internal load bearing, Stacoff et al. concluded that shoe design should focus more on the control of rearfoot movement than on shock attenuation. This may be true, but other explanations of adaptation are also possible. Perhaps the increased pronation that occurred in softer shoes posed no serious threat to the runners, and no changes were necessary to be able to run safely. Perhaps the harder shoes initially caused increased loads to be applied to the body, but because these loads caused internal stresses greater than desirable, the runners showed an immediate adaptation in running style to minimize the internal stresses. The adaptations could cause loads to return to values near those found with the softer shoes. Although such a scenario is hypothetical and perhaps unlikely, the potential influence that adaptation could have on measurements should be considered.

Running Economy

The amount of oxygen consumed during distance running, commonly referred to as *running economy*, is one of several measures often used to judge the effects of variations in running mechanics. Although economy does not provide information about the mechanism of how different segmental movement patterns influence energy cost, the logical assumption that reducing oxygen consumption by optimizing running mechanics will improve running performance is reasonable and appealing (Frederick, 1983; Williams, 1990). Using running performance as a criterion measure for studies involving running mechanics creates several difficulties because physiological state, psychological factors, and race strategy often are confounding influences on performance times. Experimental conditions are much easier to control in a laboratory situation than in a performance setting, and measures of economy provide an appropriate, though somewhat indirect, insight into the potential influences biomechanical factors may have on performance.

Intuitively it seems straightforward that alterations in running movement patterns influence economy: Changes in segmental movements result from changes in muscle activation, and the majority of the energy costs associated with running result directly from skeletal muscle metabolism. It is difficult, however, to know how changing a specific aspect of running mechanics affects energy costs and whether such changes are universal for all runners. We might expect that trained runners have optimized most aspects of their running mechanics such that major changes to running style would be detrimental (Cavanagh & Williams, 1982); however, the practical experiences of coaches and runners indicate that there is still considerable tinkering with subtle aspects of running style even in successful elite runners.

Stride length is a good example of a measure that most individuals are likely to have successfully optimized (Cavanagh & Williams, 1982) to minimize energy expenditure, as illustrated in Figure 1.5 for a single runner. There do not appear to be strong relationships between either absolute or relative (to leg length or stature) stride length and oxygen consumption (Cavanagh & Kram, 1989; Williams, 1990). This suggests that whereas there will be an optimal stride length for an individual running at a particular speed, predictive data relating $\dot{V}O_2$, stride length, and leg length from groups of runners is not likely to help select the most appropriate stride length for the individual.

Similar reasoning suggests that results from group data cannot be effectively used to predict individual relationships with oxygen consumption for a variety of biomechanical measures of running, including vertical oscillation, maximal angle of knee flexion during swing, and position of the lower extremity at footstrike. Although moderate correlations have been shown between biomechanical variables and measures of economy (Williams & Cavanagh, 1986; Williams & Cavanagh, 1987; Williams, Cavanagh, & Ziff, 1987), the correlating variables in different studies are not always the same. The question of which set of biomechanical variables are most important to monitor or change when attempting to improve economy has not been convincingly answered, nor has the magnitude of expected changes been well defined. T.A. Miller, Milliron, and Cavanagh (1990) trained four runners with poor economy for 10 days using real-time feedback, giving each information on one style variable thought to be contributing to their uneconomical state. The modifications in running style did tend to improve economy (approximately 1.0 $ml \cdot kg^{-1} \cdot min^{-1}$ lower compared to changes for a control group), but the results were nonsignificant, perhaps influenced by the small sample size. Egbuonu and Cavanagh (1990) manipulated running mechanics in a group of runners, either by having them put their arms behind their backs or by increasing

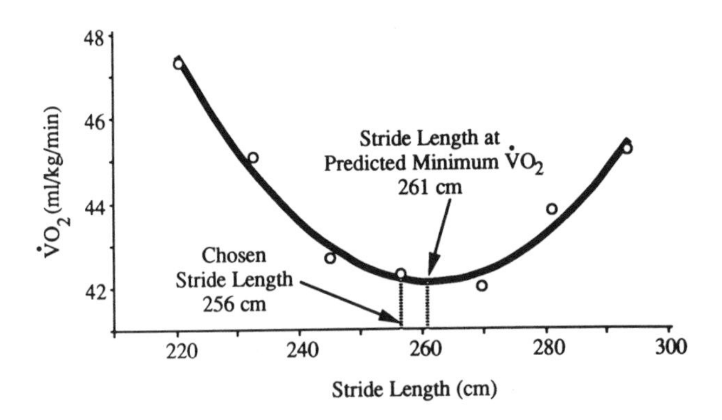

Figure 1.5 Changes in oxygen consumption ($\dot{V}O_2$) for a subject running at 3.83 $m \cdot s^{-1}$ with different stride lengths (the distance between ipsilateral foot contacts) showing that the freely chosen length is very close to the stride length predicted to minimize $\dot{V}O_2$.

vertical oscillation, and found increases in $\dot{V}O_2$ of only 1.6 and 1.9 $ml \cdot kg^{-1} \cdot min^{-1}$ for the two interventions, respectively. Egbuonu and Cavanagh suggest that this might be indicative of the order of magnitude of the maximal change in economy when drastic alterations to mechanics are undertaken.

A pertinent question concerns how much change is needed in $\dot{V}O_2$ to be important. Most well-trained runners are not likely to make extreme changes to mechanics, so it might be questioned whether it is possible to make substantial changes in the typical runner's style that lead to a marked improvement in economy. Frederick (1983) and Williams (1990) reasoned that even reductions in oxygen consumption as small as a percent or two could lead to meaningful improvements in performance times. Also, perhaps some individuals can make improvements in mechanics that result in substantial changes in economy. In a recent pilot study in my laboratory, one individual decreased oxygen consumption during running at 3.83 $m \cdot s^{-1}$ by 3.0 $ml \cdot kg^{-1} \cdot min^{-1}$ (5.6%) after training for 3 weeks at a stride length longer than his freely chosen one. Preliminary testing of this individual at various stride lengths indicated that running at longer strides should be more economical, and the desired results were achieved at the end of a 3-week training period. Similar reductions for another subject were only 0.8 $ml \cdot kg^{-1} \cdot min^{-1}$ and 1.7%, respectively.

To date, studies investigating relationships between economy and distance running mechanics for groups of runners have not identified strong specific relationships that can be used to optimize training for individuals, nor have they provided insight into the mechanisms for why a given segmental movement pattern might increase or decrease energy demands compared with a slightly different pattern. For example, it has yet to be shown why a given individual adopts a particular stride length (or alternatively, any other specific measure of running mechanics)—whether it is a function of body structure, strength, flexibility, muscle efficiency, or a multitude of other factors—nor are the full consequences of permanently lengthening or shortening a runner's stride (or other measure of mechanics) well understood.

Biomechanical Relationships With Injury

The Mechanism of Injury

The number of studies directly investigating the relationship between injury and running mechanics remains relatively small. Instead, relationships between biomechanical factors and injury are primarily correlational. The majority of running injuries have been attributed to training errors; anatomical factors, muscle imbalances, shoes, and surfaces are other important contributors (James & D.C. Jones, 1990; Lysholm & Wilkander, 1987; McKenzie, Clement, & Taunton, 1985; Micheli, 1986). Experimental studies demonstrating the specific mechanism for overuse injuries in distance runners have not been reported, but high stresses related to both impact (Denoth, 1986; Nigg, 1985, 1987; Valiant, 1990) and excessive motion (Gehlsen & Seger, 1980; James & D.C. Jones, 1990; McKenzie, Clement, & Taunton; Nigg, 1987; Viitasalo & Kvist, 1983) have been implicated. Lysholm and Wilkander (1987) found that two thirds of 39 runners followed over one year sustained at least one injury, and they reported a significant correlation between injury rate and distance covered during the preceding month. Jacobs and Berson (1986) reported similar results in a study of 451 self-reporting race participants. Forty-seven percent had been injured during the previous 2 years, with training intensity in terms of training mileage, pace, and days run per week significantly related to incidence of injury. Studies involving animals have suggested that repeated impact loading can lead to osteoarthritis (Radin, Orr, Kelman, Paul, & Rose, 1982), though similar direct evidence for humans has not been reported. Sohn and Micheli (1985) surveyed former collegiate distance runners and swimmers to see if long distance running could be implicated as a factor in the future development of osteoarthritis in the hips and knees. They found no association between moderate levels of running and osteoarthritis and suggested that neither heavy mileage nor the number of years of running contributed to osteoarthritis.

A Model for Injury Development

Overuse injuries result from stresses applied to tissues at intensities that cause a gradual breakdown and eventual failure of the tissues. These unacceptably high stresses may come from impact-related sources, muscle loading, or excessive movement at a joint or within a body segment. Structural deficiencies or misalignments may contribute to atypical stress levels. Figure 1.6 suggests a model for the mechanism of overuse injuries. If training intensity increases, whether as a result of higher mileage or faster running speeds, mechanical stress to the involved tissues increases. Similar increases in stress might be caused by alterations in footwear or surface conditions or other factors. The increase in stress leads to microscopic damage at the cellular level that stimulates remodeling of affected tissues (Radin et al., 1982). The normal process of adaptation to training would follow the path to the left in Figure 1.6. As long as tissue remodeling occurs at a rate faster than continued damage to the tissues, stronger tissue will result that is then able to withstand further increases in training intensity. A typical progression of training might be seen as continuous loops of this path.

If, however, tissue damage continues at a rate faster than remodeling of the tissue, the possibility of an overuse injury increases, and the sequence to the right in Figure 1.6 could take place. As the symptoms of an injury occur, the athlete is forced to reduce the intensity of training until the body is able to remodel the tissues sufficiently to be able to handle the applied stresses. This might be a matter of days if the injury is caught early enough, or it might be weeks or months if the proper intervention is delayed. The model described would be equally appropriate for well-trained and untrained individuals. At any level of fitness, a

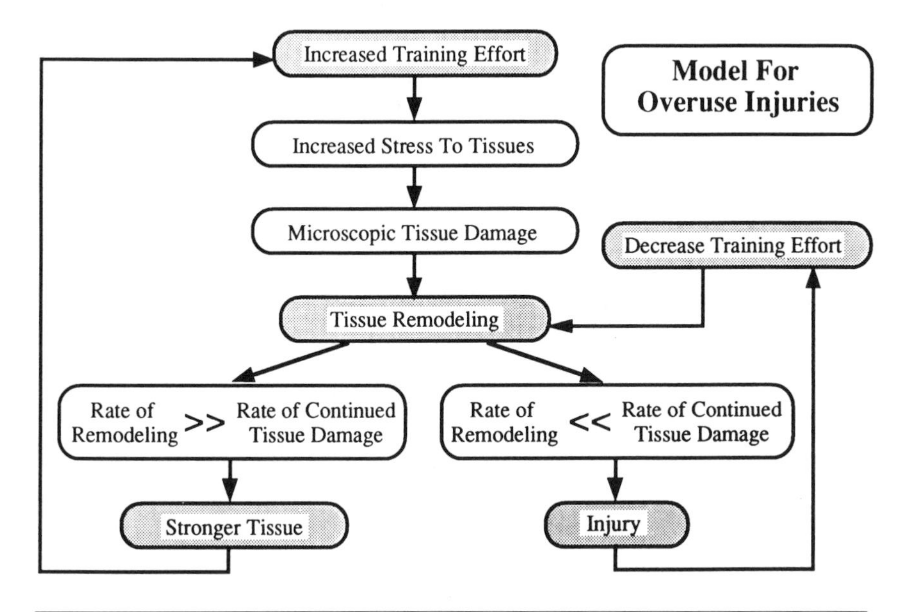

Figure 1.6 Model for the remodeling process by which tissues might become stronger or develop an overuse injury in response to increased stress.

person who increases training intensity beyond some threshold level (which is different for each individual and depends on previous training history) risks developing an overuse injury. This model is also consistent with the hypothesis that some individuals are more susceptible to injury because of structural or functional characteristics of the body. For example, when someone with a malalignment of the patella increases training effort, the increases in stress to tissues in the knee might be greater than would be found for an individual with normal alignment. This could result in greater tissue damage and greater susceptibility to injury.

Impact Stress

Although the exact mechanism of injury due to impact stress has not been clearly explained, many factors have been associated indirectly with injury. Ground reaction force, leg acceleration, and pedal pressure have all been used to monitor the loads applied to the body. Whereas high applied loads are likely to increase the risk of injury, factors such as anatomical structure, physiological and mechanical state of the tissues, and segmental movement patterns can all modify the transmission of those loads to various parts of the body (James & D.C. Jones, 1990; McKenzie et al., 1985). The critical factors determining injury susceptibility are stress to the tissues and the body's response to the stress. It is not unusual to find two runners with very similar ground reaction force patterns yet vastly different injury histories related to loads applied to the body.

It is unclear which ground reaction force measures are most appropriate to monitor in relation to injury. Measures often considered from force data include the magnitude of the vertical impact (first or initial) peak and thrust maximum (second or active) peak, the rate of vertical loading (the slope of the initial part of the vertical ground reaction force curve), average vertical force, and peak magnitudes and impulses from the braking and propulsive anteroposterior forces and mediolateral forces (Cavanagh & Lafortune, 1980; D.I. Miller, 1990). Attempts have been made to identify subsets of variables that have some commonality (Bates, Osternig, Sawhill, & J. Hamill, 1983), but confirmation that a given variable subset is better associated with injury than are other measures is still lacking. Nigg, Stacoff, and Segesser (1984) found subjects with lower extremity pain to have greater maximal vertical forces and greater differences in center of pressure movements than did runners with no symptoms.

Vertical impact peak values and loading rate are both likely to be related to the shock wave transmitted through the body at footstrike and may be sources of impact-related injuries. Tibial accelerometer measures have been used to provide a measure of this shock wave in other parts of the lower extremity (Valiant, 1990). In addition to rapidly applied impulsive loads, forces applied during midsupport may be involved in injury mechanisms. Although applied less rapidly, these loads are often of higher magnitude and longer duration than impact forces are. Measures of thrust maximum, average vertical force, or vertical impulse might be appropriate measures to evaluate this type of load. Although vertically applied forces are most often examined because of their high magnitude—in the range of two to five times body weight—it is also likely that shear forces, either in the anteroposterior or mediolateral region, can influence injury. The magnitudes of these forces are typically less than one body weight (D.I. Miller, 1990), but the direction and point of application of these forces give them longer moment arms about lower extremity joints, and they may cause torques that cause high local stresses.

Within any group of runners, there is a wide distribution of the magnitude in various ground reaction force measures (see Tables 1.1 and 1.2). Runners with relatively high values are likely to be more susceptible to impact-related injuries than are runners with lower forces. In my work with elite distance runners, those who have a history of stress-related injuries have often, but not always, shown higher than average maximal forces. Although the impact peak is often absent in midfoot strikers at slow or moderate running speeds (Cavanagh & Lafortune, 1980; D.I. Miller, 1990) some midfoot strikers show very high vertical impact peaks at faster speeds. Figure 1.7, a and b, shows vertical and anteroposterior ground reaction forces for two midfoot strikers running at 5.36 $m \cdot s^{-1}$. One runner shows the lack of an initial spike typical of midfoot strikers at slower speeds, but the other shows a very high initial peak typically seen only in rearfoot strikers at slower speeds. Almost all ground reaction force measures have been shown to increase in magnitude as running speed increases (Munro, D.I. Miller, & Fuglevand, 1987), implying that risk for injury may be higher when running faster.

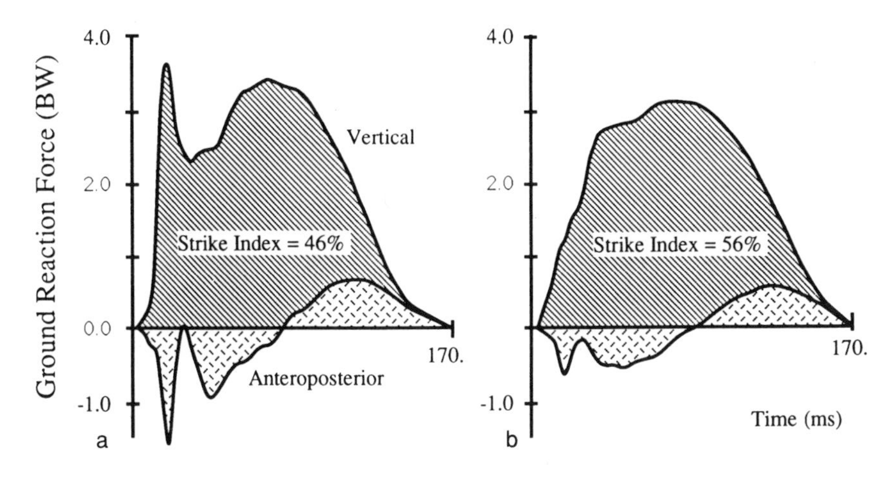

Figure 1.7 Vertical and anteroposterior ground reaction force curves for two midfoot strikers at 5.36 $m \cdot s^{-1}$. A pattern typical of midfoot landings at slower speeds (a) and a pattern typical of rearfoot strikers at slower speeds (b).

Footwear has been shown to modify the pattern of ground reaction forces and peak magnitudes to some extent, but reports differ about the exact nature of this relationship. For example, T.E. Clarke, Frederick, and Cooper (1983) reported that when wearing softer shoes runners showed slower rise times to peak vertical force than when wearing harder shoes, but there were no significant differences in maximal vertical forces. In contrast, Nigg and Bahlsen (1988) found harder shoes to lead to slower loading rates and lower peak vertical impact forces compared with softer shoes. A similar study from the same group (Nigg et al., 1987) showed no relationship between midsole hardness and loading magnitudes.

Another method of analyzing the loads borne by the body that has not received much attention is modeling of the internal moments and forces in the lower extremity joints. Burdett (1982) predicted peak resultant ankle joint forces in the range of 9.0 to 13.3 times body weight and Achilles tendon forces ranging from 5.3 to 10.0 times body weight, and Winter (1983) provided joint moment data for slow jogging. Burdett found that small changes in some of the assumptions used in his model had marked influence on estimated forces, highlighting that care must be taken in the exact procedures used in such predictive models. Stacoff et al. (1988) predicted subtalar joint forces and muscle moments for running in shoes with different constructions and found that joint forces were reduced by approximately 7% in soft shoes (shore 20) compared with hard shoes (shore 50) and that medial muscle forces were markedly influenced by aspects of shoe construction that alter the effective lever arm at footstrike. These types of data are important in establishing relationships between applied forces and injury, but the modeling procedures involved are more complicated than simple kinematic or ground reaction force measurements and also rely on the determination of appropriate body segment parameters. Of additional interest is linking measures of joint moments and forces with EMG data from the muscles. Whereas several studies have described EMG patterns during running (see McClay, Lake, & Cavanagh, 1990), relatively few have attempted to join together joint dynamics and muscle activity measures (Elliot & Blanksby, 1979). Figure 1.1 shows approximate EMG activity in relation to lower extremity movements.

Accelerometers firmly attached to the skin (Valiant, 1990) or pins embedded in bone (Hennig & Lafortune, 1988) also have been used to assess loads to the lower extremity. This procedure can provide information specific to the body area upon which the accelerometer is mounted. Care must be taken with skin-mounted accelerometers because they overestimate peak accelerations to a degree, depending on the mass of the accelerometer and the firmness and location of the attachment (Hennig & Lafortune; Valiant). Peak acceleration measures in the lower extremity have been shown to increase at faster speeds of running (T.E. Clarke, Cooper, D.E. Clarke, & C.L. Hamill, 1985), to be higher for downhill running and lower for uphill running (C.L. Hamill, T.E. Clarke, Frederick, Goodyear, & Howley, 1984), to be higher at slower stride rates (and thus longer stride lengths) when speed is kept constant (T.E. Clarke, Cooper, C.L. Hamill, & D.E. Clarke, 1985), and to be attenuated to some degree by footwear (Valiant). The exact relationships between ground reaction forces, lower extremity accelerations, and injury have not been clearly delineated. Data presented by T.E. Clarke, Frederick, and Cooper (1983) showed no consistent relationship between tibial shank deceleration, vertical force impact peak, and shoe cushioning properties as measured by drop impact test scores.

Movement Stress

A second category thought to cause overuse injuries involves stress resulting from body segment movements. The most commonly cited movements are those in the lower extremity associated with rearfoot pronation (Buchbinder, Napora, & Biggs, 1979; Gehlsen & Seger, 1980; James & D.C. Jones, 1990; McKenzie, Clement, & Taunton, 1985; Nigg, Stacoff, & Segesser, 1984; Viitasalo & Kvist, 1983). Clinically, relationships between excessive rearfoot motion and injury have been readily accepted, but only a few research studies have actually shown direct relationships between biomechanical or anatomical factors and incidence of injury. Gehlsen and Seger reported that asymptomatic runners with a history of shin splints showed greater rearfoot pronation than a control group, and similar results were found by Viitasalo and Kvist. Nigg, Stacoff, and Segesser found subjects with Achilles tendinitis, tibial tendinitis, and fibular ligament pain to have greater maximal pronation.

Most references to lower extremity injury linked to rearfoot pronation suggest that runners with excessive or prolonged pronation are the most susceptible, yet a substantial number of runners exhibit what might be termed excessive pronation but have had few or no related injuries. Figure 1.8, a and b, shows angle versus time curves for the lower leg, heel, and rearfoot angle for two elite runners. The runner with excessive pronation (21°) has never had a lower extremity injury, whereas the one with limited pronation (2°) has had chronic problems with Achilles tendinitis and plantar fasciitis. It would be equally easy to find a runner with an excessive rearfoot motion pattern who has had a string of related injuries, but this example illustrates that the amount of pronation is not the only factor influencing injury. An understanding of the mechanism of injury involving excessive or rapid pronation is needed. Ultimately, the mechanism has to be related to stress in the tissues, but we do not yet understand how to determine an injury threshold for a given runner; that is, why one runner is injured by a certain movement pattern or stress level and another is unaffected. Whereas changes in the angle of the rearfoot have been most often examined, researchers need to focus as well on changes in the heel, lower leg, hip, and knee during the support phase. Structural features of the foot have been associated with susceptibility to pronation-related injuries (James & D.C. Jones, 1990), but the relationship between lower extremity structure and pronation needs to be explored further. By looking at the lower extremity as a whole, we may be able to better understand how movements starting at the foot affect the entire leg.

If a runner suffers from pronation-related symptoms, how much reduction in pronation is needed to relieve symptoms? Shoe company advertisements may suggest that the necessary changes are massive, but there is little evidence to demonstrate how much change is needed. Considerable effort has been made in footwear design toward controlling rearfoot motion, and several factors have been shown to influence the amount or velocity of pronation during running (T.C. Clarke, Frederick, & C.L. Hamill, 1983; Frederick, 1986; Nigg, 1986, 1987; Nigg & Bahlsen, 1988). Although it has been clearly demonstrated that design factors can influence rearfoot control, and implications have been made about

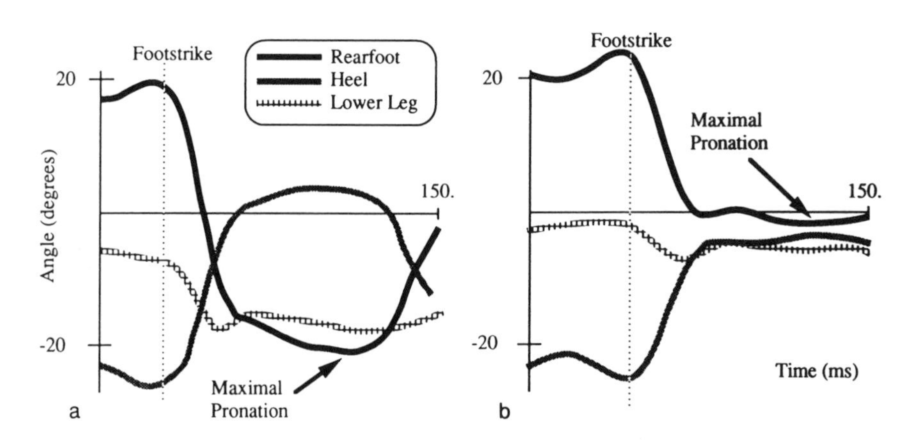

Figure 1.8 Changes in the lower leg, heel, and rearfoot angles during pronation following initial contact with the ground for a runner with excessive rearfoot pronation (a) and a runner with limited rearfoot pronation (b).

controlling pronation and reducing injury, it has not been scientifically demonstrated that selecting appropriate footwear will reduce or eliminate symptoms.

Studies involving orthotics can provide some insight as to how much change in rearfoot pronation is needed to reduce the risk of injury. Many runners have been markedly helped by wearing orthotics, and changes in rearfoot pronation with and without orthotics might suggest the magnitude of change needed to relieve injury symptoms. Surprisingly, the typical change is relatively small, on the order of 1° to 2°, although a few studies have shown greater changes (see Smith, T. Clark, C. Hamill, & Santopietro, 1983; Williams, 1985a). There is limited evidence that the rate of rearfoot pronation is reduced through the use of orthotics (Smith et al.), suggesting that both the magnitude and rate of pronation may be of importance to injury. It does not appear that changes in rearfoot kinematics need to be exceptionally large to bring about a marked alteration in internal stresses causing injury; however, more direct research evidence would be useful. Lafortune (1984) provided some insight into the internal functioning of the knee joint during the pronation movement by inserting intracortical pins into the tibia, femur, and patella and monitoring directly the 3-D movements at the tibiofemoral and the patellofemoral joints. Despite changes in footwear that ranged from 10° varus to 10° valgus, running in the valgus shoe caused the tibia to rotate internally on average only 3.5° more than with the varus shoe. Based on the successes that orthotic devices have brought about, Lafortune concluded that these joints of the knee are probably highly sensitive to minor intrinsic derangements caused by the footwear conditions.

Muscle Forces

During the running movement, there are several situations in which muscle activity might influence tissue stress levels (Radin, 1986). Whereas the muscle and tendon system is directly affected by muscle activation and might be strained at various times during the running motion, the muscle forces also have indirect influences on other tissues: loading of the bones at a joint, stress in ligaments or cartilage, and stress due to the changing influence of gravity as muscles move the body segments. Little is known about the direct influence of muscle forces on injury. While muscle force will sometimes increase stress to tissues, it can also help to relieve certain types of stresses by countering forces applied from other sources, such as gravity or the ground. As with other measures related to injury, direct information about mechanisms is lacking. For example, whereas it can be shown that the hamstring muscles undergo a phase of eccentric contraction late in the swing phase and it has been implied that excessive force during this phase of movement might lead to injury (Agre, 1985), no direct evidence shows that tissue stresses become excessive and result in hamstring tears.

Conclusion

In the idealized (and unrealistic?) biomechanics laboratory of the future, there may be a device called a *biomechani-scanner*, a computerized tomography scannerlike instrument with multiple analysis capabilities that can be used to evaluate a running subject. Its holographic output would show areas of localized stress in the body through a running cycle, a metabolic profile for specific muscle locations throughout the body, and complete kinematic and kinetic information for the body segments. Programmable self-adapting footwear would be easily adjusted to give the right combination of cushioning and support during the run, and mathematical models would provide templates for segmental movements that would minimize stress to tissues and metabolic energy expenditure and maximize efficiency and performance. In order to learn these optimized movement patterns, runners might be given real-time feedback of tissue stress levels and of muscle metabolism that would allow them to develop their most efficient neuromuscular sequencing. Upon leaving the laboratory, runners might be given a shoe-mounted miniature biomechani-scanner that would monitor and record the most important features of their training until they returned to the laboratory for a major checkup.

Whereas the technological aspects of the biomechani-scanner still need work, unless biomechanists adopt a more deliberate and multidisciplinary approach to running research, the conceptual and mechanistic basis needed to understand the complex interactions of physiological, psychological, and mechanical systems will be lacking. Relatively simple studies using a single technology or investigating a specific question can still provide needed information, but meaningful progress depends on multidimensional approaches to analyzing distance running. Biomechanical information is still needed in several relatively unexplored areas not mentioned in detail here: gender differences in distance running (Atwater, 1990), youth running, handicapped or amputee running (Enoka, D.I. Miller, & Burgess, 1982), relationships between anthropometric factors and running mechanics, symmetry in running. Biomechanists need to integrate various technologies so that they can look at the body as a system and monitor a more complete set of factors involved in running. Aspects of physiology and neurobiology must be included if the potential of biomechanics research is to be fully realized. Running serves as a good model for investigating the functioning of the human body because the physical, physiological, and psychological systems are maximally stressed and complex interactions of these systems can be studied.

References

Agre, J.C. (1985). Hamstring injuries: Proposed aetiological factors, prevention, and treatment. *Sports Medicine,* **2**, 21-33.

Atwater, A.E. (1990). Gender differences in distance running. In P.R. Cavanagh (Ed.), *Biomechanics of distance running* (pp. 321-362). Champaign, IL: Human Kinetics.

Bates, B.T., Osternig, L.R., Mason, B.R., & James, S.L. (1979). Functional variability of the lower extremity during the support phase of running. *Medicine and Science in Sports,* **11**(4), 328-331.

Bates, B.T., Osternig, L.R., Sawhill, J.A., & Hamill, J. (1983). Identification of critical variables describing ground reaction forces during running. In H. Matsui & K. Kobayashi (Eds.), *Biomechanics VIII-B* (pp. 633-640). Champaign, IL: Human Kinetics.

Bates, B.T., Osternig, L.R., Sawhill, J.A., & James, S.L. (1983). An assesssment of subject variability, subject-shoe interaction, and the evaluation of running shoes using ground reaction force data. *Journal of Biomechanics,* **16**(3), 181-191.

Buchbinder, M.R., Napora, N.J., & Biggs, E.W. (1979). The relationship of abnormal pronation to chondromalacia of the patella in distance runners. *Journal of the American Podiatry Association,* **69**(2), 159-162.

Burdett, R.G. (1982). Forces predicted at the ankle during running. *Medicine and Science in Sports and Exercise,* **14**(4), 308-316.

Cavanagh, P.R. (Ed.) (1990). *Biomechanics of distance running*. Champaign, IL: Human Kinetics.

Cavanagh, P.R., & Kram, R. (1989). Stride length in distance running: Velocity, body dimensions and added mass effects. *Medicine and Science in Sports and Exercise,* **21**(4), 467-479.

Cavanagh, P.R., & Lafortune, M.A. (1980). Ground reaction forces in distance running. *Journal of Biomechanics,* **13**, 397-406.

Cavanagh, P.R., & Williams, K.R. (1982). The effect of stride length variation on oxygen uptake during distance running. *Medicine and Science in Sports and Exercise,* **14**(1), 30-35.

Clarke, T.E., Cooper, L.B., Clarke, D.E., & Hamill, C.L. (1985). The effect of increased running speed upon peak shank deceleration during ground contact. In D. Winter, R. Norman, R. Wells, K. Hayes, & A. Patla (Eds.), *Biomechanics IX-B* (pp. 101-105). Champaign, IL: Human Kinetics.

Clarke, T.E., Cooper, L.B., Hamill, C.L., & Clarke, D.E. (1985). The effect of varied stride rate upon shank deceleration in running. *Journal of Sports Sciences,* **3**, 41-49.

Clarke, T.E., Frederick, E.C., & Cooper, L.B. (1983). Biomechanical measurement of running shoe cushioning properties. In B.M. Nigg & B.A. Kerr (Eds.), *Biomechanical aspects of sport shoes and playing surfaces* (pp. 25-33). Calgary, AB: University of Calgary.

Clarke, T.E., Frederick, E.C., & Hamill, C.L. (1983). The effects of shoe design parameters on rearfoot control in running. *Medicine and Science in Sports and Exercise,* **15**(5), 367-381.

Dainty, D.A., & Norman, R.W. (Eds.) (1987). *Standardizing biomechanical testing in sport*. Champaign, IL: Human Kinetics.

Denoth, J. (1986). Load on the locomotor system and modelling. In B.M. Nigg (Ed.), *Biomechanics of running shoes* (pp. 63-116). Champaign, IL: Human Kinetics.

Egbuonu, M.E., & Cavanagh, P.R. (1990). Degradation of running economy through changes in running mechanics. *Medicine and Science in Sports and Exercise,* **22**(2), S17.

Elliot, B.C., & Blanskby, B.A. (1979). The synchronization of muscle activity and body segment movements during a running cycle. *Medicine and Science in Sports,* **11**(4), 322-327.

Engsberg, J.R., & Andrews, J.G. (1987). Kinematic analysis of the talocalcaneal/talocrural joint during running support. *Medicine and Science in Sports and Exercise,* **19**(3), 275-284.

Enoka, R.M., Miller, D.I., & Burgess, E.M. (1982). Below-knee amputee running gait. *American Journal of Physical Medicine,* **61**(2), 66-84.

Frederick, E.C. (1983). Extrinsic biomechanical aids. In M. Williams (Ed.), *Ergogenic aids in sport* (pp. 323-339). Champaign, IL: Human Kinetics.

Frederick, E.C. (1986). Biomechanical consequences of sport shoe design. *Exercise and Sport Science Reviews,* **14**, 375-399.

Gehlsen, G.M., & Seger, A. (1980). Selected measures of angular displacement, strength, and flexibility in subjects with and without shin splints. *Research Quarterly for Exercise and Sport,* **51**(3), 478-485.

Hamill, C.L., Clarke, T.E., Frederick, E.C., Goodyear, L.J., & Howley, E.T. (1984). Effects of grade running on kinematics and impact force. *Medicine and Science in Sports and Exercise,* **16**, 185.

Hennig, E.M., & Lafortune, M.A. (1988). Tibial bone and skin accelerations during running. *Proceedings of the Fifth Biennial Conference and Human Locomotion Symposium of the Canadian Society for Biomechanics* (pp. 74-75). London, ON: Spodym.

Hinrichs, R.N. (1982). *Upper extremity function in running*. Unpublished doctoral dissertation, The Pennsylvania State University, University Park.

Jackson, K.M. (1979). Fitting of mathematical functions to biomechanical data. *IEEE Transactions of Biomedical Engineering,* **BME-26**, 122-124.

Jacobs, S.J., & Berson, B.L. (1986). Injuries to runners: A study of entrants to a 10,000 meter race. *American Journal of Sports Medicine,* **14**(2), 151-155.

James, S.L., & Jones, D.C. (1990). Biomechanical aspects of distance running injuries. In P.R. Cavanagh (Ed.), *Biomechanics of Distance Running* (pp. 249-269). Champaign, IL: Human Kinetics.

Jones, J. (1989). *The effects of shoe wear on the kinetics and kinematics of running*. Unpublished master's thesis, The University of California, Davis.

Lafortune, M.A. (1984). *The use of intra-cortical pins to measure the motion of the knee joint during walking*. Unpublished doctoral thesis, The Pennsylvania State University, University Park.

Lysholm, J., & Wilkander, J. (1987). Injuries in runners. *American Journal of Sports Medicine,* **15**(2), 168-171.

McClay, I.S., Lake, M.J., & Cavanagh, P.R. (1990). Muscle activity in running. In P.R. Cavanagh (Ed.), *Biomechanics of Distance Running* (pp. 165-166). Champaign, IL: Human Kinetics.

McKenzie, D.C., Clement, D.B., & Taunton, J.E. (1985). Running shoes, orthotics, and injuries. *Sports Medicine,* **2**, 334-347.

Micheli, L.J. (1986). Lower extremity overuse injuries. *Acta Medica Scandinavica,* (Suppl. 711), 171-177.

Miller, D.I. (1990). Ground reaction forces in distance running. In P.R. Cavanagh (Ed.), *Biomechanics of Distance Running* (pp. 203-224), Champaign, IL: Human Kinetics.

Miller, T.A., Milliron, M.J., & Cavanagh, P.R. (1990). The effect of running mechanics feedback training on running economy. *Medicine and Science in Sports and Exercise,* **22**(2), S17.

Milliron, M.J., & Cavanagh, P.R. (1990). Sagittal plane kinematics of the lower extremity during distance running. In P.R. Cavanagh (Ed.), *Biomechanics of Distance Running*, (pp. 65-106). Champaign, IL: Human Kinetics.

Munro, C.F., Miller, D.I., & Fuglevand, A.J. (1987). Ground reaction forces in running: A reexamination. *Journal of Biomechanics,* **20**(2), 147-155.

Nigg, B.M. (1985). Biomechanics, load analysis and sports injuries in the lower extremities. *Sports Medicine,* **2**, 367-379.

Nigg, B.M. (Ed.) (1986). *Biomechanics of running shoes.* Champaign, IL: Human Kinetics.

Nigg, B.M. (1987). Biomechanical analysis of ankle and foot movement. In E. Jokl & M. Hebbelinck (Eds.). *Medicine and Sports Science,* **23**, 22-29.

Nigg, B.M., & Bahlsen, H.A. (1988). Influence of heel flare and midsole construction on pronation, supination, and impact forces for heel-toe running. *International Journal of Sport Biomechanics,* **4**, 205-219.

Nigg, B.M., Bahlsen, H.A., Luethi, L.M., & Stokes, S. (1987). The influence of running velocity and midsole hardness on external impact forces in heel-toe running. *Journal of Biomechanics,* **20**, 951-959.

Nigg, B.M., Stacoff, A., & Segesser, B. (1984). Biomechanical effects of pain and sportshoe corrections. *Australian Journal of Science and Medicine in Sport,* **16**(1), 10-16.

Radin, E.L. (1986). Role of muscles in protecting athletes from injury. *Acta Medica Scandinavica,* (Supplement, 711), 143-147.

Radin, E.L., Orr, R.B., Kelman, J.L., Paul, I.L., & Rose, R.M. (1982). Effects of prolonged walking on concrete on the knees of sheep. *Journal of Biomechanics,* **15**, 487-492.

Smith, L., Clarke, T., Hamill, C., & Santopietro, F. (1983). The effects of soft and semi-rigid orthoses upon rearfoot movement in running. *Medicine and Science in Sports and Exercise,* **15**(2), 171.

Sohn, R.S., & Micheli, L.J. (1985). The effect of running on the pathogenesis of osteoarthritis of the hips and knees. *Clinical Orthopaedics and Related Research,* **198**, 106-109.

Soutas-Little, R.W., Beavis, G.C., Verstraete, M.C., & Markus, T.L. (1987). Analysis of foot motion during running using a joint coordinate system. *Medicine and Science in Sports and Exercise,* **19**(3), 285-293.

Stacoff, A., Denoth, J., Kaelin, X., & Stuessi, E. (1988). Running injuries and shoe construction: Some possible relationships. *International Journal of Sport Biomechanics,* **4**(4), 342-357.

Valiant, G.A. (1990). Transmission and attenuation of heelstrike accelerations. In P.R. Cavanagh (Ed.), *Biomechanics of Distance Running* (pp. 225-247). Champaign, IL: Human Kinetics.

Vaughan, C.L. (1985). Biomechanics of running gait. *CRC Critical Reviews in Biomedical Engineering,* **12**(1), 1-48.

Viitasalo, J.T., & Kvist, M. (1983). Some biomechanical aspects of the foot and ankle in athletes with and without shin splints. *American Journal of Sports Medicine,* **11**, 125-130.

Wells, R.P. (1988). Mechanical energy costs of human movement: An approach to evaluating the transfer possibilities of two-joint muscles. *Journal of Biomechanics,* **21**(11), 955-964.

Williams, K.R. (1985a). Biomechanics of running. *Exercise and Sport Science Reviews,* **13**, 389-441.

Williams, K.R. (1985b). The relationship between mechanical and physiological energy estimates. *Medicine and Science in Sports and Exercise,* **17**(3), 317-325.

Williams, K.R. (1990). Relationships between distance running biomechanics and running economy. In P.R. Cavanagh (Ed.), *Biomechanics of Distance Running* (pp. 271-305). Champaign, IL: Human Kinetics.

Williams, K.R., & Cavanagh, P.R. (1983). A model for the calculation of mechanical power during distance running. *Journal of Biomechanics,* **16**(2), 115-128.

Williams, K.R., & Cavanagh, P.R. (1986). Biomechanical correlates with running economy in elite distance runners. In *Human Locomotion IV (Proceedings of the North American Congress on Biomechanics* (pp. 287-288). Montreal: The organizing committee.

Williams, K.R., & Cavanagh, P.R. (1987). The relationship between distance running mechanics, economy, and performance. *Journal of Applied Physiology,* **63**(3), 1236-1245.

Williams, K.R., Cavanagh, P.R., & Ziff, J.L. (1987). Biomechanical studies of elite female distance runners. *International Journal of Sports Medicine,* **8**, 107-118.

Winter, D.A. (1978). Calculation and interpretation of mechanical energy of movement. *Exercise and Sport Science Reviews,* **6**, 183-201.

Winter, D.A. (1983). Moments of force and mechanical power in jogging. *Journal of Biomechanics,* **16**(1), 91-97.

Wood, G.A. (1982). Data smoothing and differentiation procedures in biomechanics. *Exercise and Sport Science Reviews,* **10**, 308-362.

Chapter 2

Biomechanics of the Foot During Locomotion

Mary M. Rodgers

Mary Rodgers presents a topic related to the biomechanics of distance running, yet by comparison a less examined topic—foot biomechanics during locomotion. All impact forces associated with locomotion are transmitted through the foot. Therefore, a greater understanding of how the foot functions and influences proximal segments enhances the understanding of other locomotion questions. However, the complexity of the foot (33 joints, 26 bones) provides just one example of the difficulty of solving even "simple" biomechanical questions. Computing technology eases the difficulty of laborious, CPU-intensive calculations associated with complex mechanical models. Similarly, measurement technology facilitates analysis of pressures beneath the plantar surface of the foot. Foot biomechanics is an area of great interest clinically, where orthopedists and others who work with foot pathologies are now hitting their strides, so to speak.

The foot, as a result of its location, forms a vital, dynamic connection between the human body and the ground. The foot is essential to all upright human locomotion, continually adapting to facilitate a compatible coupling between the body and its surroundings for effective movement. The dynamic attributes of the foot have been traditionally inferred from cadaveric study and qualitative clinical evaluation. Advancements in biomechanical methods for dynamic analysis have facilitated a more quantitative and precise description of foot function during movement, especially during walking.

This chapter provides a selected summary of quantitative information relevant to the dynamic function of the foot. Although results from biomechanical studies have often verified accepted anatomical assumptions regarding foot function, they have also contradicted long-accepted theories in certain cases. For example, pressure distribution studies have disproved the classical tripod theory of weight bearing during standing and have shown a wide variety of pressure distribution patterns. This case and others are addressed in this chapter.

The most frequently performed foot movements for healthy people occur during walking. Considerable research has been conducted in the analysis of walking, and most of this chapter concentrates on the dynamic biomechanics of the foot during this activity. A classical description of the biomechanics of gait as found in clinical literature is followed by an overview of quantitative findings that document kinematic and kinetic characteristics of walking.

As interest in physical fitness continues to grow, health professionals are treating an increasing number of runners, both recreational and competitive. The foot kinematics and kinetics that occur during running are briefly presented in the final section.

Foot Kinematics During Walking

Although the foot traditionally has been viewed as a static tripod or a semirigid support for body weight (BW), it has evolved primarily for walking and is, therefore, a dynamic mechanism. The body requires a flexible foot to accommodate the variations in the external environment, a semirigid foot that can act as a spring and lever arm for the push-off during gait, and a rigid foot to enable BW to be carried with adequate stability. The dynamic biomechanics of the foot complex that allow successful performance of all these requirements can only be understood when studied in relation to the biomechanics of the lower limb during walking.

The gait cycle, or stride period, provides a standardized frame of reference for the various events that occur during walking (see Figure 2.1). The gait cycle is the period of time for two steps and is measured from initial contact of one foot to the next initial contact of the same foot. The gait cycle consists of two phases:

- Stance (when the foot is in contact with the supporting surface)
- Swing (when the limb is swinging forward, out of contact with the supporting surface)

Along with providing forward leg momentum, the swing phase also prepares and aligns the foot for heelstrike and ensures that the swinging foot clears the floor. Stance comprises approximately 60% of the total gait cycle at freely chosen speeds and functions to allow weight bearing and to provide body stability. Five distinct events occur during the stance phase:

- Heelstrike (HS)
- Foot flat (FF)
- Midstance (MS)
- Heel rise (HR)
- Toeoff (TO)

Generalized Description

An understanding of the various joint axes of the lower limb is essential to the discussion of walking kinematics. Table 2.1 summarizes these joint motions as they relate to different phases of gait. Figure 2.2, a, b, and c, illustrates foot and ankle anatomy. Numerous authors have contributed to a clinical description of walking kinematics based primarily on observation. In order to understand the movements of the foot during walking, one must consider other portions of the lower extremity. During walking, rotation of the pelvis causes the femur, fibula, and tibia to rotate about the long axis of the limb (Inman & Mann, 1973). The magnitude of this rotational motion increases progressively from pelvis to tibia. For example, during normal walking on level ground, the pelvis undergoes a maximum rotation in each gait cycle of about 6°, while the tibia undergoes a rotation of about 18° in the same period (Inman, Ralston, & Todd, 1981). Generally, the limb rotates internally during the swing phase and early stance phase and then externally until the stance phase is complete and TO has occurred (Manley, 1980).

At HS, the tibia is rotated internally about 5° from its neutral position, and the ankle joint is either in its neutral position or in slight plantar flexion (Soderberg, 1986). According to Perry (1983), compression of the heel pad occurs at HS and is followed by traction on both anterior and posterior calcaneal attachments during terminal stance. Immediately following HS, the foot continues toward the floor, with the dorsiflexors controlling this plantar motion to prevent the foot from slapping down to the FF position (foot drop). From HS to just prior to FF, the increasing inward rotation of the tibia and fibula is transmitted through the ankle

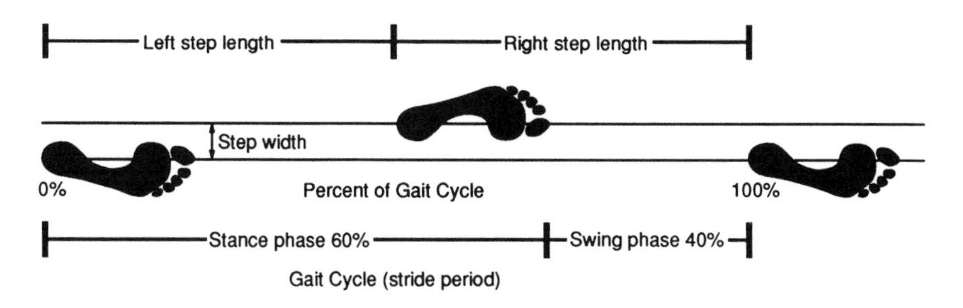

Figure 2.1 Summary of the temporal components of the gait cycle. (Adapted by permission from Vaughan, Davis, & O'Connor, 1992.)

Table 2.1 Joint Movements Occurring During Walking

Limb component	Motion	% of gait cycle
Lower limb	Internal rotation	0-18
	External rotation	18-60
	Internal rotation	60-100
Ankle joint	Plantar flexion	0-10
	Dorsiflexion	10-50
	Plantar flexion	50-60
	Dorsiflexion	60-100
Subtalar joint	Pronation	0-18
	Supination	18-60
	Pronation	60-100
Transverse tarsal joint	Free motion	0-18
	Increasingly restricted	18-60
	Free motion	60-100

mortise to the talus (Mann, 1985). The inward rotation of the mortise combined with the plantar-flexed position of the ankle tends to shift the forefoot medially from its neutral, toe-out position. The heel contact with the ground is lateral to the center of the ankle joint, where BW is transmitted to the talus, creating a pronatory moment at the subtalar joint that, in turn, stresses the structures of the medial arch. The talus rotates medially on the calcaneus about the subtalar axis, forcing the calcaneus into pronation. According to Wright, Desai, and Henderson (1964), the foot quickly pronates, approximately 10° within the first 8% of stance at an average walking speed. In this pronated position, free motion is available at the transverse tarsal joint so that the foot remains flexible, distal to the navicular and cuboid, and can bend into close contact with the supporting surface.

At the FF position, the lower limb begins to rotate externally. Since the forefoot is now fixed on the ground, the entire external rotation of the ankle mortise is transmitted to the talus. As external rotation continues, the foot supinates, increasing stability at the transverse tarsal joint and along the longitudinal arch of the foot. The stability of the transverse tarsal joint is further improved by the increasing body load being carried and by the firm fit of the convex head of the talus into the concave face of the navicular bone (Mann, 1985).

When the leg has passed over the foot, the ankle begins dorsiflexion. After HR, the ankle joint moves back into plantar flexion, forcing the metatarsophalangeal joints to dorsiflex. Since the plantar aponeurosis wraps around the metatarsal heads, a "windlass" effect takes place that increases tension across the longitudinal arch, further elevating the arch and increasing foot stability. Just before TO, the combination of weight bearing, windlass effect, and supination

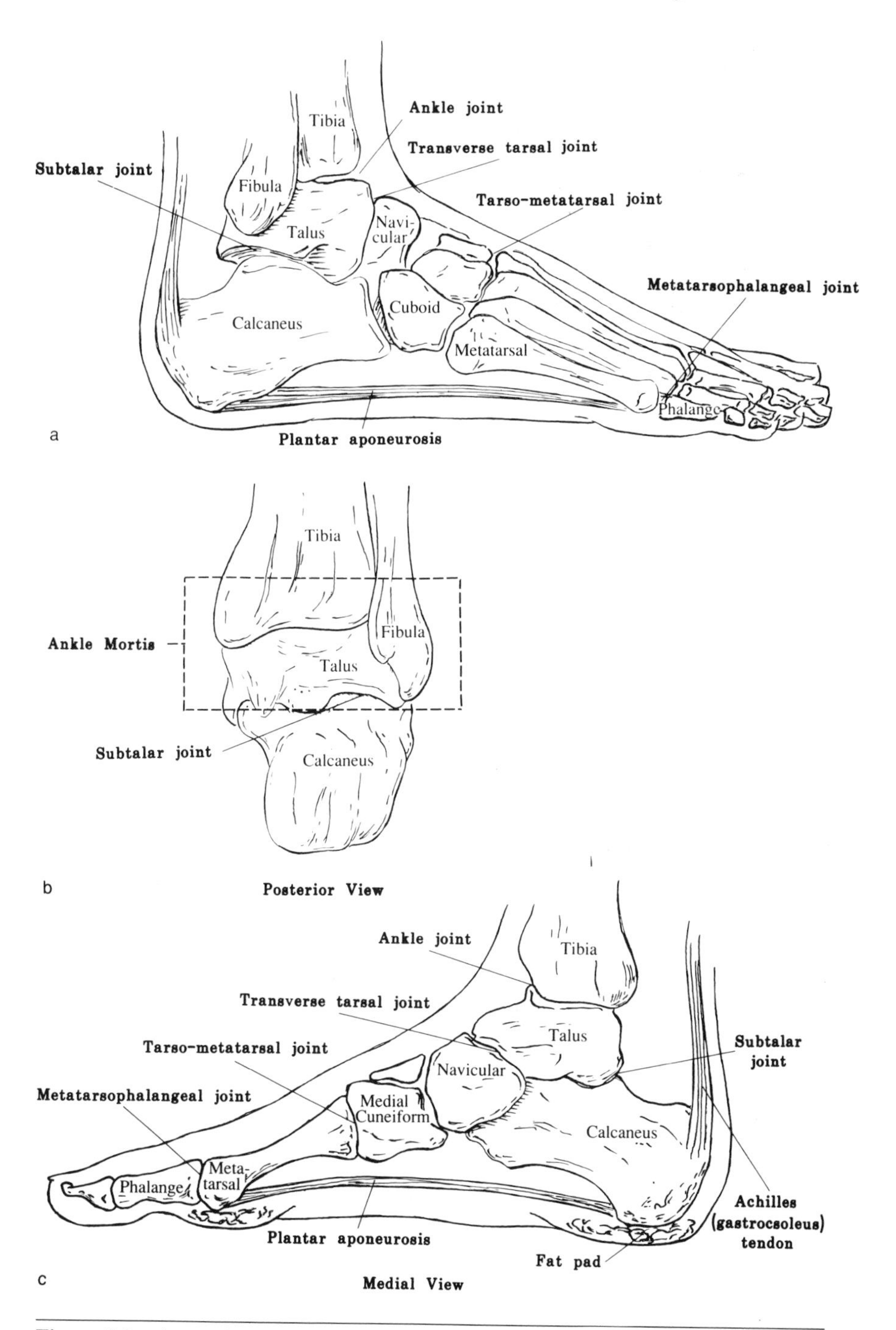

Figure 2.2 Anatomical diagrams showing the bony structures and joints of the foot and ankle from lateral (a), posterior (b), and medial (c) views.

ensures that the foot is in a maximally stable position for lift-off. After TO, the leg rotates medially, once again pronating the foot and unlocking the transverse tarsal joint so that the foot returns to its flexible state for the swing phase of gait.

Kinematic Studies

Kinematics refers to the description of motion, independent of the forces that cause the movement. Linear and angular displacements, velocities, accelerations, center of rotation for joints, and joint angles are examples of kinematics (Rodgers & Cavanagh, 1984). Kinematic information can be collected using direct measurement techniques (i.e., goniometers, accelerometers) or with indirect measurement using imaging techniques (i.e., cinematography, high-speed video, stroboscopy). The advantages and disadvantages of each technique have been described by several authors (Winter, 1979; Yack, 1984) and are not discussed in detail in this chapter. Instead, the results of selected studies relevant to dynamic biomechanics of the foot during walking and running are presented.

Walking Cadence and Velocity. Many factors affect foot biomechanics during walking, including the velocity of gait and anthropometric characteristics (e.g., limb length). Winter (1987) defined natural or free cadence as the number of steps per minute (step/min) when a subject walks as naturally as possible and reported an average natural cadence range of 101 to 122 step/min. In general, the natural cadence for females is 6 to 9 step/min higher than that of males; this is probably related to differences in limb length. Numerous studies have documented the changes in foot biomechanics that occur with increasing walking speed (Andriacchi, Ogle, & Galante, 1977; Winter, 1984). For this reason, walking velocity must be considered when comparing biomechanical findings.

Displacements: Paths of Movement. Motion of the heel in walking has been reported by Winter (1987) in a study with 14 subjects walking at their natural cadences. Vertical displacement of the heel begins well before TO and reaches maximum upward velocity just prior to TO. The heel reaches its highest displacement shortly after TO. Horizontal velocity builds up gradually after HR, reaching a maximum late in the swing phase, and then rapidly decreases just prior to HS. Vertical velocity of the heel slows abruptly at about 1 cm above ground level, after which the heel is lowered very gently to the ground.

The path of the forefoot differs from that of the heel. An initial rise in the forefoot during late push-off and early swing has been reported by Winter (1987). As the leg and the foot are swung forward,the forefoot just clears the ground and then rises to a second peak just prior to HS. Since the toe is the last to leave the ground, and because of the accompanying leg and foot angles, the toe rises to no more than 2.5 cm above the ground and then drops to only 0.87 cm clearance at midswing. As the knee extends and the foot dorsiflexes, the toe rises to a maximum of 13 cm just prior to HS.

Ankle Range of Motion, Foot Placement, and Arch Movement. Ankle joint angles, foot placement angles, and arch movement are other kinematic characteristics that have been investigated. Winter (1987) reported mean ankle joint ranges of motion during walking for 19 subjects of a maximum of 9.6° dorsiflexion and 19.8° plantar flexion. Murray, Kory, and Sepic (1970) found that foot placement angle showed high variability on successive steps of the same foot. A mean value of 6.8° foot abduction was reported, with the average difference between successive foot angles being 2.4°.

Dynamic arch movement was studied by Kayano (1986) using an *electro arch gauge*. He found that the medial longitudinal arch lengthens from the vertical force of BW during the period of early stance to FF. It then shortens with the decrease in BW and activation of the arch supporting muscles. As the posterior calf muscles activate for push-off, the arch lengthens again. It finally shortens rapidly because of the windlass action of the plantar aponeurosis as the toes dorsiflex for TO.

Foot Kinetics During Walking

Kinetics is the study of the forces that cause movement, both internally (muscle activity, ligaments, friction in muscles and joints) and externally (from the ground, active bodies, or passive bodies). Many researchers have analyzed muscular activity and ground reaction forces (GRF) during gait. Joint moments, segmental energy, joint reaction, and pressure distribution beneath the foot during walking have received less attention. The findings from electromyography (EMG) studies of the foot muscles during walking are presented in the first subsection, followed by findings from force plate and pressure distribution studies. Calculated kinetic parameters, such as ankle joint moments and joint reaction forces, are included in the final subsection.

EMG of Foot Muscles During Walking

Many researchers have investigated the electrical activity of lower extremity muscles during walking, and Basmajian (1985) presented a review of their findings. In general, studies have shown that many of the changes in levels of muscular activity occur at 15% to 20% of the cycle (FF), when the foot adapts to the supporting surface.

Winter and Yack (1987) have contributed extensively to the literature on EMG during walking. The tibialis anterior has its major activity at the end of swing to keep the foot in a dorsiflexed position. Immediately after HS, the tibialis anterior peaks and generates forces to lower the foot to the ground in opposition to the plantar-flexing ground reaction forces. The tibialis anterior is the only inverting muscle active during the period of maximum everting stress, when BW is completely on the heel. In some individuals, the tibialis anterior plays a minor role in pulling the leg forward over the foot shortly after FF. A second burst of tibialis anterior activity commences at TO and results in dorsiflexion for foot clearance during midswing.

The extensor digitorum has activity almost identical to that of the tibialis anterior. It functions to lower the foot after HS and to dorsiflex the foot and toes for clearance during swing. A minor third phase is seen during push-off and appears to be a co-contraction to stabilize the ankle joint.

The gastrocnemius and soleus show a significant phase of activity evident throughout the single limb support period. It begins just prior to HS and rises during stance, reaching peak just before mid-push-off (50% stride). From FF to 40% stride, the muscles lengthen as the leg rotates forward about the ankle under its control. During push-off, the posterior calf muscles shorten to actively plantarflex the foot and to generate an explosive push-off (estimated at 250% of BW in tension). Activity rapidly drops until TO, when low-level gastrocnemius activity continues into swing, probably showing the gastrocnemius acting as a knee flexor to cause adequate knee flexion prior to swing-through.

The peroneus longus has a small burst of activity during weight acceptance (10% of stride) that appears to stabilize the ankle (possibly as a co-contraction to the tibialis anterior). A larger burst during push-off (50%) shows the peroneus longus acting as a plantar flexor. Low-level peroneus longus activity during early swing is likely a co-contraction to the tibialis anterior to control the amount of foot dorsiflexion and supination.

Other investigators have reported their findings of intrinsic muscle activity in the foot during walking (Basmajian & Stecko, 1963; Mann & Inman, 1964). The group of intrinsic muscles covered by the plantar fascia (flexor digitorum brevis, abductor hallucis, and abductor digiti minimi) were shown to be active at 35% of the gait cycle. This part of the gait cycle includes the onset of HR, the concentration of BW on the forefoot, and the beginning of foot resupination. The intrinsic muscle activity assists the foot in becoming more rigid as it resupinates.

Force Plate Studies

Force platforms are commonly found in gait laboratories, and GRF are among the most commonly measured biomechanical parameters. GRF show the magnitude and direction of loading directly applied to the foot structures during locomotion. Because the feet are the only parts of the body to contact the ground during walking, they must be able to withstand and transmit these GRF. GRF data also provide information necessary for the calculation of ankle joint reaction forces, which is discussed later in the chapter.

Figure 2.3 shows a graph of typical vertical GRF during walking. The magnitude of vertical GRF has been reported by Cavanagh, Williams, and Clarke (1981) to range from 1.1 to 1.3 times BW, depending on walking speed. Footwear with rubber soles (such as running shoes) has been shown to attenuate the peak vertical GRF values. A rapid loading rate, often seen in vertical GRF during the first 25 ms after contact, has been described by Radin et al. (1986) as a possible contributing factor in knee joint degeneration because of transmission of impulse through the tibia to the knee joint.

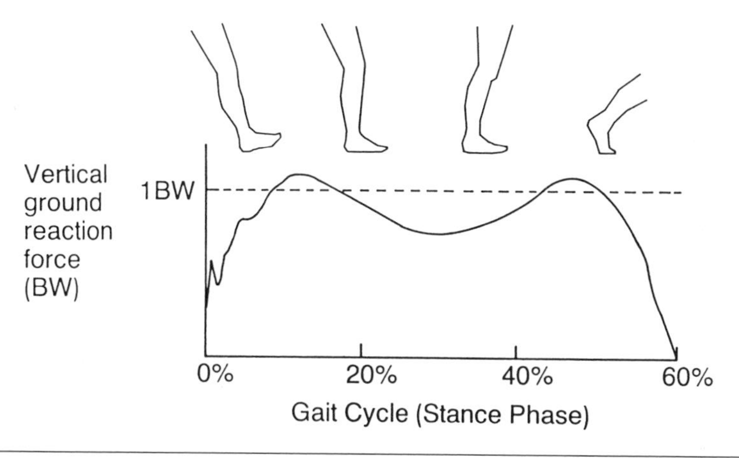

Figure 2.3 Typical vertical ground reaction forces beneath the foot during walking (BW = body weight).

The force plate provides only one instantaneous measure of force distribution. This measure, called the *center of pressure* (COP), identifies the geometric centroid of the applied force distribution. The path of the COP is created by plotting the instantaneous COP at regular time intervals during the entire stance phase of gait. Studies of the COP show a normal progression of the path from just slightly lateral to the midline of the heel, along the midline of the foot, up to the metatarsal heads (Cavanagh et al., 1981; Katoh, Chao, Laughman, Schneider, & Morrey, 1983). At this point medial migration occurs so that by TO the COP lies under the first or second toe. This medial migration aspect of the COP path has been described as the most variable among subjects. The COP path is altered by different footwear and foot positions (i.e., primarily supinated or pronated during stance phase).

Pressure Distribution Studies

Force plate systems are limited in the analysis of foot movement because the force information is not specific to foot anatomical locations. For example, the forces recorded may occur underneath both the fore and rear part of the foot simultaneously, so the COP may fall at some intermediate point that may not actually be loaded. Pressure distribution devices provide the specific location of pressures as they occur beneath the moving foot. Recent studies in pressure distribution have revealed new information regarding dynamic foot function during walking.

Although considerable variability exists in foot pressures during walking, the usual location of peak pressure is beneath the heel. A comparison of mean regional peak pressures found by several different investigators is shown in Table 2.2. Differences in values reported result from the variety of techniques and subject samples used. These pressure studies have shown that all metatarsal heads are loaded during the stance phase of gait. This finding negates the classical concept of tripod stance, which would not allow pressure beneath the middle metatarsal heads. Pressures during the first step have been shown to be similar to midgait pressures. Figure 2.4 shows the pressures recorded for 60 male subjects, 40 to 81 years of age, during first step and during midgait walking. Significant differences were found under the toes, under the first metatarsal head, and in the heel regions between the two conditions. Thus, for most of the foot, the pressures that occur during walking are similar to those present at the first step. The implications of this work are that patients who are unable to walk the distances required for normal pressure measurement, such as people with balance problems, can still be measured using their first step if regional conversions are used.

Many variables have been identified that directly influence pressure distribution beneath the foot. Clarke (1980) found that, with increasing speed, pressures increase and shift medially. The toes contribute more (by assuming more loading) as the walking speed increases. Walking barefoot alters both kinetic and kinematic parameters, compared with walking in shoes. Structural characteristics of the foot, such as arch type, also affect pressure distribution. The more rigid high-arched foot tends to concentrate pressure beneath the heel and forefoot, with minimal pressure beneath the midfoot (Cavanagh & Rodgers, 1985). This absence of midfoot pressure is present even in the higher loading conditions that occur

Table 2.2 Comparison of Mean (and Standard Deviations) of Regional Peak Pressures

Region	Mean pressure (*SD*) (kPa)
Hallux	307.3 (127.4)
Medial toes	286.0 (99.7)
Lateral toes	174.3 (22.3)
First metatarsal	320.0 (133.6)
Second metatarsal	353.8 (109.0)
Lateral metatarsals	323.6 (144.0)
Medial midfoot	57.0 (12.8)
Lateral midfoot	88.5 (60.1)
Medial heel	426.2 (215.0)
Lateral heel	349.0 (89.8)

Note. Reprinted by permission from Rodgers, 1985.

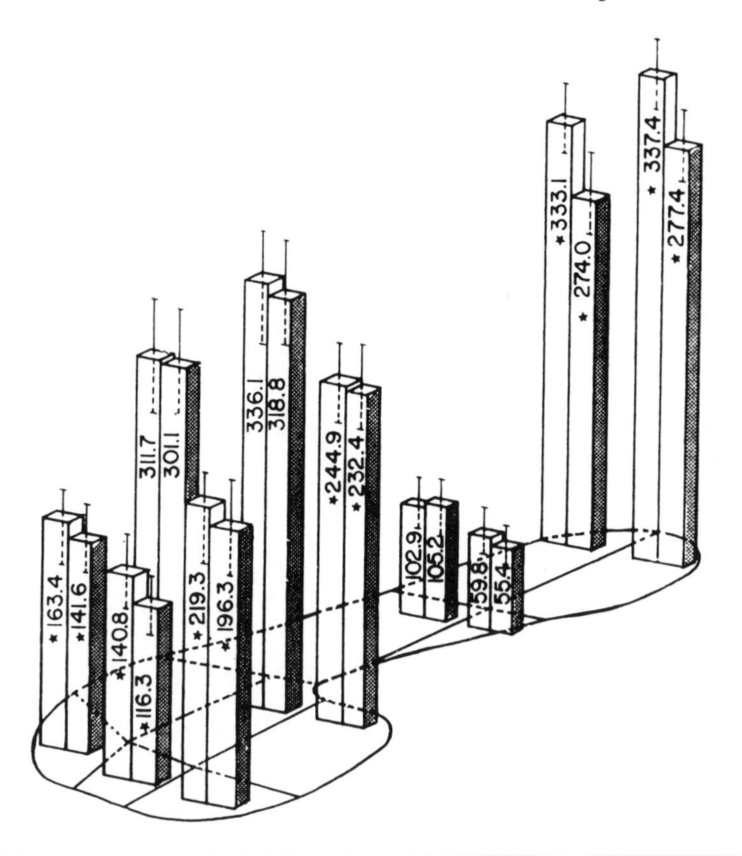

Figure 2.4 Regional pressure means and standard deviations during first step (shaded columns) and midgait walking in a group of males, ages 40 to 81 years (Rodgers, 1985). Regions where the difference between the two conditions was significant at the .05 level are indicated with an asterisk. (Reprinted by permission from Rodgers, 1985.)

with increasing speed of locomotion. The flexible flat-arched foot shows more spreading of pressure, including the area beneath the midfoot.

The classic Morton foot structure, characterized by a second metatarsal head that is more distally placed than the first, has also been shown to influence pressure distribution. Rodgers and Cavanagh (1989) reported that second metatarsal head pressures are significantly higher in Morton foot subjects than in non-Morton control subjects (see Figure 2.5). This finding suggests that people with a Morton foot structure may be more prone to second metatarsal pressure problems (e.g., stress fractures or inflammation) than are those with other foot structures. Pressure studies have also been useful in identifying areas of concentrated pressure that may lead to pressure ulcers for individuals with insensitive feet (Cavanagh, Hennig, Rodgers, & Sanderson, 1985). The use of pressure studies to screen patients at risk because of foot structure or disease presents a direct clinical application of the biomechanical technique.

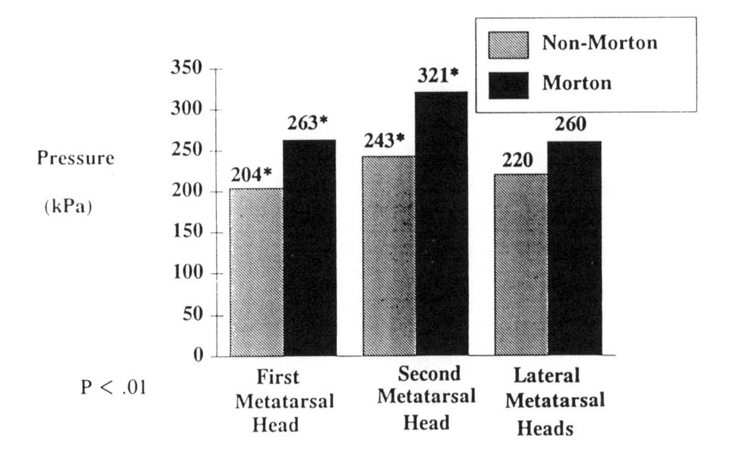

Figure 2.5 Regional pressure (kPa = kilopascal) means and standard deviations for Morton's foot compared with normal foot structure. (Adapted by permission from Rodgers & Cavanagh, 1989.)

Joint Moments and Joint Reaction Forces

Indirect methods have been used to calculate gait kinetics when direct methods are not feasible. These methods are necessary to calculate forces within the joint because force transducers cannot presently be safely used in subjects. Winter and colleagues (Winter, 1979, 1987; Winter & Robertson, 1978) have made significant contributions in the calculation of joint moments of force and energy patterns of walking. The mean maximum ankle joint moment (normalized to body mass) generated during walking was found to be a plantar moment of 1.6 N·m/kg, occurring between 40% and 60% of the gait cycle. Plantar flexors were found to absorb energy during the early and midstance phase as the leg rotates over foot. Late in stance, these same muscles plantar-flex rapidly (producing the plantar moment) and generate an explosive burst of energy (push-off).

As mentioned in the section on force plate studies, the GRF during gait are transmitted proximally to the rest of the body through the foot, compressing each joint along the way. These compressive forces have been shown (Radin, Boyd, Martin, Burr, Caterson, & Goodwin, 1985; Radin et al., 1986) to contribute to the formation of osteoarthrosis. Joint reaction studies of the ankle have been few, probably because this joint demonstrates osteoarthritic changes less often than the hip and knee joints do. Stauffer, Chao, and Brewster (1977) have shown ankle joint compressive forces of approximately 3 times BW from HS to FF. A further rise to a peak value of 4.5 to 5.5 times BW occurs during heel-off when the plantar flexors are undergoing strong contraction. Seireg and Arvikar (1975) have derived maximal ankle joint reaction forces of 5.2 times BW from mathematical models. Procter and Paul (1982) found a peak of 3.9 times BW for ankle joint reaction force during walking.

Stauffer et al. (1977) also reported ankle shear forces of 0.6 times BW in a posterior direction. After HR, talocrural shear was anterior and reduced to less than half of the previous posterior forces. Subtalar joint reaction forces have been calculated by Seireg and Arvikar (1975). The peak resultant force in the anterior facet of the talocalcaneonavicular joint was 2.4 times BW and for the posterior facet, 2.8 times BW. Peaks for both locations occurred in the late stance phase of the gait cycle.

Foot Kinematics During Running

Considerable research has been conducted in running biomechanics and is reviewed in detail by Williams (1985). The position of other body parts and the timing of their movements is basic to an understanding of foot motion. Although other body parts (primarily the hip and knee) have received most of the attention, several investigators have contributed to a functional description specific to foot motions during running at moderate speeds (Bates, Osternig, & Mason, 1978; Mann, Baxter, & Lutter, 1981).

Generalized Description

For the running gait in which HS occurs, initial contact is at the lateral heel with the foot slightly supinated. This position results from swinging the leg toward the line of progression. Slight plantar flexion of the subtalar joint occurs along with supination of the forefoot and calcaneus. The subtalar joint passes from a supinated to a pronated position between HS and 20% into support phase. The foot remains pronated between 55% and 85% of the support phase. Maximum pronation occurs between 35% and 40% of the support phase, approximately when total body center of gravity passes over the base of support. Full pronation marks the end of the absorbing and braking period of support as the foot begins its propulsive period. Maximum ankle dorsiflexion occurs 50% to 55% into the support phase when the center of gravity is forward of the support leg. The foot begins to supinate and returns to neutral position from 70% to 90% of the support phase. The foot then assumes a supinated position for push-off.

Kinematic Studies

Several stride variables that directly affect running kinematics and kinetics have been described by Cavanagh (1987). These include stride length at different speeds, optimal stride length, timing of the phases of running gait, and foot placement. Timing of the biomechanical events in running varies because it depends on running speed, type of shoe, and individual anatomic variations. For example, Kaelin, Unold, Stussi, and Stacoff (1985) reported the interindividual (n = 70) and intraindividual variabilities (20 repetitions each for 6 subjects) for several parameters of running. The maximum pronation angle during foot-ground contact showed a range of 20° among the 70 subjects, but only 7° to 12° within one individual. Vertical touchdown velocity of the foot during running varied between 0.64 and 2.3 m/s for the 70 subjects. Scranton, Rutkowski, and Brown (1980) reported an average duration of the support phase for jogging of 0.2 s and for sprinting, 0.1 s.

Clinical evaluations have suggested a relationship between foot pronation during running and lower extremity problems such as shin splints and knee pain. At present, quantitative data do not support the relationship, although this may result from inadequate analytical techniques. For example, studies of rearfoot motion have been conducted in two dimensions, but pronation occurs in more than one plane. Clarke, Frederick, and Hamill (1984) reviewed several studies of rearfoot movement in running. They reported an average maximum pronation angle of 9.4° over all studies. The authors suggest that a maximum pronation angle of 13° and total rearfoot motion greater than 19° during running would be considered excessive. However, at present no single parameter reliably predicts safe rearfoot movement during running.

Foot Kinetics During Running

Direct measurement of running kinetics poses more difficult technical problems than does measurement of the slower speeds of walking gait. Targeting a force plate, without altering the normal running gait patterns, is more difficult at the higher speeds. The faster motion requires more distance for running, and therefore longer cables or telemetry systems must be used for EMG data collection. Treadmill running has been used for EMG data collection although the pattern of running is different from that seen over natural terrain or on a track. Because of these problems, few researchers have directly measured foot muscle activity during running. More research has been conducted in GRF and pressure distribution during running. Indirect calculations of foot muscle forces, segmental moments, and joint reaction forces during running have been performed by a few researchers.

EMG of the Foot Muscles During Running

Studies have shown that EMG activity increases with running compared with walking. M. Miyashita, Matsui, and Miura (1971) reported that integrated EMG activity of the tibialis anterior and gastrocnemius increases exponentially with increasing speed. Ito, Fuchimoto, and Kaneko (1985) reported that with increasing running speed, the integrated EMG increased during swing but remained the same during the contact phase.

Force Plate Studies

Several authors have suggested a link between common running injuries and the impact forces at footstrike that can occur thousands of times during running (James, Bates, & Osternig, 1978; Mann et al., 1981). Force plate analysis has shown that peak vertical loading force during running is more than twice that of walking and occurs at least twice as fast. Perry (1983) extrapolated that the forces imposed on the supporting tissues would reflect a fourfold increase in strain. Microtrauma is cumulative; therefore, running creates symptoms that do not arise with ordinary walking.

Force plate data for jogging and running are much more variable from step to step than are such data for walking. The pattern and magnitude of the vertical GRF during running also differ significantly from those that occur during walking. Variables that affect vertical GRF data include touchdown velocity of the heel, position of the foot and lower leg before contact, and movement of these structures during impact (Nigg, 1987). The vertical GRF curve for heel-toe running (heel strikers) usually shows two distinct peaks: the impact force peak and the active force peak (see Figure 2.6) (Cavanagh & Lafortune, 1980; Frederick, Hagy, & Mann, 1981). Typical peak vertical GRF values for distance running speeds are 2.5 to 3.0 times BW.

The pattern of force depends on the orientation of the foot at initial contact, which is determined by whether the runner is a forefoot, midfoot, or rearfoot striker (Cavanagh & Lafortune, 1980). Most runners initially contact the ground with the outside border of the shoe, some with the rear lateral (rearfoot strikers) and some with the middle lateral (midfoot strikers). Harrison, Lees, McCullagh,

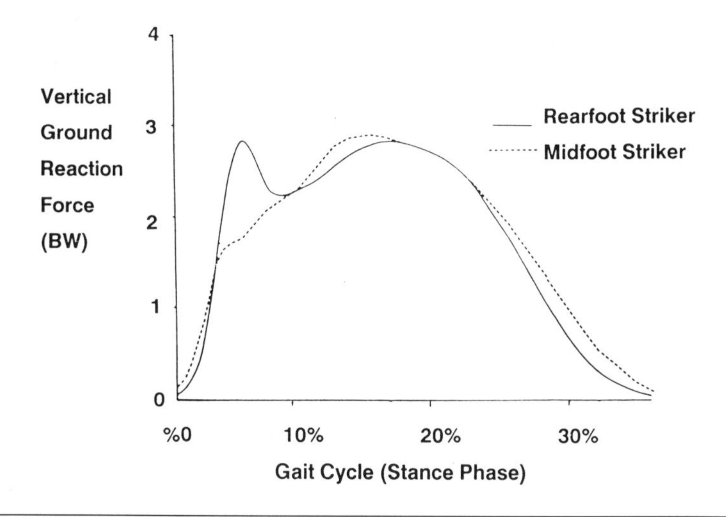

Figure 2.6 Typical vertical ground reaction forces (BW = body weight) for a rearfoot striker and a midfoot striker running at a velocity of 4.5 m/s. (Adapted by permission from Cavanagh & Lafortune, 1980.)

and Rowe (1987) reported that mean foot contact time is reduced in forefoot strikers (0.19 s) compared with that of rearfoot strikers (0.20 s). Cavanagh and Lafortune (1980) also found slightly shorter contact times for midfoot strikers than for rearfoot strikers.

Additional differences in GRF patterns have been described by Cavanagh and Lafortune (1980). Rearfoot strikers demonstrate a sharp initial spike in vertical GRF generally absent from the midfoot striker patterns. Midfoot strikers produced two positive peaks in the anteroposterior force during the braking phase. The mean peak-to-peak amplitude for mediolateral GRF was three times greater for the midfoot strikers than for the rearfoot strikers (0.35 BW and 0.12 BW respectively). These findings indicate that the loading rates within the muscles and joints are affected by the type of initial foot contact during running.

The COP path also depends on the type of initial foot contact. Cavanagh and Lafortune (1980) found that the COP path for rearfoot strikers followed from the rear lateral border to the midline within 15 ms of contact. The COP path then continued along the midline to the center of the forefoot, where it remained for almost two thirds of the entire 200-ms contact phase. Midfoot strikers running at the same speed made initial contact at 50% of shoe length. The COP path then migrated posteriorly as the rear part of the shoe made contact with the ground. This posterior movement coincided with a drop in the anteroposterior GRF. When the end of posterior migration was reached, the COP rapidly moved to the forefoot, where it remained for most of the contact phase.

Pressure Distribution Studies

Very little information is available about pressure distribution under the foot during running. Cavanagh and Rodgers (1985) have shown that pressure patterns vary with foot type (e.g., different arch heights or metatarsal lengths). The increased loading that occurs with running remains concentrated under the heel and forefoot in the more rigid high-arched foot. In the more flexible flat-arched foot, the increased load is spread beneath the entire foot, including the midfoot region. Cavanagh and Hennig (1983) found that the average peak pressure during the contact phase of running (868 kPa) occurred under the heel for a sample of 10 rearfoot strikers. Although pressures were much higher beneath the heel of these rearfoot strikers, more of the contact time was spend on the forefoot.

Muscle Forces, Segmental Impulse, and Joint Reaction Forces

Several investigators have developed mathematical models to predict muscle forces during running. Forces generated by the dorsiflexors and the gastrocnemius have been calculated by Harrison et al. (1987). They reported peak forces in the dorsiflexors of 0.52 times BW that are active only during the first 10% of the stance phase. The gastrocnemius generated a substantially greater peak force of 7.53 times BW. Calculations by Burdett (1982) revealed that the gastrocsoleus group had the highest predicted force (5.3 to 10.0 times BW) of the ankle muscle groups. Predicted forces in the tibialis posterior, flexor digitorum longus, and flexor hallucis longus group ranged from 4.0 to 5.3 times BW. The peroneus tertius and extensor digitorum longus did not show any predicted force during the stance phase of running.

Impulse, which is the effect of a force acting over a period of time, is determined mathematically as the integral of the force-time curve. Ae, K. Miyashita, Yokoi, and Hashihara (1987) calculated the impulse generated by different body segments during running and found that the foot generated the largest mean impulse. This impulse increased with faster running, suggesting that the foot plays an important role in projecting the body and increasing running velocity.

Ankle joint reaction forces during running have also been calculated by several investigators. Harrison et al. (1987) reported maximum ankle joint reactions of 8.97 and 4.15 times BW for the compressive and shear components, respectively. Burdett (1982) predicted that compressive forces on the foot along the longitudinal axis of the leg reached peak values from 3.3 to 5.5 times BW during running. In addition, he reported mediolateral shear forces that ranged from a medial force of 0.8 times BW to a lateral force of 0.5 times BW. Furthermore, the vertical reaction forces and other calculated forces were determined to be approximately 2.5 times larger in running (at a 4.47 m/s pace) than in walking.

Conclusion

This chapter has described current findings related to the dynamic biomechanics of the asymptomatic foot during walking and running. Functional descriptions of walking and running biomechanics have been provided along with quantitative findings from current biomechanical studies. Extensive data bases are still not available for most of the biomechanical parameters that affect dynamic foot motion. However, as advances in biomechanical methods continue and more clinicians include quantitative techniques in their routine evaluations, more insight into dynamic foot function will be provided. As our understanding improves, we can provide more effective therapeutic approaches to lower extremity injury and can better design protective shoewear. We may eventually see computer-aided design of shoes and therapeutic exercise programs based on clinical biomechanical measurements.

References

Ae, M., Miyashita, K., Yokoi, T., & Hashihara, Y. (1987). Mechanical power and work done by the muscles of the lower limb during running at different speeds. In B. Jonsson (Ed.), *Biomechanics X-B* (pp. 895-899). Champaign, IL: Human Kinetics.

Andriacchi, T.P., Ogle, J.A., & Galante, J.O. (1977). Walking speed as a basis for normal and abnormal gait measurements. *Journal of Biomechanics* **10**, 261-268.

Basmajian, J.V. (1985). *Muscles alive* (5th ed.). Baltimore: Williams & Wilkins.

Basmajian, J.V., & Stecko, G. (1963). The role of muscles in arch support of the foot. *Journal of Bone and Joint Surgery* [American] **45**, 1184-1190.

Bates, B.T., Osternig, L.R., & Mason, B. (1978). Lower extremity function during the support phase of running. In E. Asmussen & K. Jorgensen (Eds.), *Biomechanics VI* (pp. 31-39). Baltimore: University Park Press.

Betts, R.P., Franks, C.I., & Duckworth, T. (1980). Analysis of pressures and loads under the foot: 2. Quantification of the dynamic distribution. *Journal of Clinical Physics and Physiological Measurement*, **1**(2), 113-124.

Burdett, R.G. (1982). Forces predicted at the ankle during running. *Medicine and Science in Sports and Exercise*, **14**, 308-316.

Cavanagh, P.R. (1987). The biomechanics of lower extremity action in distance running. *Foot and Ankle*, **7**, 197-217.

Cavanagh, P.R., & Hennig, E.M. (1983). Pressure distribution measurement: A review and some new observations on the effect of shoe foam materials during running. In B.M. Nigg & B.A. Kerr (Eds.), *Biomechanical aspects of sports shoes and playing surfaces* (pp. 187-190). Calgary, AB: University of Calgary Press.

Cavanagh, P.R., Hennig, E.M., Rodgers, M.M., & Sanderson, D.J. (1985). The measurement of pressure distribution on the plantar surface of diabetic feet. In M. Whittle & D. Harris (Eds.), *Biomechanical measurement in orthopaedic practice* (pp. 159-166). Oxford, UK: Clarendon Press.

Cavanagh, P.R., & Lafortune, M.A. (1980). Ground reaction forces in distance running. *Journal of Biomechanics*, **13**, 397-406.

Cavanagh, P.R., & Rodgers, M.M. (1985). Pressure distribution underneath the human foot. In S.M. Perren & E. Schneider (Eds.), *Biomechanics: Current interdisciplinary research* (pp. 85-95). Dordrecht: Martinus Nijhoff.

Cavanagh, P.R., Williams, K.R., & Clarke, T.E. (1981). A comparison of ground reaction forces during walking barefoot and in shoes. In A. Morecki, K. Fidelus, K. Kedzior, & A. Wit (Eds.): *Biomechanics VII* (pp. 151-156). Baltimore: University Park Press.

Clarke, T.E. (1980). *The pressure distribution under the foot during barefoot walking*. Unpublished doctoral dissertation, The Pennsylvania State University, University Park.

Clarke, T.E., Frederick, E.C., & Hamill, C.L. (1984). The study of rearfoot movement in running. In E.C. Frederick (Ed.), *Sport shoes and playing surfaces* (pp. 166-189). Champaign, IL: Human Kinetics.

Frederick, E.C., Hagy, J.L., & Mann, R.A. (1981). Prediction of vertical impact force during running. *Journal of Biomechanics*, **14**, 498.

Grieve, D.W., & Rashdi, T. (1984). Pressures under normal feet in standing and walking as measured by foil pedobarography. *Annals of Rheumatic Diseases*, **43**, 816-818.

Harrison, R.N., Lees, A., McCullagh, P.J.J., & Rowe, W.B. (1987). Bioengineering analysis of muscle and joint forces acting in the human leg during running. In B. Jonsson (Ed.), *Biomechanics X-B* (pp. 855-861). Champaign, IL: Human Kinetics.

Inman, V.T., & Mann, R.A. (1973). Biomechanics of the foot and ankle. In V.T. Inman & H.L. Du Vries (Eds.), *Surgery of the foot* (pp. 3-22). St. Louis: Mosby.

Inman, V.T., Ralston, H.J., & Todd, F. (1981). *Human walking*. Baltimore: Williams & Wilkins.

Ito, A., Fuchimoto, T., & Kaneko, M. (1985). Quantitative analysis of EMG during various speeds of running. In D.A. Winter, R.W. Norman, R.P. Wells, K.C. Hayes, & A.E. Patla (Eds.), *Biomechanics IX-B* (pp. 301-306). Champaign, IL: Human Kinetics.

James, S.L., Bates, B.T., & Osternig, L.R. (1978). Injuries to runners. *American Journal of Sports Medicine*, **6**, 40-50.

Kaelin, X., Unold, E., Stüssi, E., & Stacoff, A. (1985). Interindividual and intraindividual variabilities in running. In D.A. Winter, R.W. Norman, R.P. Wells, K.C. Hayes, & A.E. Patla (Eds.), *Biomechanics IX-B* (pp. 356-360). Champaign, IL: Human Kinetics.

Katoh, Y., Chao, E.Y.S., Laughman, R.K., Schneider, E., & Morrey, B.F. (1983). Biomechanical analysis of foot function during gait and clinical applications. *Clinical Orthopedics*, **177**, 23-33.

Kayano, J. (1986). Dynamic function of medial foot arch. *Journal of the Japanese Orthopedic Association*, **60**, 1147-1156.

Manley, M.T. (1980). Biomechanics of the foot. In A.J. Helfet & D.M.G. Lee (Eds.), *Disorders of the foot* (pp. 21-30). Philadelphia: Lippincott.

Mann, R.A. (1985). Biomechanics of the foot. In American Academy of Orthopedic Surgeons (Eds.), *Atlas of orthotics: Biomechanical principles and application* (pp. 112-125). St. Louis: Mosby.

Mann, R.A., Baxter, D.E., & Lutter, L.D. (1981). Running symposium. *Foot and Ankle*, **1**, 190-224.

Mann, R.A., & Inman, V.T. (1964). Phasic activity of intrinsic muscles of the foot. *Journal of Bone and Joint Surgery* [Am], **46**, 469-481.

Miyashita, M., Matsui, H., & Miura, M. (1971). The relation between electrical activity in muscle and speed of walking and running. In J. Vredenbregt & J.W. Wartenweiler (Eds.), *Biomechanics II* (pp. 192-196). Baltimore: University Park Press.

Murray, M.P., Kory, R.C., & Sepic, S. (1970). Walking patterns of normal women. *Archives of Physical Medicine and Rehabilitation*, **51**, 637-650.

Nigg, B.M. (1987). Biomechanical analysis of ankle and foot movement. *Medicine and Science in Sports and Exercise*, **23**, 22-29.

Perry, J. (1983). Anatomy and biomechanics of the hindfoot. *Clinical Orthopedics*, **177**, 9-15.

Procter, P., & Paul, J.P.L. (1982). Ankle joint biomechanics. *Journal of Biomechanics*, **15**, 627-634.

Radin, E., Boyd, R., Martin, B., Burr, D.B., Caterson, B., & Goodwin, C. (1985). Mechanical factors influencing articular cartilage damage. In J.G. Peyron (Ed.), *Etiology of osteoarthrosis* (pp. 92-101). Paris: CIBA-Geigy.

Radin, E., Whittle, M., Yang, K.H., Jefferson, R.J., Rodgers, M.M., Kish, V.L., & O'Conner, J.J. (1986). The heel strike transient, its relationship with the angular velocity of the shank, and the effects of quadriceps paralysis. *Proceedings of the American Society of Mechanical Engineers Annual Conference*, **2**, 121-123.

Rodgers, M.M. (1985). *Plantar pressure distribution measurement during barefoot walking: Normal values and predictive equations.* Unpublished doctoral dissertation, The Pennsylvania State University, University Park.

Rodgers, M.M., & Cavanagh, P.R. (1984). Glossary of biomechanical terms, concepts, and units. *Physical Therapy*, **64**, 1886-1902.

Rodgers, M.M., & Cavanagh, P.R. (1989). Pressure distribution in Morton's foot structure. *Medicine and Science in Sport and Exercise*, **21**(1), 23-28.

Scranton, P.E., Rutkowski, R., & Brown, T.D. (1980). Support phase kinematics of the foot. In J.E. Bateman & A. Trott (Eds.), *The foot and ankle* (pp. 195-205). New York: Thieme-Stratton.

Seireg, A., & Arvikar, R.J. (1975). The prediction of muscular load sharing and joint forces in the lower extremities during walking. *Journal of Biomechanics*, **8**, 89-102.

Soames, R.W. (1985). Foot pressure patterns during gait. *Journal of Biomedical Engineering*, **7**, 120-126.

Soderberg, G.L. (1986). *Kinesiology: Application to pathological motion.* Baltimore: Williams & Wilkins.

Stauffer, R.N., Chao, E.Y.S., & Brewster, R.C. (1977). Force and motion analysis of the normal, diseased, and prosthetic ankle joint. *Clinical Orthopedics*, **127**, 189-196.

Williams, K.R. (1985). Biomechanics of running. In R.L. Terjung (Ed.), *Exercise and sport science reviews* (pp. 389-442). New York: Macmillan.

Winter, D.A. (1979). *Biomechanics of human movement.* New York: Wiley.

Winter, D.A. (1984). Kinematic and kinetic patterns in human gait: Variability and compensating effects. *Human Movement Science*, **3**, 51-76.

Winter, D.A. (1987). *The biomechanics and motor control of human gait.* Waterloo, ON: University of Waterloo Press.

Winter, D.A., & Robertson, D.G.E. (1978). Joint torque and energy patterns in normal gait. *Biological Cybernetics*, **29**, 137-142.

Winter, D.A., & Yack, H.J. (1987). EMG profiles during normal human walking: Stride-to-stride and inter-subject variability. *Electroencephalography Clinics in Neurophysiology*, **67**, 402-411.

Wright, D.G., Desai, M.E., & Henderson, B.S. (1964). Action of the subtalar and ankle-joint complex during the stance phase of walking. *Journal of Bone and Joint Surgery* [American] **46**, 361-382.

Yack, H.J. (1984). Techniques for clinical assessment of human movement. *Physical Therapy,* **64**, 1821-1830.

Chapter 3

Human Gait: From Clinical Interpretation to Computer Simulation

Christopher L. Vaughan and Michael D. Sussman

Christopher "Kit" Vaughan and Michael Sussman approach the biomechanical analysis of human walking gait from a point of view quite removed from the athletic arena—that of the clinical orthopedics environment. Whereas the athlete turns to the biomechanist to improve an already sound neuromuscular system, the cerebral palsy patient may seek medical help for a mildly to severely impaired gait. Vaughan and Sussman demonstrate how the biomechanical tools and techniques described in previous chapters are used to enhance the quality of a child's life by providing quantitative information to the surgeon. They also discuss the potential promise of computer simulation in gait analysis. Finally, Vaughan and Sussman challenge present and future biomechanists to address the theoretical and clinical issues of gait analysis using integrated, interdisciplinary methods.

Almost 40 years ago, Saunders and his colleagues in San Francisco proposed a theory in which they defined six major determinants in normal and pathological gait (Saunders, Inman, & Eberhart, 1953). They based their theory on the hypothesis that "fundamentally locomotion is the translation of the center of gravity throughout space along a pathway requiring the least expenditure of energy." They started with the concept that human walking could be represented, in a minimal configuration, by a pelvis and two stiff legs that allowed only flexion and extension at the hip joints (*compass gait*). This model obviously produced a jerky, undulating gait. The addition of six determinants—pelvic rotation, pelvic tilt, knee flexion during stance, knee and ankle interactions, and lateral displacement of the pelvis—converted this gait into a smooth, sinusoidal pattern. They further postulated that the loss of one of these determinants could be compensated for by the other determinants; however, to compensate for the loss of two required a threefold increase in energy consumption. This theory, based on simple kinematic arguments, has a certain logical appeal. It is, therefore, not surprising that the theory has been repeated, almost as proven fact, in more recent publications (Inman, Ralston, & Todd, 1981; Pandy & Berme, 1988b; Yamaguchi, 1990). In the sense that this theory recognized the three-dimensional nature of human gait, and the importance of the interaction between the pelvis and the distal segments, it was certainly ahead of its time. With the notable exception of the research by Pandy and Berme (1989a, 1989b), very little work has tested this theory rigorously, and this would be an important project for future researchers to undertake.

A recent, and far more radical, theory has been proposed by Gracovetsky (1985, 1990). Gracovetsky's theory of the spinal engine posits that locomotion was first achieved by the motion of the spine. Gracovetsky justifies this point of view by tracing the evolution of the vertebrate spine. He believes that the extremities came later on as an improvement, not as a substitute, for locomotion. He argues that the musculoskeletal system of the lumbar spine is the key structure to land-based locomotion, with the pelvis being driven by the spine (cf. Figure 3.1, a and b). Optimum control of motion dictates that the stresses at the intervertebral joints should be minimized, with the central nervous system modulating the torques at these joints. Gracovetsky (1985, 1990) invokes the research of other investigators to support this theory, challenging traditional approaches to bipedal gait that have concentrated exclusively on the contributions of the legs. He speculates that the spinal engine theory would be more successful in predicting the outcome of certain disabilities (e.g., loss of control of the psoas muscle) than are the classical theories of locomotion. We suggest that researchers of human gait pick up the gauntlet thrown down by Gracovetsky and test his hypothesis both experimentally and theoretically.

Role of Gait Analysis in Orthopedics

The motion analysis laboratory can provide information to orthopedists in a number of areas. First, gait analysis may provide insight into the basic mechanism underlying a disease process. For example, the underlying neuropathology in cerebral palsy is poorly understood (Bleck, 1987). Gait analysis may help to answer the question, Is the dysphasic muscle activity that is seen in cerebral palsy secondary to poor central coordination, or is the prolonged phasic activity due to an accentuated stretch reflex that prolongs the activity of the muscle? In other situations, where there is a better understanding of the etiology of a disease (such

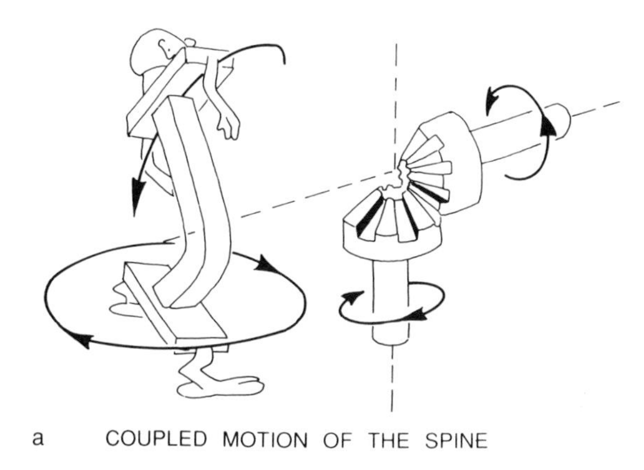

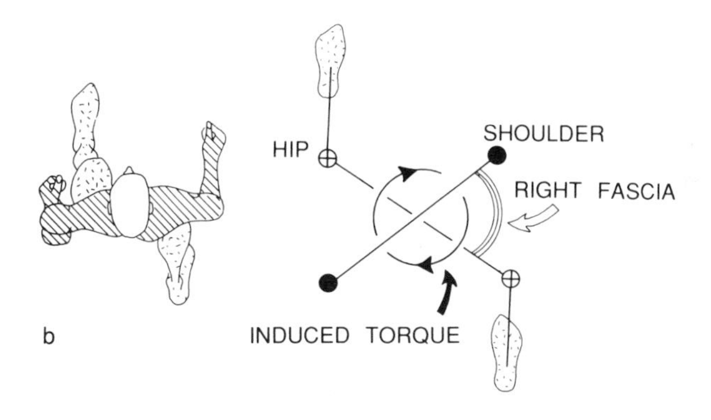

Figure 3.1 In the spinal engine theory, a lateral bend to the left produces an axial torque that forces the pelvis to rotate in a clockwise direction (a); a top view showing counter rotation of the pelvis and shoulders (b). (Adapted by permission from Gracovetsky, 1985.)

as juvenile rheumatoid arthritis or Duchenne muscular dystrophy), gait analysis allows us to understand the impact of the disease process on the function of a patient and to assess the progression or regression of the problem.

Second, gait analysis allows us to localize deficits more precisely and assists us in planning interventions. Gait analysis alone cannot provide absolute information to determine an intervention, but along with standard assessment techniques, it can assist us in decision making regarding the choice of surgical procedures or the choice of orthotic support for a particular patient (Wang, Kuo, Andriacchi, & Galante, 1990).

Third, when used appropriately, gait analysis allows us to assess accurately the efficacy of a specific intervention. If this intervention is reversible (such as the use of an orthosis), then we can decide whether to continue its use. If it is an irreversible intervention (such as surgery), we can decide whether further interventions need to be done. Perhaps more importantly, we can assess the efficacy of consistent interventions in similar situations, which allows us to develop reliable treatment plans. For example, Segal, Thomas, Mazur, and Manterer (1989) have shown that 30% of patients who underwent lengthening of the Achilles tendon exhibited evidence of too much weakness of triceps surae muscle function as demonstrated by calcaneal gait. This is much higher than had previously been recognized by clinical examination alone.

Application of Gait Analysis to Cerebral Palsy Children

Cerebral palsy is a condition wherein there has been damage to the motor cortex of the brain with resultant alteration of muscle function. Spastic muscles are set at a constantly higher tension and contract readily to passive stretch. In addition, there may be impairment in the processing of sensory information from proprioceptive, vestibular, and visual systems, thereby preventing the proper feedback needed for coordinated motion (Bleck, 1987).

Hip flexion deformity is manifested by the inability to extend the hip during stance phase, which results in shortened stride length and possibly also increased lumbar lordosis. In the series of patients reported by Gage, Fabian, Hicks, and Tashman (1984), increased stride length was shown following release of the iliopsoas, although dynamic gait electromyograms (EMGs) were not used to separate rectus femoris activity or other hip flexors from the iliopsoas. Children with cerebral palsy often manifest an internal rotation deformity of the lower extremity with gait that may be due to increased internal torsion of the femur but may also be caused by increased activity of the medial hamstrings.

The predominant problem about the knee in patients with cerebral palsy is increased knee flexion during the stance phase. Static studies by Perry, Antonelli, and Ford (1975) demonstrated that for each degree of knee flexion, the quadriceps must develop an amount of force equal to approximately 6% of body weight to maintain the upright position. When further challenged by the dynamics of walking, the quadriceps force required to prevent a flexed knee from collapsing becomes quite large. In a study by Thometz, Simon, and Rosenthal (1989), it was found that patients who had only mild hamstring activity during early stance phase were prone to the development of knee hyperextension following hamstring lengthening.

Following lengthening of the hamstring tendons, patients improve knee extension during stance phase but also diminish knee flexion during swing phase. In these patients, significant co-contraction of the rectus femoris with the hamstrings during swing phase has also been found (Gage, Perry, Hicks, Koop, & Werntz, 1987; Thometz et al., 1989), as illustrated in Figure 3.2. Based on this observation, Perry suggested a surgical procedure wherein the rectus femoris was transferred to the sartorius insertion simultaneously or subsequently to hamstring tendon lengthening. Gage et al. (1987) found that the patients maintained a greater arc of motion at the knee and presumably improved their gait by developing greater swing phase knee flexion (cf. Figure 3.3, a and b). This procedure probably would

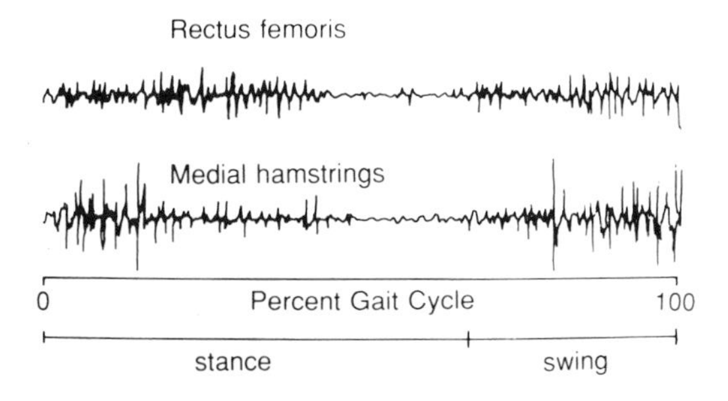

Figure 3.2 EMG recording of a patient with cerebral palsy shows coactivation of the rectus femoris and medial hamstrings. (Adapted by permission from Gage, Perry, Hicks, Koop, & Werntz, 1987.)

not have been developed on the basis of pure clinical observation alone: The operation was designed to derive the desired kinematic result based on the evidence of muscle dysfunction acquired from dynamic EMG studies.

Foot deformity in the cerebral palsy child may occur in all planes. Varus deformity is more common in patients with hemiplegia and causes them to walk on the lateral side of the foot, thereby concentrating the weight bearing over a smaller surface area. This causes pain and instability and problems with shoe fitting and shoe wear. A varus deformity can be due to overactivity of the tibialis anterior, the tibialis posterior, or both muscles. Although Kling, Kaufer, and Hensinger (1985) felt that a trained observer could distinguish whether the varus-producing force was the tibialis anterior or posterior purely on the basis of clinical examination, Wills, Hoffer, and Perry (1988) concluded that clinical examination was inadequate. In this latter group's series of 51 patients, they found that the tibialis anterior was the sole deforming muscle in 23 patients' feet and the tibialis posterior was the sole deforming force in 13 patients' feet. They had previously shown that when a tibialis anterior transfer was done in cases of continuous tibialis anterior activity, this resulted in an improvement for 4 out of 4 patients (Perry & Hoffer, 1977). Even though the muscle still exhibited continuous activity, alteration of its insertion converted the functional effects of this activity so that the foot remained in a balanced position.

The major limitation of all of these studies on cerebral palsy children is that although the authors used the EMG to demonstrate specific muscle dysfunction and then chose the surgical procedure based on this specific muscle dysfunction, they did not demonstrate that this approach has any advantage over clinical decision making without the use of this technique. In general, muscles do not change phase but change their activity due to weakening or transfer, although in some cases changes in phasic activity may occur.

Limitations of Gait Analysis in Clinical Practice

Although gait analysis does provide objective information, there are some limitations to its use. These include the following:

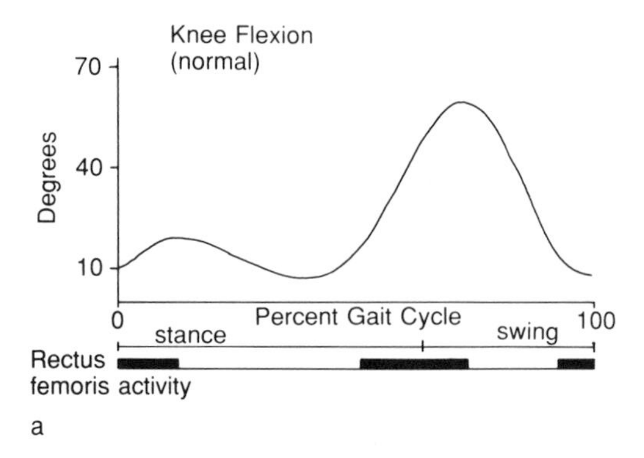

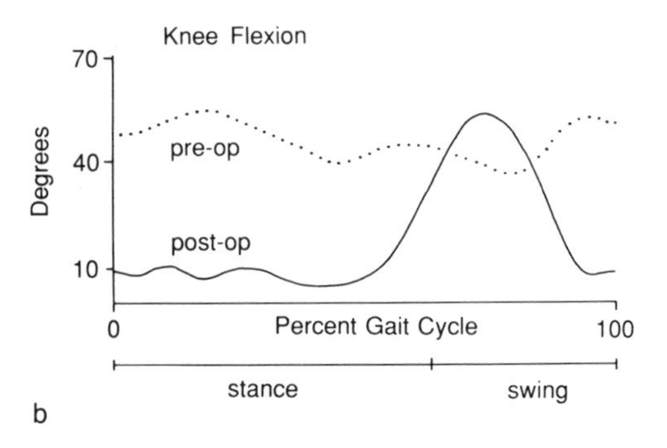

Figure 3.3 Average knee flexion motion for 23 normal adults (a); equivalent graph for a 13-year-old patient following medial and lateral hamstring lengthening and transfer of rectus femoris to the iliotibial band (b). (Adapted by permission from Gage, Perry, Hicks, Koop, & Werntz, 1987.)

- If patients require assistive devices such as walkers or canes for ambulation, these devices will interfere with data collection, although in many cases data can still be acquired.

- Because the markers are placed on skin rather than on bone, they may not relate to the underlying bony landmarks as accurately as we would like. Three problems are associated with markers attached to the skin: Prediction equations, based on very limited data, are used to estimate the internal landmarks (e.g., joint centers) from external markers (Vaughan, Davis, & O'Connor, 1992); a small change in the position of a marker can lead to a relatively large error in joint angle measurements (Kadaba, Ramakrishnan, & Wooten, 1990); and the skin, with the marker attached to it, can move significantly relative to the underlying skeletal landmark (Macleod & Morris, 1987; van Weeren, van den Bogert, & Barneveld, 1988).

- Some variation is observed in the way subjects walk at different times and on different days. Therefore, multiple observations may be required. There is also tremendous variability in the output of the EMG signals (Kadaba, Ramakrishnan, Wootten, Gainey, Gorton, & Cochran, 1989). The signal may vary in intensity depending on where the electrode is located within the muscle. Therefore, EMGs are most likely to be useful for only qualitative information about muscle activity, although recent work by Wootten, Kadaba, & Cochran (1990) suggests some promise for standard characterization.

- Although we try to relax the patients maximally, the gait laboratory is not their natural environment and patients try to overcome their deficits and demonstrate their best walking pattern. This phenomenon, well known to orthopedic surgeons, is referred to as *clinic walk.* This problem can be reduced by multiple walks in a nonthreatening environment but cannot be completely eliminated. In addition, encumbrance of the patient (with markers, electrode wires, etc.) will also distract the patient and alter the gait pattern.

These limitations should not deter us from pursuing gait analyses, but we should recognize that there are inherent limitations of the measurement system. In spite of these limitations, computerized gait analysis remains the most effective technique for objectively quantifying dynamic function and will undoubtedly grow in importance over the next decade.

Computer Modeling and Simulation of Gait

Although the terms *computer modeling* and *computer simulation* are often used synonymously, they have different meanings. The following definitions, taken from Vaughan (1984), are suggested for this chapter:

Computer modeling refers to the setting up of mathematical equations to describe the system of interest, the gathering of appropriate input data, and the incorporation of these equations and data into a computer program.

Computer simulation is restricted to mean the use of a validated computer model to carry out ''experiments,'' under carefully controlled conditions, on the real-world system that has been modeled.

These definitions are presented in diagrammatic form in Figure 3.4, a and b. As you can see, both from the definitions and Figure 3.4, computer simulation cannot be performed without first developing a computer model. It is, however, feasible to build a computer model without performing computer simulation. Most studies in human gait that have used a computer have dealt only with modeling and have stopped short at linking point 1 in Figure 3.4. The advantages and disadvantages of computer simulation are reviewed briefly next (Vaughan, 1984).

It is the exploratory nature of computer simulation that offers the greatest potential benefit—it can answer the what-if questions. These could include questions such as ''What would happen to the patient's gait if I transferred tibialis

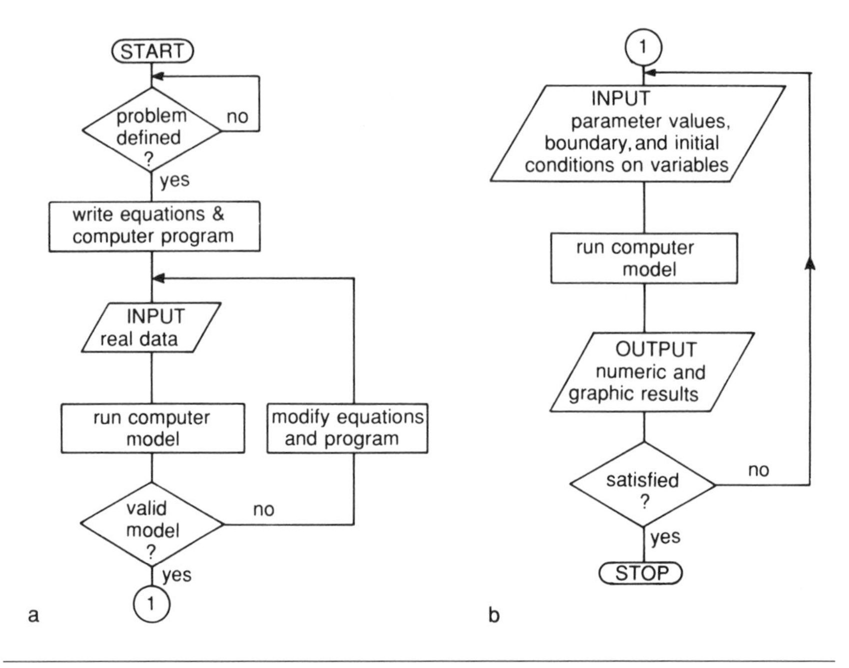

Figure 3.4 Flow diagrams for computer modeling (a) and computer simulation (b). (Adapted by permission from Vaughan, 1984.)

anterior instead of tibialis posterior?'' or ''Would this cerebral palsy child achieve a true heelstrike if I gave her an ankle foot orthosis?'' The predictive aspects of gait simulation models have the greatest potential when compared with the traditional diagnostic application of gait analysis (Yamaguchi, 1990). The advantages of computer simulation may be summarized as follows:

- It is safe (no hazardous experiments are performed on the patient).
- It saves time (many different simulations can be performed relatively quickly).
- It can be used to predict optimal function.
- It saves money (no need to build different devices or perform unnecessary operations).
- It can be performed on microcomputers, which are easy to use and have powerful graphic displays (the advent of microcomputers has been crucial to the development of simulation models; see Vaughan, 1984).

Perhaps the greatest drawback to computer simulation in human gait is the difficulty of validating the computer model, a point argued so succinctly by Panjabi (1979). The disadvantages may be summarized as follows:

- Validation is difficult, and incorrect models prejudice results.
- Advanced mathematics and computer skills are often required of the user.

- Results are often difficult to interpret and to translate to practicality.
- Computers are still not sufficiently powerful to perform multisegment, many-degrees-of-freedom simulations in a reasonable period of time.

It is vitally important that one integrate all the biomechanical parameters when modeling human walking so that the causes of the movement can be studied. In the past few years, investigators have implemented true three-dimensional (3-D) models to measure the joint moments (or torques) in normal subjects (Apkarian, Naumann, & Cairns, 1989; Kadaba et al., 1989) and also in patients (Winter, Olney, Conrad, White, Ounpuu, & Gage, 1990). Furthermore, efforts have been made to relate these joint moments to muscular activity (EMG), although there are still some discrepancies between measurement and prediction (Hof, 1990).

One problem with joint moments is that they are a measurement of the resultant vector acting about a joint. This parameter cannot distinguish between agonist and antagonist muscle activity, nor can it reveal anything about the contribution from soft tissues such as the ligaments. Various investigators have recognized this shortcoming with joint moments and have begun to develop accurate anatomical models (Hoy, Zajac, & Gordon, 1990; White, Yack, & Winter, 1989; Yamaguchi, Sawa, Moran, Fessler, & Winters, 1990).

One area of computer modeling that has gained in popularity over the past decade is mathematical optimization. With this approach, the problem of indeterminacy (fewer equations than the number of unknown force actuators) can be overcome by minimizing some cost function. Crowninshield and Brand (1981) used a model with 47 muscle elements and postulated that, during gait, the person moves in such a way as to minimize the sum of muscle stresses raised to the third power (they argued that this was equivalent to maximizing muscle endurance time). The validation of their model, using EMG data, is shown in Figure 3.5. However, Crowninshield and Brand emphasized that caution should be exercised when drawing conclusions about the validity of an optimization criterion based solely on muscle activity predictions. The cost function to be minimized cannot be known a priori and may vary from person to person depending on the demands of the task (Vaughan, Hay, & Andrews, 1982). Some of the early optimization algorithms—for example, gradient projection used by Crowninshield and Brand (1981) and Vaughan et al. (1982)—were based on a quasi-static approach with the assumption that the force actuators were time-independent. Dynamic optimization, on the other hand, which allows both activation and musculotendon mechanics to be modeled simultaneously, has been the approach of choice in the past few years (Davy & Audu, 1987; Pandy & Berme, 1988a, 1988b, 1989a, 1989b; Yamaguchi, 1990). This algorithm, however, is computationally very intensive, so researchers have had to study models with a limited number of degrees of freedom (Yamaguchi, 1990).

As mentioned briefly in the previous section on theories of human gait, Pandy and Berme (1989a, 1989b) have performed some interesting simulations with their 3-D model. They explored a few of the gait determinants postulated by Saunders et al. (1953) and concluded that their models were "useful implements

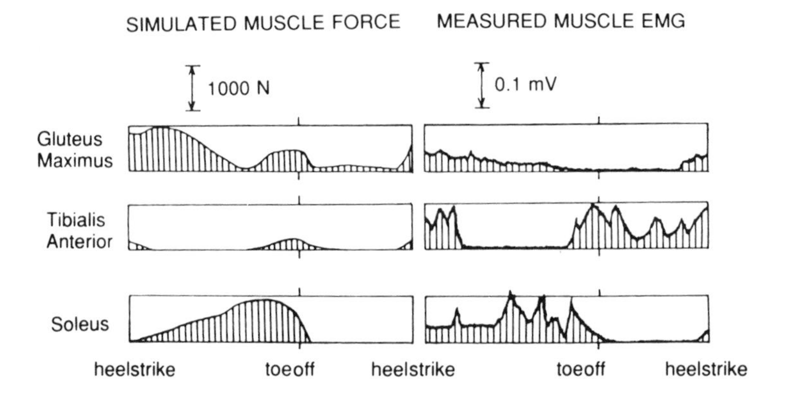

Figure 3.5 Comparison of simulated muscle forces using optimization techniques and measured muscle electromyography for the gait cycle. (Adapted by permission from Crowninshield & Brand, 1981.)

for assessing both the effect of joint motion loss, as well as the degree of compensatory action adopted.'' Another development that bodes well for simulation research in the future is the availability of commercial software packages such as DADS (cf. van den Bogert, Schamhardt, & Crowe, 1989) and AUTOLEV (cf. Yamaguchi, 1990) that automatically generate, and solve, the equations of motion for a multilink system. Although we must be careful that we do not use these packages as ''black boxes,'' simulation packages do enable the interested researcher to concentrate on simulation, that is, asking the what-if questions.

A classical challenge in musculoskeletal biomechanics is the distribution problem: How do we apportion the loads to the many muscles crossing a joint? A relatively new algorithm, known as *artificial neural networks*, has been gaining widespread recognition and application (Rumelhart, Hinton, & Williams, 1986). We first used this approach to model a highly nonlinear sensory motor transformation, studying a two-link system with five muscle actuators and three rotational degrees of freedom (Wells & Vaughan, 1989). More recently, we proposed that EMGs can be used to predict joint forces and moments in human walking and that a neural network model based on generalized back propagation has considerable promise (Sepulveda, Wells, & Vaughan, 1992). We chose a three-layer model, as shown in Figure 3.6. Using data on normal subjects (Winter, 1987), we demonstrated that the model could generalize the vector transformation from EMG to joint moments very successfully. Perhaps of greater interest, however, was the ability of the network to predict outcome based on a perturbation of the input. Figure 3.7 shows a simulation in which the activity of the soleus muscle was decreased by 30% throughout the cycle. Note the significant decrease in the plantar-flexor (extensor) ankle moment, which is consistent with the action of this muscle. This example—and there are other simulations that are similarly successful—demonstrates the potential of a neural network model to help us understand the connections between the locomotor ‘‘program’’ and the musculoskeletal effector system.

Conclusion

Clearly, there is no shortage of research publications on the mechanics of human gait (Vaughan, Besser, Bowsher, & Sussman, 1992). The challenge for future researchers, however, is to identify areas of inquiry that may ultimately lead to fruitful results. In this final section, we suggest what some of these areas might be.

Few theories on human gait have been proposed. Those theories that have been proposed (Gracovetsky, 1985; Saunders et al., 1953) have not been subjected to close scrutiny. We suggest that a fertile area of research might be to test these theories both experimentally and using computer simulation techniques (cf. Pandy & Berme, 1989b). Thereafter, it would probably be appropriate to propose, and evaluate, further theories of human gait. We believe that such theories should not look at the musculoskeletal effector system in isolation but should try to integrate the neural control system. A concept such as the central pattern generator, which has been applied in quadrupedal gait, may have some potential.

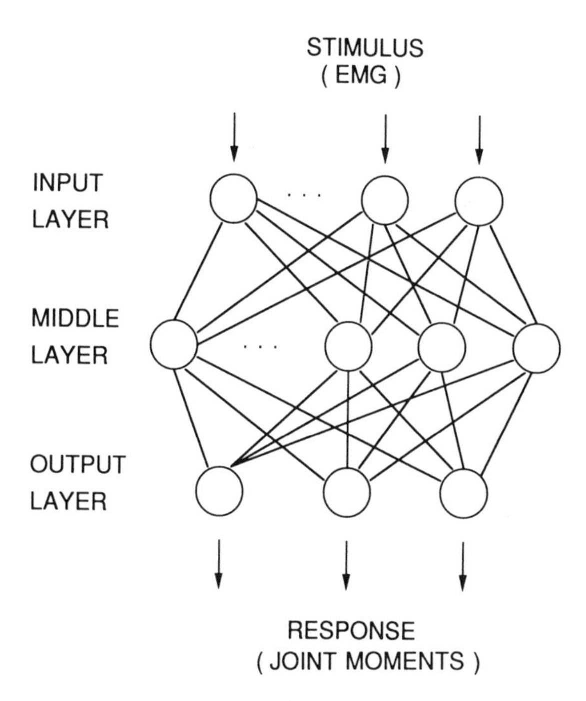

Figure 3.6 Basic neural network architecture for a model of human gait. (Adapted by permission from Sepulveda, Wells, & Vaughan, 1992.)

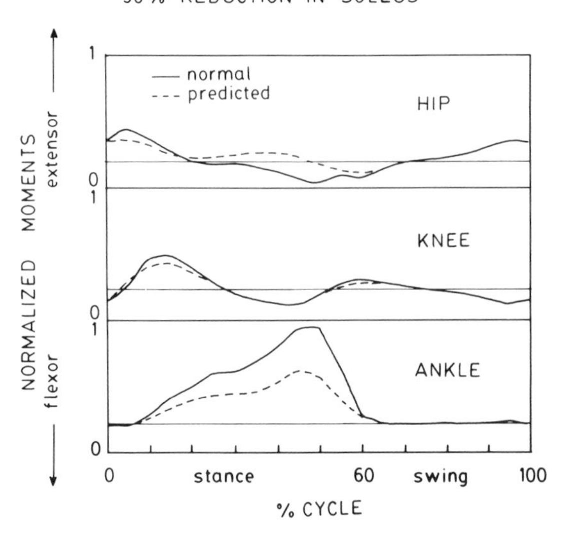

Figure 3.7 Neural network prediction of joint moments for a simulated 30% reduction in the electromyogram (EMG) signal of the soleus during the entire gait cycle. (Adapted by permission from Sepulveda, Wells, & Vaughan, 1992.)

Biomechanics researchers in the past have been concerned with ''building a better mousetrap.'' Although some might suggest that this preoccupation has been to the detriment of real progress toward understanding human gait, we argue that our field desperately needs individuals and companies who are committed to developing new and better technologies. Miniaturized EMG transmitters, weighing less than 20 g, are now routinely used in clinical practice (cf. Vaughan et al., 1992). Future work in this area should concentrate on further miniaturization and should address the problems of needle insertion for fine-wire electrodes. The insertion can be uncomfortable—not to mention the psychological impact, which can make a visit to the gait laboratory a traumatic experience for a young child. Despite recent advances in the field (cf. Walton, 1990), video-based motion systems with their passive markers are still unable to produce 3-D coordinates in real time. Whereas active markers have, at least potentially, much higher sampling rates and the data are available in real time, active markers still have the significant problem of patient encumbrance. During the next decade we hope motion systems will be developed that have the benefits of both these approaches (real-time operation, lack of encumbrance, high sampling rate) without the disadvantages (Lanshammar, 1985). Many of the force-transducing systems developed in the past few years are designed to measure forces at discrete points under the person's foot. All these systems, however, measure only normal loads; we hope transducers in the future will also be sensitive to shear loads.

Future clinical applications of gait analysis should move toward integrating kinematic, dynamic (joint moments, ground reaction forces), and electromyographic data (Vaughan et al., 1992). Despite the views of some (Winter et al.,

1990), we believe that a greater number of future studies will be three-dimensional. Despite the objectivity of gait data, clinical decisions are still made on a subjective basis. Expert systems may have some potential in this regard, but we think that education of clinicians—where they have ready access to 3-D gait data, such as the software in Vaughan et al. (1992) — also needs to be emphasized. Finally, clinical gait laboratories must commit themselves to long-term studies and to the development of large data bases (Vaughan, Berman, Peacock, & Eldridge, 1989).

Despite the promise of computer simulation as an adjunct to clinical decision making, this has yet to be accomplished. However, as models become more realistic, algorithms easier to implement, computers faster and more powerful, this promise may yet be realized before the turn of the century. Computer simulation will play a pivotal role in our understanding the theories of human locomotion, and emerging algorithms such as neutral networks may be at the forefront of future modeling endeavors. It seems clear that in the future researchers of human gait will be limited only by their own imaginations.

References

Apkarian, J., Naumann, S., & Cairns, B. (1989). A three-dimensional kinematic and dynamic model of the lower limb. *Journal of Biomechanics,* **22**, 143-155.

Bleck, E.E. (1987). *Orthopaedic management in cerebral palsy.* Oxford, UK: MacKeith Press.

Crowninshield, R.D., & Brand, R.A. (1981). A physiologically based criterion of muscle force prediction in locomotion. *Journal of Biomechanics,* **14**, 793-801.

Davy, D.T., & Audu, M.L. (1987). A dynamic optimization technique for predicting muscle forces in the swing phase of gait. *Journal of Biomechanics,* **20**, 187-201.

Gage, J.R., Fabian, D., Hicks, R., & Tashman, S. (1984). Pre- and postoperative gait analysis in patients with spastic diplegia: A preliminary report. *Journal of Pediatric Orthopedics,* **4**, 715-725.

Gage, J.R., Perry, J., Hicks, R.R., Koop, S., & Werntz, J.R. (1987). Rectus femoris transfer to improve knee function of children with cerebral palsy. *Developmental Medicine and Child Neurology,* **29**, 159-166.

Gracovetsky, S. (1985). An hypothesis for the role of the spine in human locomotion: A challenge to current thinking. *Journal of Biomedical Engineering,* **7**, 205-216.

Gracovetsky, S. (1990). Musculoskeletal function of the spine. In J.M. Winters & S.L.Y. Woo (Eds.), *Multiple muscle systems* (pp. 410-437). New York: Springer-Verlag.

Hof, A.L. (1990). Effects of muscle elasticity in walking and running. In J.M. Winters & S.L.Y. Woo (Eds.), *Multiple muscle systems* (pp. 591-607). New York: Springer-Verlag.

Hoy, M.G., Zajac, F.E., & Gordon, M.E. (1990). A musculoskeletal model of the human lower extremity: The effect of muscle, tendon, and moment arm on the moment-angle relationship of musculotendon actuators at the hip, knee and ankle. *Journal of Biomechanics,* **23**, 157-169.

Inman, V.T., Ralston, H.J., & Todd, F. (1981). *Human walking.* Baltimore: Williams & Wilkins.

Kadaba, M.P., Ramakrishnan, H.K., & Wootten, M.E. (1990). Measurement of lower extremity kinematics during level walking. *Journal of Orthopaedic Research,* **8**, 383-392.

Kadaba, M.P., Ramakrishnan, H.K., Wootten, M.E., Gainey, J., Gorton, G., & Cochran, G.V.B. (1989). Repeatability of kinematic, kinetic and electromyographic data in normal adult gait. *Journal of Orthopaedic Research,* **7**, 849-860.

Kling, T.F., Kaufer, H., & Hensinger, R.N. (1985). Split posterior tibial-tendon transfers in children with cerebral spastic paralysis and equinovarus deformity. *Journal of Bone and Joint Surgery,* **67A**, 186-194.

Lanshammar, H. (1985). Measurement and analysis of displacement. In H. Lanshammar (Ed.), *Gait analysis in theory and practice* (pp. 29-45). Uppsala, Sweden: Uppsala Universitet.

Macleod, A., & Morris, J.R.W. (1987). Investigation of inherent experimental noise in kinematic experiments using superficial markers. In B. Jonsson (Ed.), *Biomechanics X-B* (pp. 1035-1039). Champaign, IL: Human Kinetics.

Pandy, M.G., & Berme, N. (1988a). A numerical method for simulating the dynamics of human walking. *Journal of Biomechanics,* **21**, 1043-1051.

Pandy, M.G., & Berme, N. (1988b). Synthesis of human walking: A planar model for single support. *Journal of Biomechanics,* **21**, 1053-1060.

Pandy, M.G., & Berme, N. (1989a). Quantitative assessment of gait determinants during single stance via a three-dimensional model: 1. Normal gait. *Journal of Biomechanics,* **22**, 717-724.

Pandy, M.G., & Berme, N. (1989b). Quantitative assessment of gait determinants during single stance via a three-dimensional model: 2. Pathological gait. *Journal of Biomechanics,* **22**, 725-733.

Panjabi, M. (1979). Validation of mathematical models. *Journal of Biomechanics,* **12**, 238.

Perry, J., Antonelli, D., & Ford, W. (1975). Analysis of knee joint forces during flexed knee stance. *Journal of Bone and Joint Surgery,* **57A**, 961-967.

Perry, J., & Hoffer, M.M. (1977). Preoperative and postoperative dynamic electromyography as an aid in planning tendon transfers in children with cerebral palsy. *Journal of Bone and Joint Surgery,* **59A**, 531-537.

Rumelhart, D.E., Hinton, G.E., & Williams, R.J. (1986). Learning representations by back propagation of errors. *Nature,* **323**, 533-536.

Saunders, J.B., Inman, V.T., & Eberhart, H.D. (1953). The major determinants in normal and pathological gait. *Journal of Bone and Joint Surgery,* **35A**, 543-559.

Segal, L.S., Thomas, S.E.S., Mazur, J.M., & Manterer, M. (1989). Calcaneal gait in spastic diplegia after heel cord lengthening: A study with gait analysis. *Journal of Pediatric Orthopedics,* **9**, 697-701.

Sepulveda, F.S., Wells, D.M., & Vaughan, C.L. (1992). A neural network representation of electromyography and joint dynamics in human gait. *Journal of Biomechanics,* (in press).

Thometz, J., Simon, S., & Rosenthal, R. (1989). The effect on gait of lengthening of the medial hamstrings in cerebral palsy. *Journal of Bone and Joint Surgery,* **71A**, 345-353.

van den Bogert, A.J., Schamhardt, H.C., & Crowe, A. (1989). Simulation of quadrupedal locomotion using a rigid body model. *Journal of Biomechanics,* **22**, 33-41.

van Weeren, P.R., van den Bogert, A.J., & Barneveld, A. (1988). Quantifying skin displacement near the carpal, tarsal and fetlock joints of the walking horse. *Equine Veterinary Journal,* **20**, 203-208.

Vaughan, C.L. (1984). Computer simulation of human motion in sports biomechanics. *Exercise and Sports Science Reviews,* **12**, 373-411.

Vaughan, C.L., Berman, B., Peacock, W.J., & Eldridge, N.E. (1989). Gait analysis and rhizotomy: Past experience and future considerations. *Neurosurgery: State of the Art Reviews,* **4**, 445-458.

Vaughan, C.L., Davis, B.L., & O'Connor, J.C. (1992). *Dynamics of human gait.* Champaign, IL: Human Kinetics.

Vaughan, C.L., Hay, J.G., & Andrews, J.G. (1982). Closed loop problems in biomechanics: 2. An optimization approach. *Journal of Biomechanics,* **15**, 201-210.

Vaughan, C.L., Besser, M.P., Bowsher, K., & Sussman, M.D. (1992). *Biomechanics of human gait: An electronic bibliography* (3rd ed.). Champaign, IL: Human Kinetics.

Walton, J.S. (1990). *Image-based motion measurement* (Society of Photo-Optical Instrumentation Engineers Vol. 1356). Bellingham, Washington: International Society for Optical Engineering.

Wang, J.W., Kuo, K.N., Andriacchi, T.P., & Galante, J.O. (1990). The influence of walking mechanics and time on the results of proximal tibial osteotomy. *Journal of Bone and Joint Surgery,* **72A**, 905-909.

Wells, D.M., & Vaughan, C.L. (1989). A 3D transformation of a rigid link system using back propagation. *Proceedings of the International Joint Conference on Neural Networks,* **2**, 630A.

White, S.C., Yack, H.J., & Winter, D.A. (1989). A three-dimensional musculoskeletal model for gait analysis: Anatomical variability estimates. *Journal of Biomechanics,* **22**, 885-893.

Wills, C.A., Hoffer, M.M., & Perry, J. (1988). A comparison of foot-switch and EMG analysis of varus deformities of the feet of children with cerebral palsy. *Developmental Medicine and Child Neurology,* **30**, 227-231.

Winter, D.A. (1987). *The biomechanics and motor control of human gait*. Waterloo, ON: University of Waterloo Press.

Winter, D.A., Olney, S.J., Conrad, J., White, S.C., Ounpuu, S., & Gage, J.R. (1990). Adaptability of motor patterns in pathological gait. In J.M. Winters, & S.L.Y. Woo (Eds.), *Multiple muscle systems* (pp. 680-693). New York: Springer-Verlag.

Wootten, M.E., Kadaba, M.P., & Cochran, G.V.B. (1990). Dynamic electromyography: 2. Normal patterns during gait. *Journal of Orthopaedic Research*, **8**, 259-265.

Yamaguchi, G.T. (1990). Performing whole-body simulations of gait with 3-D, dynamic musculoskeletal models. In J.M. Winters, & S.L.Y. Woo (Eds.), *Multiple muscle systems* (pp. 663-679). New York: Springer-Verlag.

Yamaguchi, G.T., Sawa, A.G.U., Moran, D.W., Fessler, M.J., & Winters, J.M. (1990). A survey of human musculotendon actuator parameters. In J.M. Winter & S.L.Y. Woo (Eds.), *Multiple muscle systems* (pp. 717-773). New York: Springer-Verlag.

Chapter 4

Low Back Biomechanics in Industry: The Prevention of Injury Through Safer Lifting

Stuart M. McGill and Robert W. Norman

Stuart McGill and Robert Norman illustrate the application of the same biomechanical analysis tools used in the analysis of walking and running gait to a major health problem in the industrial work environment: lumbar injuries sustained during lifting. Low back disorders affect more than 70% of the general population at some time during their lives and cost $15 billion to $30 billion a year in lost wages, medical treatment, and lost productivity in the United States. McGill and Norman combine the classical biomechanical analysis techniques with newly acquired anatomical data, laboriously taken from cadaveric dissections and magnetic resonance images (MRI), and the electromyographical data from the trunk musculature represented in their model of the lumbar spine. They discuss concepts regarding the estimation and partitioning of vertebral loads, trunk function, and the mechanisms of injury to the low back. They offer a set of tentative guidelines for safe lifting procedures and encourage other researchers to test these guidelines.

To combat the enormous cost of disabling low back injury, industry has undertaken a major effort in the last two decades by training workers in the proper techniques of lifting. While the emphasis has been ''Bend the knees—not the back''; other widely prescribed instructions have included ''Hold objects close,'' ''Don't jerk loads—move slowly,'' and ''Turn feet to avoid twisting the trunk.'' Very few tasks in industry can be conducted in a manner that adheres to these instructions.

Many workers prefer to stooplift, possibly because there is an increased physiological cost in squatting (Garg & Herrin, 1979). Some studies attempted to evaluate the issue of the stoop versus squat posture (e.g., Garg & Herrin, 1979; Park & Chaffin, 1974; Troup, 1977) but were unable to uncover a clear biomechanical rationale for the promotion of either. These studies led to more recent investigations of low back mechanics that together have produced data and improved ways to evaluate the risk of injury to the low back, information essential to planning and implementing injury avoidance strategies. However, present instructions remain insufficient, some are inaccurate, and many are unworkable in many job tasks. Despite widespread educational efforts, compensation costs for low back injury continue to escalate. The reasons for this include problems of public policy and of pedagogy, but they also include the biomechanical accuracy of the information transferred to workers. One must ask what scientific evidence supports the ''rules'' of safe lifting that are frequently taught in back education programs.

Whereas the theoretical basis of estimating injury risk is quite simple, estimating the risk is much more difficult. If the load applied to a tissue exceeds the strength of the tissue, the result is failure and injury. Thus, two bodies of knowledge are vital for the risk analysis. The discipline of tissue mechanics provides information on failure tolerance that depends on tissue properties such as homogeneity, isotropy, viscoelasticity, fatigue characteristics, collagen-elastin ratios, and architectural arrangements. The other necessary component for the analysis is knowledge of the load applied to, or generated by, the specific tissue. Obtaining direct in vivo measures of tissue loads on humans remains extremely difficult, so biomechanical modeling appears to be the only tenable option for estimating individual loads in living workers.

Biomechanical models have been used to estimate loads in the low back tissues and to identify high-risk jobs for approximately three decades. In the process, the output of some very simple models has been used to interpret certain aspects of low back mechanics and subsequently to suggest strategies for workers to avoid injury. Some of these suggestions have been seriously questioned in recent years. For example, the issue of maintaining a lordotic curve in the lumbar spine while lifting, or lifting with a flexed spine, remains controversial; some have suggested that certain types of awkward lifts impose larger loads on the spine if they are performed in a slow manner than if the loads are skillfully accelerated; abdominal belts are prescribed to industry without proven biomechanical rationale. Perhaps the issues regarding low back injury in industry are more mechanically complex than were previously realized, and the answers cannot be expected from simple models. Clearly, the common instruction to bend the knees and not the back is overly simplistic and, indeed, may not always be the best advice. Moreover, it addresses only one or two of the several mechanical issues of low back mechanics related to safe lifting.

The first section of this chapter briefly examines the biomechanical literature to review methods used to assess the size of loads on tissues at risk of injury during occupational lifting. Specifically, modeling methods used to estimate loads are discussed, together with a functional interpretation of the calculated tissue load-time histories. In the second section, issues in low back mechanics specific to lifting are addressed, and scientific evidence available to assess these issues is reviewed. In the third section, this information is integrated into a set of lifting guidelines that we believe are justifiable. Some of these guidelines are based on recent research findings; therefore, they remain tentative until their efficacy is proven or disproven. Whereas some may think it prudent to wait for the definitive study, our intent in listing the tentative guidelines here is to provoke discussion and to motivate others to initiate experiments to test their viability.

Biomechanical Models for Estimating Loads

The use of mathematical models to estimate tissue loads and injury risk to the lumbar spine, and to provide insight into its function, has increased over the last 30 years. It is important to review model features because lumbar tissue loading predicted by these models forms the cornerstone for the development of lifting guidelines. The following brief review is an attempt to analyze underlying modeling philosophies and to interpret the output of a few spine models.

The 5-cm Equivalent Moment Arm Model

One of the most widely used models for analyzing industrial tasks and predicting lumbar joint compression assumes the extensor tissues of the spine to be reasonably represented with a single equivalent reaction moment force generator. This most simple model implicity assumes that a single vector representing the extensor tissues connects adjacent posterior spinous processes to generate force through a moment arm distance that is usually 5 cm from the disc center, producing an extensor moment extending the lower back (see Figure 4.1). This extensor force is necessary to counterbalance the weight of the upper body when a person is in a lifting posture and has an external load in the hands. This approach to modeling has proved to be valuable for use in industry (e.g., Chaffin, 1969; National Institute for Occupational Safety and Health, 1981). Injury statistics and low back compression forces estimated from a 5-cm model have been successfully used to set industrial lifting guidelines. The prediction of the size of lumbar disc compression imposed by lifting tasks tends to be quite high. If there is error in the prediction, it is in the direction of safety, and for this reason this approach is justifiable. However, some other researchers have worked beyond the intended limitations of the model and erroneously expected this simplified anatomical respresentation to provide deeper insight into lumbar mechanics than the model is capable of producing. In fact, some faulty lifting recommendations have resulted from a superficial understanding of low back function.

A compressive penalty is imposed on the spine from tension in the supporting extensor tissues, and this primary source of compressive load increases with a decrease in extensor force moment arm. The extensor force is considered to be parallel to the axis of compression, thus removing any shearing capability. Certain versions of this model have increased the mechanical advantage of the extensor tissue by increasing their moment arm to 6 cm (e.g., Wood & Hayes, 1974).

Even though the single equivalent model has been used for decades, its output presents a paradox to the researcher: Compressive loads are predicted that exceed vertebral end plate failure tolerance for lifts in which the individual is able to complete the task without reporting any ill effects. This seeming anomaly may be addressed from two perspectives. Perhaps the tissue tolerance data are too low. Certainly, in highly trained individuals, evidence of increased mineral density in the vertebral body has been presented by Granhed, Jonsson, and Hansson (1987), suggesting higher tolerances in active people. However, we believe an additional consideration contributes to the apparent inability of this simplified model to predict the failure load; specifically, the lumbar anatomy is not represented with sufficient detail or accuracy. Disc loads and facet joint reaction forces in the lumbar spine are the result of the force of gravity as well as of a multitude of ligament, muscle, and other force vectors that provide stability to the spine. We do not believe that a comprehensive mechanical perspective of such a complex system can be obtained when it is reduced to a single equivalent vector.

Model overestimates of compressive load on the joint were first recognized by Bartelink (1957), who expanded on a proposal by Keith (1923) that perhaps the

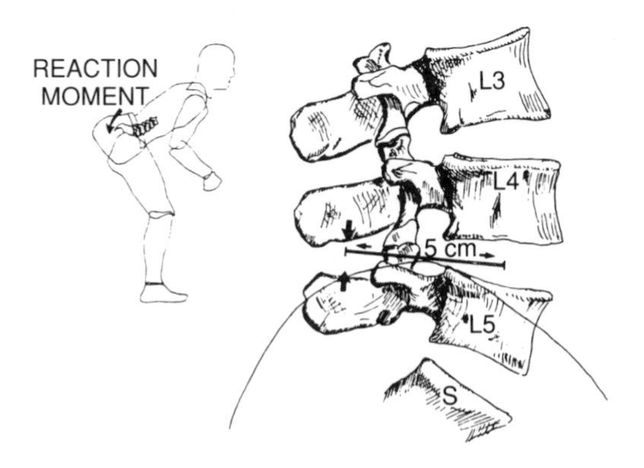

Figure 4.1 The very simple, single equivalent force vector model used to represent all of the extensor tissues to support the reaction moment of the lumbar spine. This model has been successfully used in industrial task analysis but should not be expected to provide insight into lumbar mechanics. (Reprinted by permission from McGill, 1990.)

mysteriously large predictions of compressive load from the single equivalent models could be reduced to tolerable levels by the hydraulic action of intra-abdominal pressure acting over the pelvic floor and upon the underside of the diaphragm. This proposal was mathematically operationalized by Morris, Lucas, and Bresler (1961) using simplified anatomy. They showed, theoretically, the possible dual effect of pressure to reduce disc compression directly from tensile forces on the diaphragm and indirectly from a torso extensor moment of force produced by pressure over the diaphragm. This mechanism would reduce the forces needed in the extensor tissues, and this assistance to the extensor tissues, in turn, would reduce their tensile load, hence decreasing net joint compression. Subsequently, use of intra-abdominal pressure to reduce estimates of compressive load on the spine has been widespread, and some have speculated that compression may be reduced by as much as 2,000 N (e.g., Cyron, Hutton, & Stott, 1979). However, predictions from a detailed anatomic model, and recent direct experimental evidence, have shown this claim to be extremely contentious; an argument is developed later in the chapter.

The apparent overprediction of compressive force by the single equivalent extensor model also motivated researchers to entertain other hypotheses intended to reduce the compressive load. Two thought-provoking proposals originated from Gracovetsky, Farfan, and Lamy (1977, 1981). One proposal suggested that intra-abdominal pressure exerts posterior hydraulic action on the extensor tissues, increasing the tension within them to ultimately generate extensor torque. The other proposed that abdominal forces (internal oblique and transverse abdominis) increase lateral tension in the lumbodorsal fascia, which shortens the fascia longitudinally via Poisson's effect to pull the posterior spinous processes together, resulting in extension. The crux of both proposals was that the extensor moment is generated by tissue that has the greatest mechanical advantage to minimize the compression penalty to the joint. Although attractive, these ideas were contentious and served the purpose of motivating researchers to test their viability. Scientific evidence does not appear to support these ideas, one of which has been withdrawn by the original proponents themselves. Some details of the research on this issue are presented later.

The Problem of Indeterminacy

The anatomic simplicity of the single equivalent models has been questioned for nearly a decade. As a result, more anatomically detailed models have evolved, but these models collectively suffer from mathematical indeterminacy. The number of equations of equilibrium is less than the number of unknown forces in muscles and other tissues. Therefore, a strategy must be adopted to partition the reaction moment at the level of the low back into the many components that share restorative moment-generating duties. Apart from the reductionist technique of simplifying the system into a single representative moment generator, two different partitioning strategies have been adopted for this purpose, each emanating from a different philosophy, optimization and biologically driven models.

Mathematical optimization assumes that the body's central nervous system distributes commands to the musculature to create movement in such a way as to minimize a function such as tissue stress, energy use, or fatigue. For example, the model of Gracovetsky et al. (1977, 1981) minimized intervertebral joint shear and compression; the model of Schultz et al. (Schultz, Andersson, Ortengren, Haderspeck, & Nachemson, 1982; Schultz, Haderspeck, Warwick, & Portillo, 1983) assessed loads on the spine using various strategies including minimizing muscle contraction intensity and minimizing joint compression. A major concern with the minimum compression scheme is that it cannot account for antagonist muscle co-contraction; co-contraction penalizes the spine with additional compression. Rather, tissues with the greatest moment arms, or those farthest from the spine, possess the greatest mechanical advantage and are recruited first to restore the reaction moment. Once these tissues reach a preset level of maximum stress, the next deepest force generating tissue is recruited and so on until the moment challenge is satisfied. Perhaps the use of an appropriate nonlinear cost function in the future will facilitate physiological, simultaneous recruitment of muscles. However, at present, the unequivocal choice of an optimization criterion has not been established. It is not known whether the locomotor system works to minimize joint stress, muscular work, fatigue, or some other function. In fact, the criterion may change even during the course of a simple task and is also affected by conditions such as comfort, fatigue, or presence of injury.

Despite these limitations, the Schultz et al. (1982) optimization model demonstrated reasonable correlations of predicted low back compression with directly measured intradiscal pressure, but only during a statically held position. With dynamic movement, especially complex movement, large variations in muscle co-contraction are evidenced in electromyogram (EMG) profiles. It is doubtful that a single optimization criterion could reproduce the patterns of muscle co-activation observed during both dynamic sagittal plane and complex motion. Perhaps significant developments in artificial intelligence will enable the selection of optimization criteria that are sensitive to pain, energy cost, blood pressure, and fatigue, to name a few intervening variables.

An alternate strategy (biologically driven) to partition the force-moment-generating duties to the many muscles relies on measurements of neural drive from the EMG and records of dynamic spine position. Strategic placement of electrodes makes possible monitoring of muscular co-contractions. This information, which provides temporal and amplitude information about the activation of various muscles, is lost if one of the currently available optimization procedures is selected. With appropriate processing methods, the relative contribution of an individual muscle to supporting the reaction torques is obtained from the EMG, which when coupled with spine kinematic information can provide details on muscle-ligament interplay. However, significant concerns remain with the biological method. Force estimates from EMGs remain problematic although the processing techniques of the raw waveform are improving and show promise. In addition, unless indwelling electrodes are used, deep muscle activity is unobtainable and the modeler must resort to other strategies, such as interpolation from skin surface recordings of synergists, to estimate activity. Furthermore, the biological approach rarely predicts moments about the three orthogonal axes that equal those measured. Therefore, whereas full credit is given to coactivation of muscle pairs, some errors occur in moment equilibrium for the nondominant axes.

Although both biological methods and optimization techniques are presently used to partition load-generating duties among the many tissues, future directions will demand analysis of complex dynamic motion. The EMG-based biological method is in need of basic research to better elucidate the muscle activation-force relationship. The optimization method requires development of complex algorithms that integrate mechanical and motor control variables to faithfully predict observed complex muscle patterns such as antagonist co-contraction displayed throughout the skeletal linkage. Furthermore, trial-to-trial and subject-to-subject variability in lifting tasks that have the same kinematics and reaction moment demands can be accommodated by biologically driven models but not by current optimization models.

Anatomic Anomalies of Past Models

If the purpose of a biomechanical model is to provide insight into the mechanical roles of various tissues, the model must represent the anatomic structure as closely as possible. Many inaccuracies in the description of the lumbar tissues found in the literature have been recently recognized. Because many researchers have relied solely on the literature as a source for anatomic information, these inaccuracies have been incorporated into spine models. Unfortunately, these anatomic errors have led to some erroneous conclusions regarding spinal function. The following is a list, which is by no means exhaustive, of some examples of anatomic or mechanical errors:

- Muscle areas used in some models were obtained from cadavers (e.g., Farfan, 1973). Dimensions obtained in this way are difficult to justify for use in models of healthy, young individuals because atrophy and distortion from fluids would greatly underpredict the force potential of the musculature. These underestimations of muscle area, and of force-producing potential, have been pointed out with computerized tomography (CT) scan data of younger, ambulatory adults recently presented by Nemeth and Ohlsen (1986), Reid and Costigan (1985), and McGill, Patt, and Norman (1988). An example of the muscle bulk observed in a fit 30-year-old from a transverse CT scan at the L4-L5 level can be observed in Figure 4.2.
- Muscles in the trunk do not pull in straight lines, but most models of the lumbar spine represent muscles this way. Many muscles within the trunk act around pulley systems of bone, other muscle bulk, and pressurized viscera, which alter length, force, and vector direction properties.
- Models that assume the extensor musculature can be represented by a single equivalent force vector have been introduced previously. However, cadaver dissection and CT scan measurements in our laboratory and in others cannot agree

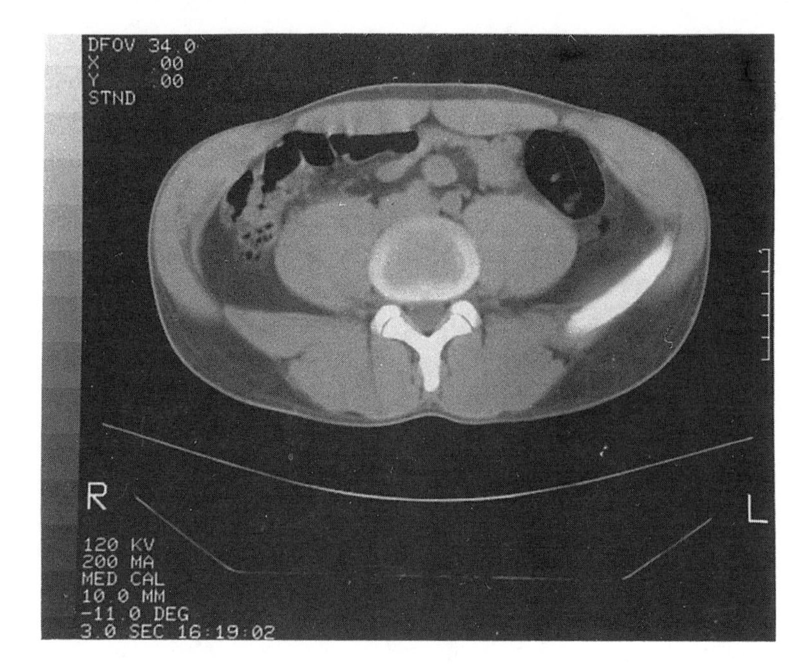

Figure 4.2 A transverse CT scan through the fourth and fifth lumbar joint (L4-L5) illustrates the magnitude and geometry of the muscle bulk of a fit 30-year-old factory worker. A very different impression would be obtained from observations of an elderly cadaver. (Reprinted by permission from McGill, 1990.)

with the anatomic or mechanical implications that result from this representation. Excellent work by Langenberg (1970) and Macintosh and Bogduk (1987) provided clear descriptions of the connections for the prime extensors of longissimus thoracis pars thoracis and pars lumborum, iliocostalis lumborum pars thoracis and pars lumborum, and multifidus. Very few of these fibers run parallel to the axis of spinal compression, demonstrating that they exert shear forces on the spine. In addition, the laminated architecture of the muscle fascicles provides for a much larger muscle cross-sectional area to contribute to extensor moment production than would be observed in a single transverse section of the abdomen. For this reason, an estimate of extensor moment potential from a single transverse scan would result in large error because only a small portion of the musculature would be measured. Bogduk (1980) recommended that any kinetic analysis must consider these muscles as a continuum of independent fibers because their action cannot be represented by a single equivalent force. The architecture of the primary extensors from a mechanical perspective in the sagittal plane is shown in Figure 4.3. The relatively large bulk of thoracic fibers (shown in Figure 4.4, a & b) is often neglected as an important contributor to extension. Thoracic fibers produce extensor forces over the full lumbar spine through a moment arm often approaching 10 cm.

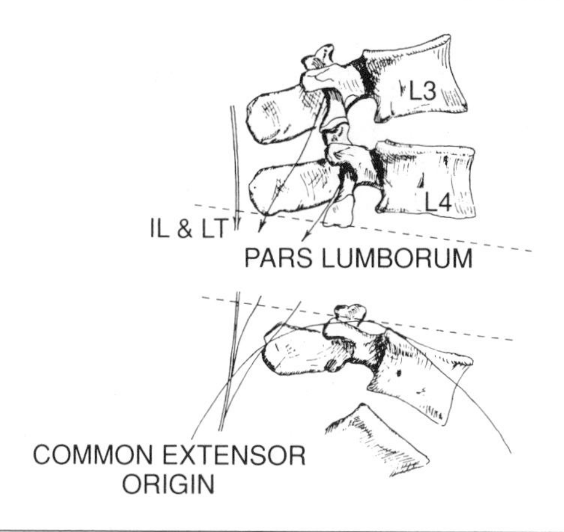

Figure 4.3 The architecture of the major extensors reveals the shearing component of the pars lumborum fibers of longissimus thoracis (LT) and iliocostalis lumborum (IL). Whereas these lumbar laminae are close to the disc (small moment arm), the tendons of the pars thoracis fibers generate extensor moments with a much greater mechanical advantage, resulting in decreased compressive force on the spine. (Reprinted by permission from McGill, 1990.)

• The extensor and torsional potential of latissimus dorsi is often neglected, yet it has the largest moment arm length of all of the posterior trunk muscles. Its association with the lumbordorsal fascia as a spine extensor is a contentious issue.

• The passive force contributions of the supraspinous and interspinous ligaments are often modeled with a single equivalent element. However, interspinous fibers run obliquely to the supraspinous ligament, thus creating nonparallel forces. In fact, the interspinous acts to generate a shear force on the joint. Farfan (1973) stated that this shear relieves facet contact force by shearing the superior vertebrae posteriorly on its inferior counterpart. However, this fiber direction, which is also depicted in any edition of *Gray's Anatomy* (e.g., Warwick & Williams, 1980), is an error, as pointed out by Heylings (1978). Instead, the interspinous has been shown to contribute significant anterior shear forces that increase facet load during large degrees of flexion (McGill, 1988; Shirazi-Adl & Drouin, 1987).

• The diaphragm area and shape are fundamental to the calculation of the potential assistance provided by intra-abdominal pressure. It is suspected that models have grossly overestimated the size of diaphragms (sizes as large as 465 cm^2 have been used); the normal surface area on which pressure is exerted is probably closer to 243 cm^2 (McGill & Norman, 1987), 276 cm^2 (Schultz et al., 1982), or 299 cm^2 (Troup, Leskinen, Stalhammear, & Kuorinka, 1983).

• The psoas complex has long been considered, and described in the textbooks, as a flexor of the lumbar spine. Recent work has shown that although it is a flexor of the hip, its lever arm to flex the spine is extremely limited (McGill et al., 1988; Thorstensson, Andersson, & Cresswell, 1989) except at the lumbosacral joint.

a

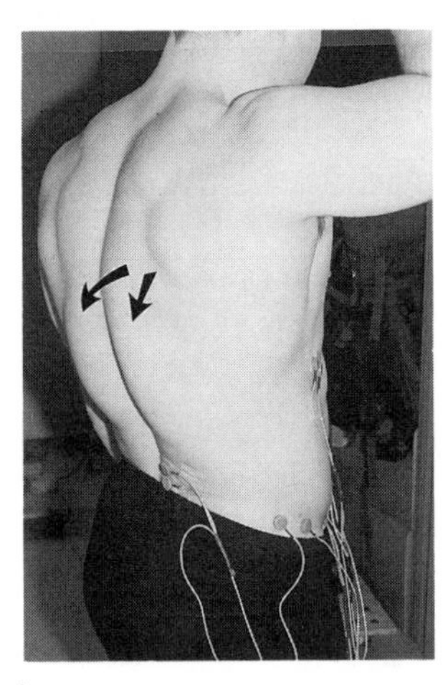

b

Figure 4.4 A bundle of fibers of longissimus thoracis has been dissected and the tendon isolated to show the insertion on the sixth and seventh thoracic (T6 and T7) ribs and sacral origin (a). Hence, these muscles create an extensor moment over the full length of the lumbar spine and minimize the compressive penalty due to their mechanical advantage. Longissimus thoracis bulk in a developed weight lifter (b). (Reprinted by permission from McGill, 1990.)

A Revised Model of the Lumbar Spine

We have attempted to incorporate detail in the anatomic structure and neurophysiological functional representation of muscle activation in a revised model of the lumbar spine. Anatomic data were obtained from cadaver dissections, CT and

MRI scans, and data-based anatomical literature. Neurological activation of muscles was monitored using EMG rather than assuming an optimization objective function to drive the model. This approach was necessary so that the muscle co-contractions and subtle trial-to-trial changes in muscle utilization, so often observed in subjects, could be accounted for. Also, previous research had shown that it was important in the model to accommodate muscle-ligament and other passive tissue interplay in the support of moments of force during load handling (e.g., Gracovetsky et al., 1977). For these reasons, we decided that the only way that objectives could be satisfied with current technology was to implement an internal load partitioning approach using biological signals; EMG and spine passive tissue loads are functions of the relative position of the vertebrae, so a scheme for monitoring the relative motion of the pelvis and the thorax was also devised.

The purpose of the model was to allow estimation of forces on individual muscles, ligaments, lumbodorsal fascia, discs, and vertebrae during dynamic load-handling tasks. The model has evolved from the ability to assess planar movements only to its current capability of analyzing full three-dimensional coupled motion.

The lifting guidelines listed in the third part of this chapter are based on interpretations of tissue loading estimated from this model and on other research in the scientific literature. Therefore, in this chapter we describe briefly some of the details of the model. For equations and more detail of the planar version of the model, refer to McGill and Norman (1986), and for the three-dimensional version refer to McGill (1992).

Estimation of the Low Back Reaction Moments About Three Orthopedic Axes. The condition of dynamic equilibrium demands that a low back reaction moment (about any axis) be calculated, as with any low back model, which is supported or restored by internal moment-generating tissues (muscle, ligament, and the disc in bending). The L4-L5 extensor reaction moments and restorative forces are shown in Figure 4.5. The reaction moment is calculated from a dynamic linked segment representation of the body that includes hands, forearms, upper arms, head-neck, thorax-abdomen, pelvis, thighs, legs, and feet and uses displacement coordinate data of the joints and forces on the hands as input (see McGill & Norman, 1985). Joint displacements are recorded on video at 30 Hz using two or more cameras to reconstruct the joint centers in three dimensions.

Skeletal Description. The three-dimensional skeleton comprised of a pelvis, rib cage, and five lumbar vertebrae was compiled from archived radiologic records and corresponded to a man of 50th-percentile dimensions as defined by Dreyfuss (1966). Features such as the intercrestal line intersecting the lower portion of the L4 body were matched to coincide with the data of MacGibbon and Farfan (1979), who used 553 living subjects. Orientation of the skeletal components was obtained from the relative rotation and translation of the rib cage with respect to the pelvis. Lumbar vertebrae were orientated between these two assumed rigid structures from which the amount of lordosis in the spine could be measured. This basic

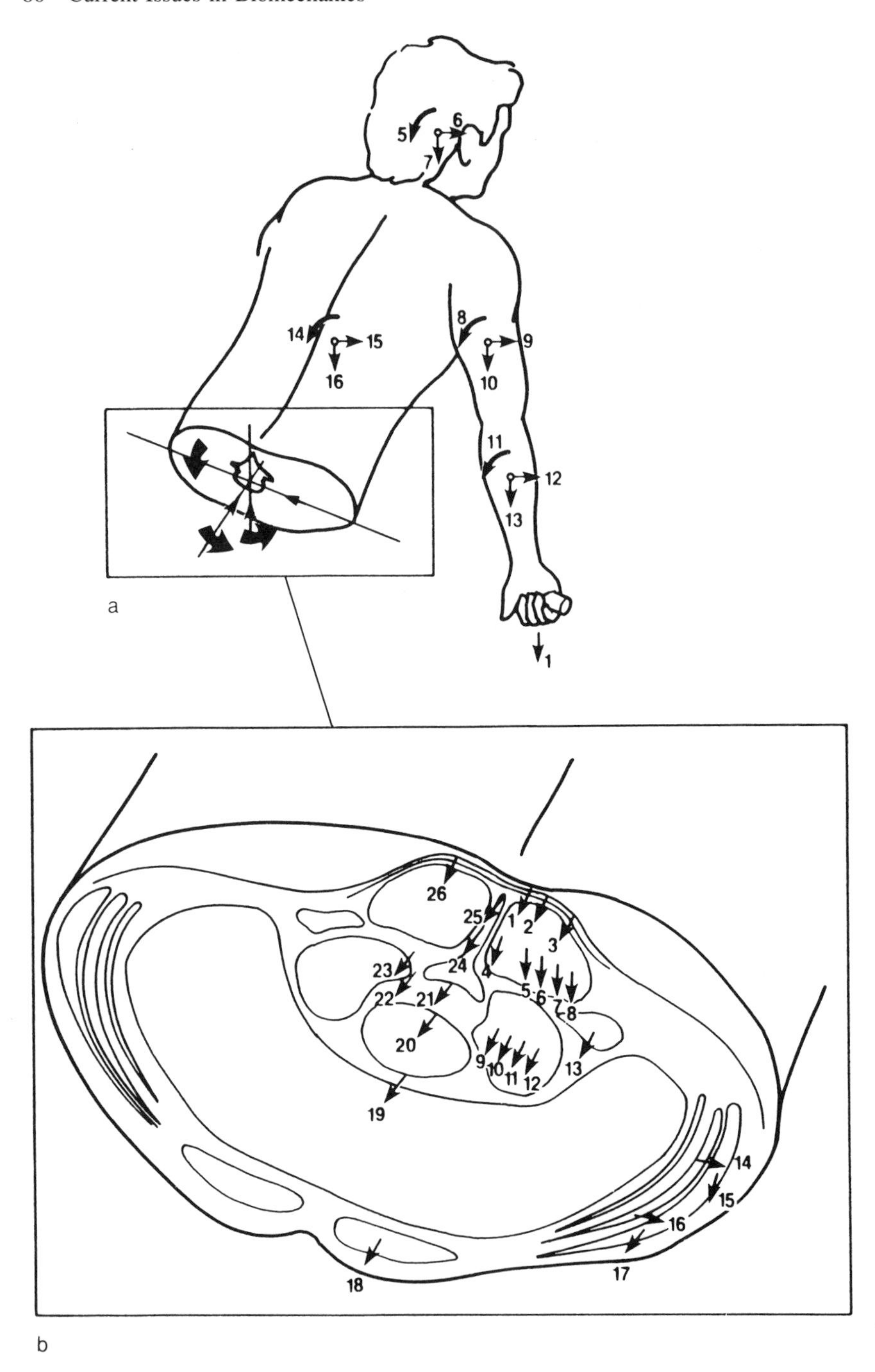

Figure 4.5 The reaction moments about three axes through L4-L5 (a) are determined from a rigid linked segment model and are partitioned into restorative components provided by muscles (forces 1-18), ligaments (forces 19-26), and the disc in bending (b).

approach to skeletal orientation in three dimensions was adapted from the two-dimensional method developed and used by Anderson, Chaffin, Herrin, and Matthews (1985).

The intervertebral discs were modeled in the direction of the compressive axis as nonlinear, elastic elements to correspond to the average of a data envelope reported by Markolf and Morris (1974).

Muscles and ligament attachment coordinates moved with the appropriate skeletal attachment points. Some muscles do not pull along a straight line between the origin and insertion. Consequently, a length and force vector correction feature was incorporated that represented the line of force, in the curved muscles, as an arc.

Components of the Restorative Moment. Because disc and ligamentous strain depends on the amount of flexion-extension, axial twist, and lateral bend, these contributions to the restorative moments were determined first in the algorithm. The remaining difference between the passive tissue moments and reaction moments were then allocated to the musculature. The disc resistance to flexion in the sagittal plane was assumed to be satisfactorily represented by the exponential equation of Anderson et al. (1985), and the data of Schultz, Warwick, Berkson, and Nachemson (1979) were adapted for lateral bending and axial twist. Ligament forces (the vectors are shown in Figure 4.6) were determined from a collection of individual ligament stress-strain data, the details of which are published elsewhere (McGill, 1988, and Cholewicki and McGill, 1992).

Muscular Contributions to Moment Generation. The task to partition the total muscular moment into the appropriate forces was aided by EMG, knowledge of geometric attachments, pennation, instantaneous muscle length and velocity, moment arms, and relationships with passive tissues. Estimates of neural activation to drive the musculature were derived from 12 EMG electrode locations

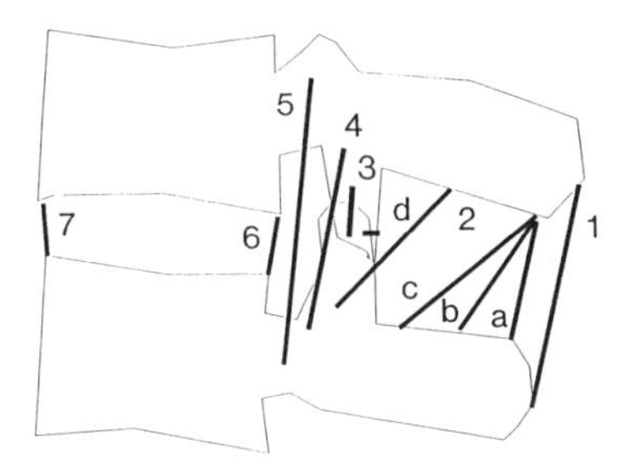

Figure 4.6 The ligament force vectors acting on the vertebrae in the sagittal plane impose both compression and shearing forces on the joint when strained. (1) Supraspinous; (2 a,b,c,d) four representative fibers of interspinous; (3) two representative fibers of the facet capsular ligaments; (4) ligament flavum; (5) intertransverse; (6) posterior longitudinal; (7) anterior longitudinal. *Note:* ligaments three and five are three-dimensional structures. (Reprinted by permission from Cholewicki & McGill, 1992.)

(6 on each side of the trunk). The EMG umbilical cord may be observed in Figure 4.7, which shows a subject lifting a laterally placed but vertically directed load. EMG was used only to guide the partitioning of the remaining restorative moment into the respective muscular components rather than to estimate absolute muscle force directly. This was done in the following way: Raw EMG was full-wave rectified and low-pass filtered at a filter cutoff value of 3 Hz (single pass to generate a time-phase lag representing electromechanical delay). This produced a rise time of 53 ms, which was consistent with the 30- to 90-ms contraction times reported by Buchthal and Schmalbruch (1970). Each raw muscle force was estimated by multiplying its physiological cross-sectional area by a force producing potential of 35 N/cm^2, and the instantaneous level of neural activation was normalized to the electrical output observed during a maximum voluntary isometric contraction. Each muscle force was modulated by coefficients reflecting instantaneous muscle length, velocity, and type of contraction (isometric, concentric, or eccentric). An activation-independent term for muscle passive elasticity based on instantaneous length was added to the modulated force. These processed muscle forces were applied to the skeleton, and the moments were calculated and summed. The sum of muscle moments was forced to equal the largest reaction moment (extensor, lateral bend, or axial twist) by multiplying each muscle force by a common gain factor. In lifting, the largest moment is nearly always the extensor moment. This gain term does not alter the relative contribution of each muscle in moment production. Thus, it must be emphasized that this EMG-based approach facilitates the partitioning of moment restoration duties among all muscles of the trunk but does not require that individual muscle forces be determined solely from EMG information.

Interpretation of Muscle Forces. No simple explanation can satisfactorily describe how a complex mechanical structure such as the spine can successfully respond to the load demands of daily living. Only knowledge of the time histories of the loads on the many individual component tissues enables evaluation of the many controversies associated with spinal mechanics.

This model has been criticized because it sometimes predicts that ligaments are not shown to be significant load-bearing structures. Most subjects that we studied did not perform lifts with the lumbar spine fully flexed; therefore, the posterior ligaments did not generate a significant extensor moment of force. This placed the full burden of moment generation on the musculature. Farfan (1973) and Gracovetsky (1986) have argued that the musculature is too small to supply the forces necessary to support the needed extensor moments; they maintain that ligaments must bear this supporting responsibility.

A muscle's ability to generate a moment is determined by the force and the perpendicular distance, or moment arm, to the fulcrum (assumed to be the center of the disc in this case). Our anatomic studies, and those of others, have shown that the moment arms of many of the low back extensors can approach 10 cm in a healthy worker (McGill et al., 1988). Farfan (1973) quoted muscle cross-sectional area data derived from only one cadaver (Eycleshymer & Shoemaker,

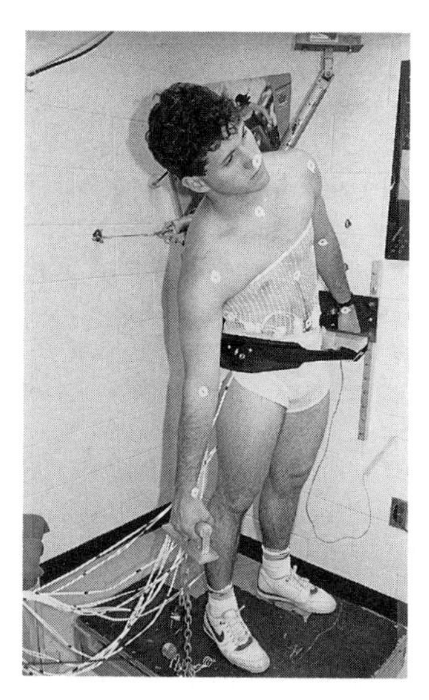

Figure 4.7 A subject lifting a laterally placed load in a lateral bending movement. The instrumented load provides the dynamic force-time history of forces applied to the body. A surgical stocking is worn to hold down EMG cables on the trunk. A linked segment representation, constructed from the joint markers, is used to estimate the lumbar reaction moments. These moments are subsequently partitioned among the supporting tissues for interpretation and analysis. (Reprinted by permission from Cholewicki & McGill, 1992.)

1911). Several CT scan studies have demonstrated much larger muscle areas of younger men who have not experienced the atrophy associated with inactivity of the elderly (e.g., McGill et al., 1988; Reid & Costigan 1985). If these areas are considered and credit is given to the full extent of the musculature that can contribute to lumbar extension, then the estimates of moments generated exclusively from muscular sources increase significantly. The muscular bulk of the thoracic components of iliocostalis lumborum and longissimus thoracis is not visible in a lumbar section, but it must not be neglected in an estimate of the moment potential of the musculature. If we assume that muscle, conservatively, can produce force at 35 N/cm^2, the average total extensor moment potential of the extensor musculature approaches 400 N·m in healthy working men. This is close to maximum back extension moments experimentally measured by Troup and Chapman (1969). Although moments of this magnitude are rare in daily activity, it would be unfair to suggest that there must be other dominant sources of extensor moment when full credit has not been given to muscular sources.

The Role of Muscles in the Controlling Joint Compression and Shear Forces. The role of the musculature in dramatically increasing disc compression in some cases and reducing it in others—and, in particular, in reducing the size of the shear force that facet joints have to support—has been underemphasized in discussions of injury mechanisms. The components of muscular moment generation are detailed in Table 4.1 for the period of peak loading in a sample sagittal plane squat lift of 27 kg that was maximally accelerated upward and produced a reaction moment in the low back of 450 N·m. The individual muscle forces, three-dimensional joint moments, and components of compression and shear (both anterior-posterior and lateral) that are imposed on the joint are very useful information. The very large magnitude of force in the pars lumborum fascicles results from the large individual cross-sectional area. These forces produce a large proportion of the extensor moment. Negative flexion-extension moments observed in Table 4.1 correspond to the flexor contributions of abdominal co-contraction. The abdominal co-contraction in this lifting example, and in most sagittal plane lifting tasks, was small at the instant of peak extensor moment.

The compression penalty from even mild abdominal activity can be observed from the column of individual muscle forces. Moreover, to meet the requirement of the net moment, additional extensor activity is necessary to offset the flexor moment produced by the abdominals. However, this creates a double contribution to joint compression: compression from abdominal activity together with compression from the additional extensor forces. Even so, when all the component forces are summed, the total predicted joint compression is less than what would have been predicted by the simple single equivalent muscle model introduced earlier.

We have observed a varying ability of subjects to reduce compression on the spine for a given extensor task that has ranged from 5% to 25% of that predicted by the 5-cm single equivalent model in sagittal plane lifts. In fact, in the complete absence of abdominal activity, a single equivalent model with a moment arm of 7.5 cm often predicts values of compression comparable with our anatomically detailed dynamic lumbar model. In most lifting cases, this reduction is large enough to reduce compression to subfailure levels of the vertebral end plate so that hypothesized, but contentious, compression-reducing mechanisms such as intra-abdominal pressure or lumbodorsal fascia activation are not necessary to produce realistic predictions from the model. An individual's ability to reduce compression appears to be determined by the degree to which he or she can reduce abdominal activity. Obviously, abdominal activity would result in a shorter equivalent moment arm. The time courses of some selected muscle forces for a sample lift are shown in Figure 4.8, a, b, and c. The EMG profiles of male subjects generating axial and lateral bend torque demonstrate the large amounts of co-contraction always observed in this type of effort (see Figure 4.9, a & b, and Figure 4.10, a-f). However, as is often observed in elite lifters, the abdominals are not completely silent and exhibit varying degrees of activity. This suggests that they are sacrificing the minimizing of compression for some other, as yet unknown, benefit. Interviews with some elite lifters as to why co-contraction is observed often reveal that the lifters feel that it stiffens the trunk to prevent buckling of the spine. This idea is addressed later in the discussion on intra-abdominal pressure.

Table 4.1 Components for Musculature Moment Generation of 450 N · m of Extension During Peak Loading for Squat Lift of 27 kg

Muscle	Force (N)	Moment (N · m)			Compression (N)	Shear (N)	
		Flexion-extension	Lateral	Twist		Anterior-posterior	Lateral
R rectus abdominis	18	−2	1	0	18	4	−3
L rectus abdominis	18	−2	−1	−0	18	4	3
R external oblique 1	51	−2	3	4	36	36	−12
L external oblique 1	51	−2	−3	−4	36	36	12
R external oblique 2	37	−1	3	1	33	12	−13
L external oblique 2	37	−1	−3	−1	33	12	13
R internal oblique 1	14	0	1	−1	10	−7	8
L internal oblique 1	14	0	−1	1	10	−7	−8
R internal oblique 2	11	−1	1	−1	7	−4	8
L internal oblique 2	11	−1	−1	−1	7	−4	−8
R longissimus thoracis pars lumborum L4	678	26	27	−6	602	−235	152
L longissimus thoracis pars lumborum L4	678	26	−27	6	602	−235	−152
R longissimus thoracis pars lumborum L3	616	32	26	2	578	−114	151
L longissimus thoracis pars lumborum L3	616	32	−26	−2	578	−114	−151
R longissimus thoracis pars lumborum L2	564	44	29	10	552	132	60
L longissimus thoracis pars lumborum L2	564	44	−29	−10	552	132	−60
R longissimus thoracis pars lumborum L1	507	40	26	9	496	119	54
L longissimus thoracis pars lumborum L1	507	40	−26	−9	496	119	−54
R iliocostalis lumborum pars thoracis	281	23	19	1	280	40	−7
L iliocostalis lumborum pars thoracis	281	23	−19	−1	280	40	7

(continued)

Table 4.1 ***(continued)***

Muscle	Force (N)	Moment (N · m)			Compression (N)	Shear (N)	
		Flexion-extension	Lateral	Twist		Anterior-posterior	Lateral
R longissimus thoracis pars thoracis	375	30	15	−1	372	54	−32
L longissimus thoracis pars thoracis	375	30	−15	1	372	54	32
R quadratus lumborum	283	10	22	5	278	68	14
L quadratus lumborum	283	10	−22	−5	278	68	−14
R latissimus dorsi L5	153	9	8	−4	137	12	−68
L latissimus dorsi L5	153	9	−8	4	137	12	68
R multifudus 1	158	8	4	3	145	22	60
L multifudus 1	158	8	−4	−3	145	22	−60
R multifudus 2	158	8	4	0	153	50	0
L multifudus 2	158	8	−4	−0	153	50	0
R psoas L1	9	−0	1	0	8	5	2
L psoas L1	9	−0	−1	−0	8	5	−2
R psoas L2	9	−0	1	0	8	5	2
L psoas L2	9	−0	−1	−0	8	5	−2
R psoas L3	9	−0	0	0	7	5	2
L psoas L3	9	−0	−0	−0	7	5	−2
R psoas L4	9	0	0	0	7	6	3
L psoas L4	9	0	−0	−0	7	6	−3

Note. Negative flexion-extension moments correspond to flexion, and negative A-P shear corresponds to L4 shearing posteriorly on L5.

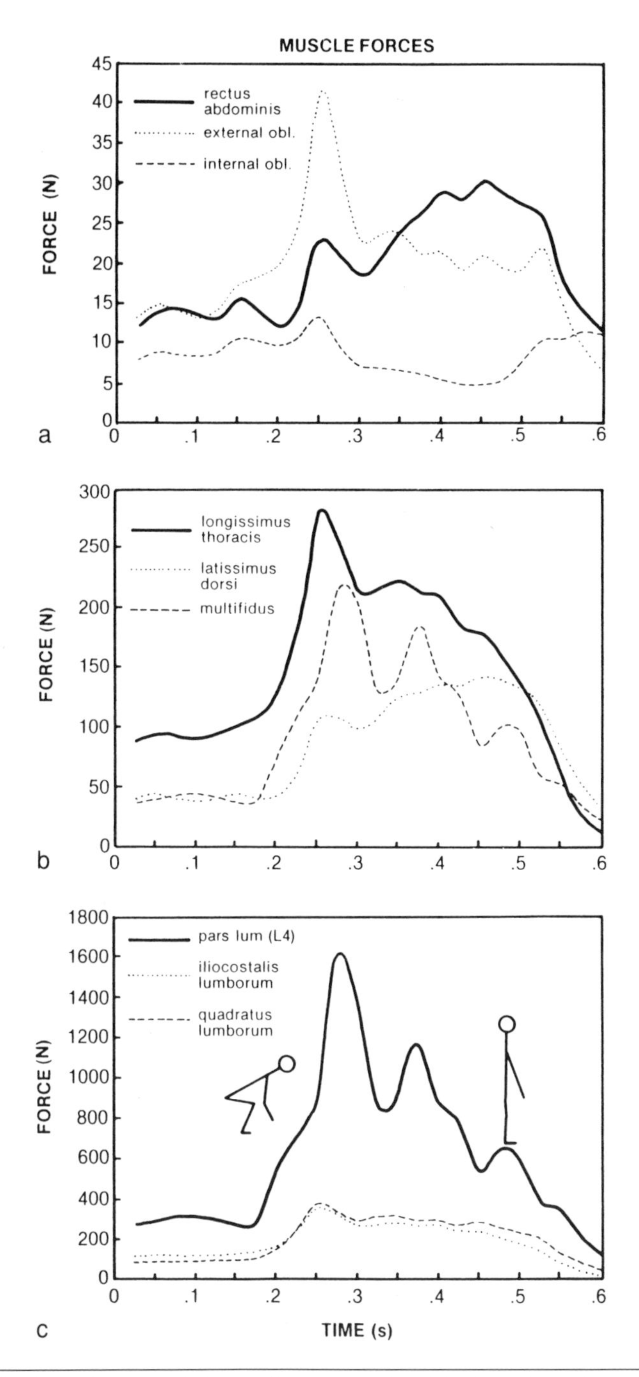

Figure 4.8 Individual muscle forces from a sample trial of a subject lifting 27 kg quite quickly in the sagittal plane. Muscle forces are shown for rectus abdominis, external oblique, and internal oblique (a); longissimus thoracis, latissimus dorsi, and multifidus (b); and pars lumborum, iliocostalis lumborum, and quadratus lumborum. (Reprinted by permission from McGill, 1990.)

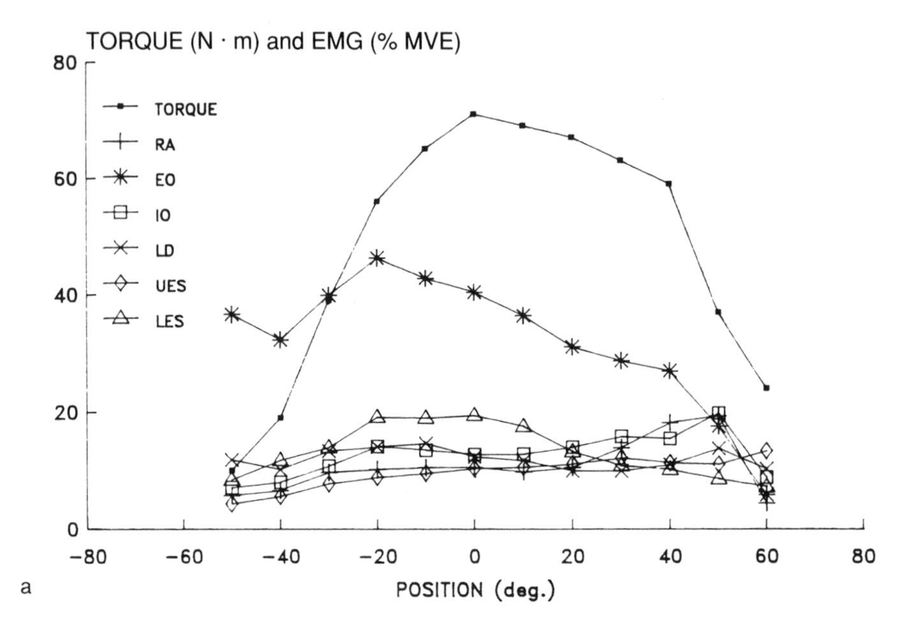

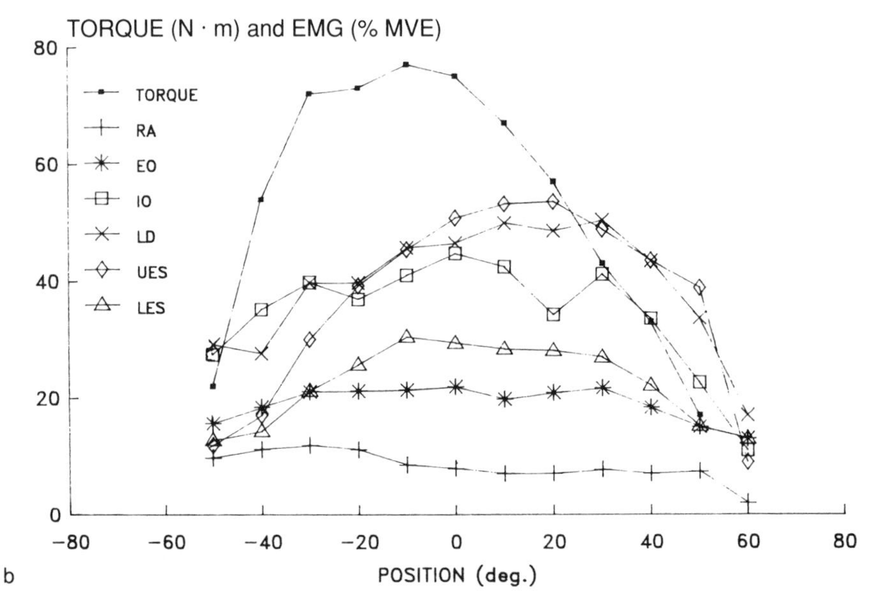

Figure 4.9 EMG profiles (averages) from the right side of the trunk of 10 male subjects generating only axial torque, twisting clockwise at a velocity of 300/second (a), and twisting counter-clockwise at the same velocity (b). The activity levels indicate co-activation of bilateral muscle pairs during the generation of twisting torque (RA = rectus abdominis, EO = external oblique, IO = internal oblique, LD = lattisimus dorsi, UES = upper erector spinae-T9, LES = lower erector spinae [L3], MVC = maximum voluntary effort). (Reprinted by permission from McGill, 1991.)

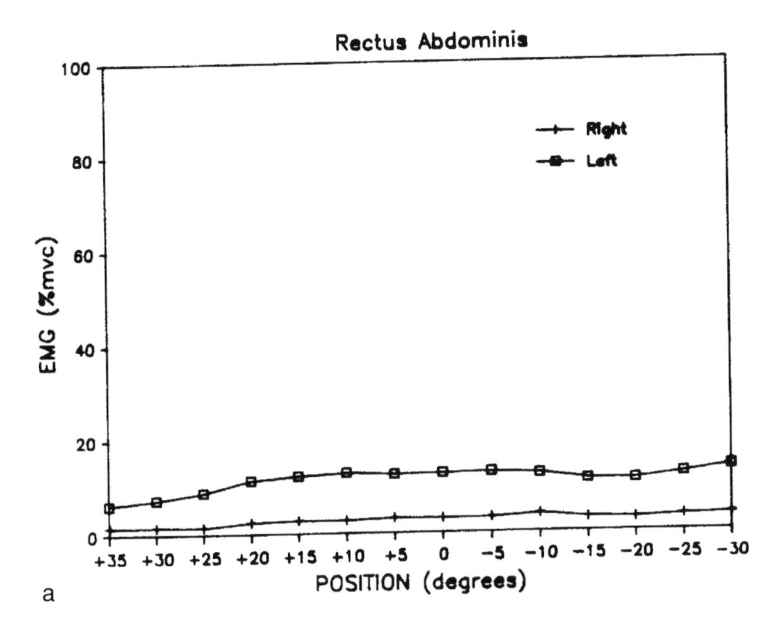

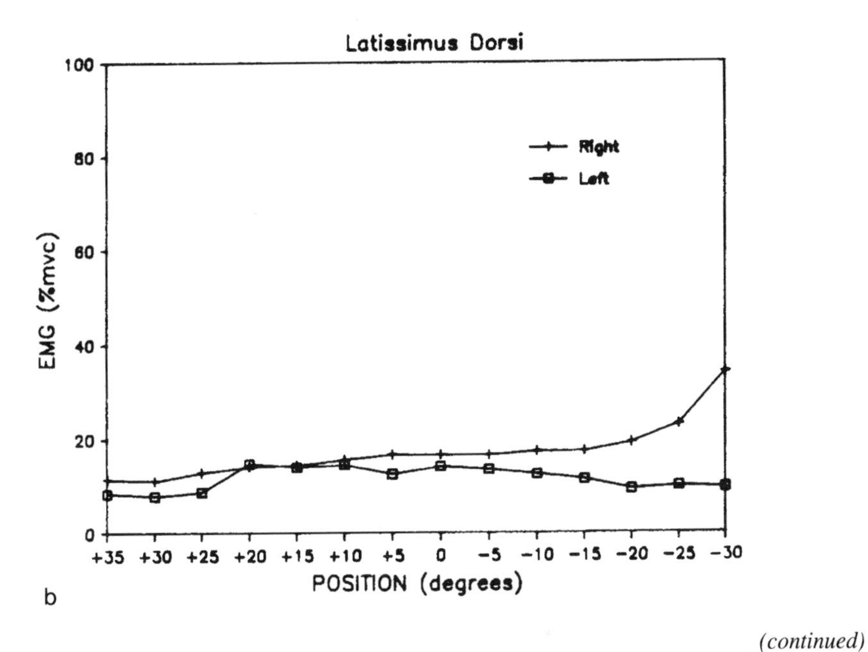

(continued)

Figure 4.10 EMG profiles (averages) of 11 male subjects generating only lateral bend. The load was held in the right hand while the subjects lifted the load and bent to the left. Profiles are shown for rectus abdominis (a), latissimus dorsi (b), external oblique (c), thoracic erector spinae (d), internal oblique (e), and lumbar erector spinae (f) (MVC = maximum voluntary effort).

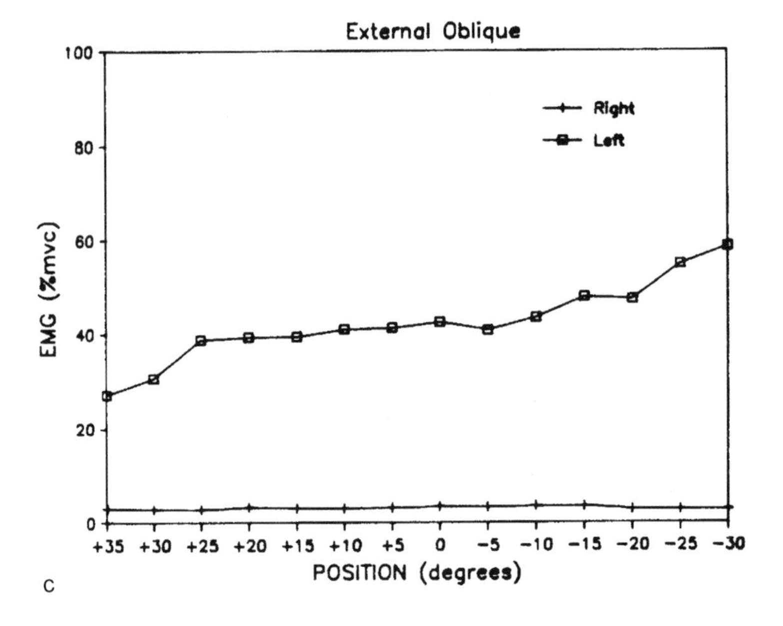

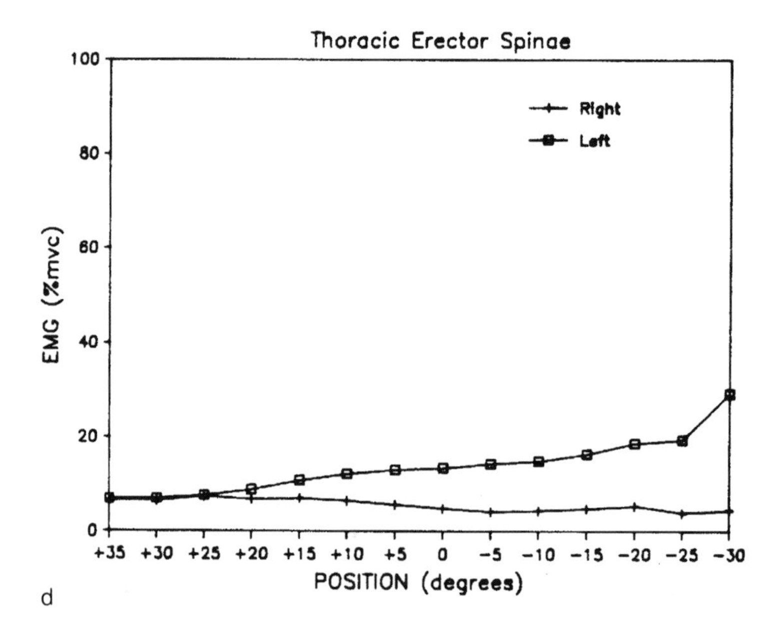

Figure 4.10 *(continued)*

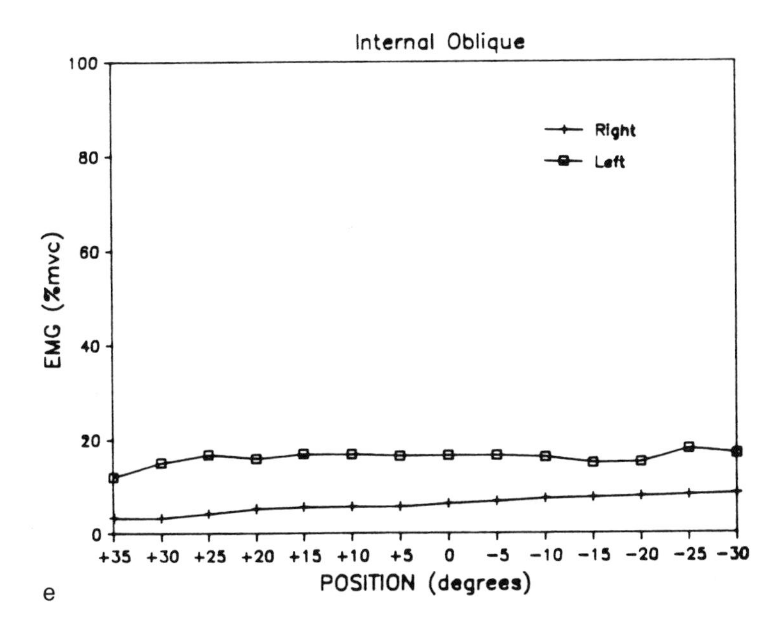

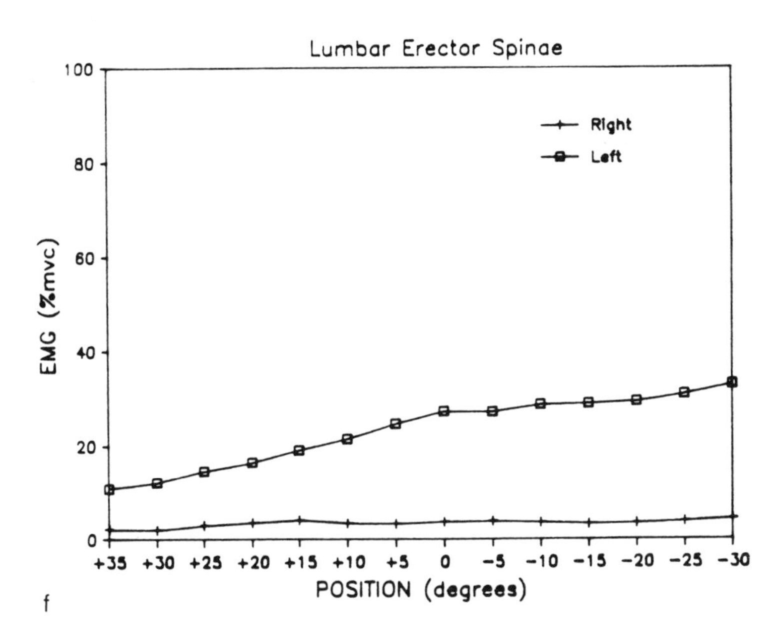

Figure 4.10 EMG profiles (averages) of 11 male subjects generating only lateral bend. The load was held in the right hand while the subjects lifted the load and bent to the left. Profiles are shown for rectus abdominis (a), latissimus dorsi (b), external oblique (c), thoracic erector spinae (d), internal oblique (e), and lumbar erector spinae (f) (MVC = maximum voluntary effort).

Negative anterior-posterior shear forces from the muscles (shown in Table 4.1) correspond to L4 shearing posteriorly on L5. Hence, a very powerful anti-anterior shear mechanism is observed in the tabulated forces due to the obliquity of the pars lumborum extensors. These muscles help to offset the anterior shear force caused by the upper body and load weight and their accelerations and also contribute to the extensor moment. These muscle forces reduce load on the facet joints and discs. Some subjects that we tested offset the reaction shear force almost completely (Potvin, McGill & Norman, 1991), depending on the forward inclination of the disc (and trunk) and on the magnitude of force in these obliquely orientated pars lumborum muscle fibers, which depends on the degree of activation of the muscles.

Many current biomechanical models of the lumbar spine represent the extensor muscles as a vector or vectors, perpendicular to the disc shear plane (parallel to the compression vector). This representation does not account for the load shear force reducing capability of the actual lines of force of the musculature. Thus, any predictions of shear from these models are only estimates of the shear produced by the body and load weights and their accelerations. They are not estimates of the size of the shear load that is supported by the facets or discs which are much lower.

The effects of lumbar spine bending and twisting on the changing alignment of muscle vectors with the orthopedic axis cause variations in the moment-producing role of some muscles, together with compressive and shearing components on the joint (McGill, 1991). For example, the effect of axial twist and lateral bending on the ability of rectus abdominis to create moments from differing lines of action is shown in Figure 4.11 (see McGill, 1991a). Muscle lengths, moment potential, and compressive and shear force contributions to the joint during some selected postures are shown in Tables 4.2, 4.3, and 4.4. The orthopedic convention is shown in Figure 4.12. This evidence demonstrates the importance of including individual muscle vectors together with posture-specific orthopedic axes for three-dimensional analyses of lumbar joint load. Furthermore, it is interesting to note the magnitude of the compressive cost imposed by the co-contracting musculature on the joint during the support of axial twist and lateral bending moments. Assuming average muscle activations (subject averages), it appears that the generation of 50 N·m of extensor torque places approximately 800 N of compression on the joint, 50 N·m of lateral bend places 1,400 N on the joint, and 50 N·m of axial twisting torque places approximately 2,500 N of compression on the joint. Capturing the forces in the co-contracting musculature in three-dimensional tasks is extremely important.

Issues in Low Back Mechanics Specific to Lifting

In the previous section a case was made for the need to represent the musculature and ligaments as accurately as possible in spine models. Such models have enabled reassessment of many mechanical issues that pertain to spine function.

Table 4.2 Muscle Lengths in cm (Including Tendon Length) Obtained From the Upright Standing Posture and From Various Extreme Postures

	Upright standing	60° flexion	25° lateral bending	10° twist	Combined[a]
R rectus abdominis	30.1	19.8[b]	27.4	30.0	17.7
L rectus abdominis	30.1	19.8[b]	32.6	30.5	22.9[b]
R external oblique 1	16.7	14.4	12.6[b]	15.5	13.3
L external oblique 1	16.7	14.4	21.3[b]	18.0	19.2
R external oblique 2	15.8	12.9	10.7[b]	15.6	9.5[b]
L external oblique 2	15.8	12.9	21.0[b]	16.2	18.2
R internal oblique 1	11.1	9.9	9.1	12.2	7.6[b]
L internal oblique 1	11.1	9.9	14.1[b]	10.0	12.2
R internal oblique 2	10.3	8.8	7.6[b]	11.2	6.9[b]
L internal oblique 2	10.3	8.8	13.2[b]	9.5	10.7
R psoas lumborum (L1)	15.1	20.0[b]	14.2	15.4	18.7[b]
L psoas lumborum (L1)	15.1	20.0	16.8	15.7	20.7[b]
R psoas lumborum (L2)	12.5	16.4[b]	11.8	12.9	15.6[b]
L psoas lumborum (L2)	12.5	16.4[b]	13.8	13.0	17.0[b]
R psoas lumborum (L3)	9.9	12.9[b]	9.5	10.0	12.5[b]
L psoas lumborum (L3)	9.9	12.9[b]	10.9	9.8	13.3[b]
R psoas lumborum (L4)	7.5	9.4[b]	7.4	7.6	9.3[b]
L psoas lumborum (L4)	7.5	9.4[b]	8.0	7.4	9.5[b]
R iliocostalis lumborum	23.0	29.7[b]	18.9	23.5	26.4
L iliocostalis lumborum	23.0	29.7[b]	25.5	22.8	31.8[b]
R longissimus thoracis	27.5	33.7[b]	25.4	27.6	31.1

(continued)

Table 4.2 ***(continued)***

	Upright standing	60° flexion	25° lateral bending	10° twist	Combined[a]
L longissimus thoracis	27.5	33.7[b]	28.8	27.4	34.8[b]
R quadratus lumborum	14.6	18.2[b]	11.9	14.4	14.9
L quadratus lumborum	14.6	18.2	17.4	15.1	20.9[b]
R latissimus dorsi (L5)	29.4	32.1	26.8	29.8	29.0
L latissimus dorsi (L5)	29.4	32.1	31.5	29.1	34.6
R multifidus 1	5.3	7.3[b]	5.2	5.1	7.1[b]
L multifidus 1	5.3	7.3[b]	5.4	5.5	7.5[b]
R mutlifidus 2	5.1	7.2[b]	5.0	5.1	7.01[b]
L multifidus 2	5.1	7.2[b]	5.2	5.1	7.2
R psoas (L1)	29.2	28.6	28.1	29.0	27.2
L psoas (L1)	29.2	28.6	30.2	29.5	29.6
R psoas (L2)	25.8	25.3	25.1	25.7	24.4
L psoas (L2)	25.8	25.3	26.5	26.0	26.1
R psoas (L3)	22.1	21.8	21.7	22.0	21.2
L Psoas (L3)	22.1	21.8	22.5	22.2	22.3
R psoas (L4)	18.7	18.6	18.6	18.7	18.4
L psoas (L4)	18.7	18.6	18.9	18.8	18.9

[a]Combinations of 60° flexion, 25° right lateral bend, 10° counterclockwise twist.

[b]Muscle lengths that differ by more than 20% from those obtained during upright standing.

Note. Reprinted by permission from ''Kinetic Potential of the Lumbar Trunk Musculature About Three Orthogonal Orthopaedic Axes in Extreme Positions,'' by S.M. McGill, 1991, *SPINE*, **16**(7), pp. 809-815.

Table 4.3 Moment Potential (N • m) of Some Representative Muscles in Various Postures

Muscle	Force (N)	Upright standing			60° flexion			25° lateral bending			10° twist			Combined[a]		
		F[b]	B[c]	T[d]	F	B	T	F	B	T	F	B	T	F	B	T
R rectus abdominis	350	–28	17	6	–35	17	12	–28	21	8	–29	15	4	–33	20	15
R external oblique (1)	315	–7	27	9	–7	24	15	–7	28	11	–8	28	6	–1	20	22
R internal oblique (1)	280	–15	19	–16	–9	8	–31	–13	18	–20	–14	17	–18	–2	2	–33
R pars lumborum (L1)	455	36	23	9	35	23	8	36	24	9	3	24	5	38	23	7
R pars lumborum (L4)	595	23	23	–7	23	25	–4	23	23	–8	23	22	–8	23	25	–6
R iliocostalis lumborum	210	18	15	–1	18	14	2	18	15	–1	18	15	–2	18	15	1
R longissimus thoracis	280	24	10	–3	22	10	0	24	13	–3	24	12	–5	24	11	–1
R quadratus lumborum	175	5	11	5	7	14	1	5	12	6	6	13	4	7	14	0
R latissimus dorsi (L5)	140	9	5	–3	8	7	–4	8	6	–5	9	5	–5	8	7	–6
R multifidus	98	5	1	3	5	1	2	5	1	3	5	1	2	5	1	2
R psoas (L1)	154	1	9	9	–5	9	2	1	9	4	1	6	3	–4	8	4

Note. Only a portion of some muscles are represented; for example, the laminae of psoas to L1 is shown rather than the whole psoas. (Reprinted by permission from "Kinetic Potential of the Lumbar Trunk Musculature About Three Orthogonal Orthopaedic Axes in Extreme Positions," by S.M. McGill, 1991, *SPINE*, **16**(7), pp. 809-815.)

[a]Combination of 60° flexion, 25° right lateral bending, and 10° counterclockwise twist in one posture.
[b]F = flexion.
[c]B = lateral bend.
[d]T = axial twist.

Table 4.4 Resultant Compressive, Lateral Shear, and Anterior-Posterior Shear Forces From Representative Muscles (Shown in Table 4.3) Imposed on the L4/L5 Joint (N)

Muscle	Maximum force (N)	Upright standing			60° flexion			25° R lateral bending			10° twist			Combined		
		C[a]	Lat[b]	A-P[c]	C	Lat	A-P	C	Lat	A-P	C	Lat	A-P	C	Lat	A-P
R rectus abdominis	350	345	–46	72	335	–70	110	341	–47	70	347	–28	62	323	–64	130
R external oblique (1)	315	290	–102	87	265	–118	139	279	–112	92	300	–91	50	207	–106	215
R internal oblique (1)	280	210	180	–28	96	225	–130	192	198	–56	198	186	–51	24	223	–159
R pars lumborum (L1)	455	446	69	89	444	41	118	445	94	90	451	17	88	446	36	117
R pars lumborum (L4)	595	486	153	–275	546	124	–173	475	169	–293	482	110	–295	537	119	–197
R iliocostalis lumborum	210	207	–7	7	206	–5	40	210	1	7	207	–26	8	208	–5	40
R longissimus thoracis	280	276	–32	57	275	–17	54	279	–22	11	276	–59	12	276	–21	54
R quadratus lumborum	175	158	10	84	175	8	16	153	–13	93	164	0	72	173	–18	27
R latissimus dorsi (L5)	140	118	–10	84	124	–56	2	98	–97	22	116	–78	14	114	–78	11
R multifidus	98	86	46	9	91	34	16	85	50	9	89	40	6	93	34	15
R psoas (L1)	154	135	33	76	132	34	79	134	34	78	136	33	73	132	36	81

Note. Negative lateral shear corresponds to L4 shearing to the left on L5 and negative A-P shear corresponds to L4 shearing posteriorly on L5. (Reprinted by permission from "Kinetic Potential of the Lumbar Trunk Musculature About Three Orthogonal Orthopaedic Axes in Extreme Positions," by S.M. McGill, 1991, *SPINE*, **16**(7), pp. 809-815.)

[a]C = compressive forces.
[b]Lat = lateral shear forces.
[c]A-P = anterior-posterior shear forces.

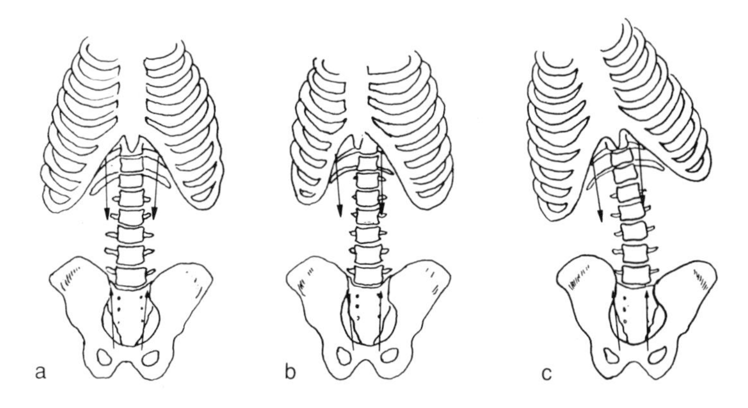

Figure 4.11 The line of action of a muscle changes as a function of lumbar position. Rectus abdominis (a) in this case changes its ability to produce torques about different axes as a function of axial twist (b) and lateral bend (c). *Note.* From ''Kinetic Potential of the Lumbar Trunk Musculature About Three Orthogonal Orthopaedic Axes in Extreme Postures'' by S.M. McGill, 1991, *SPINE*, **16**(7), pp. 809-815. Reprinted by permission of J.B. Lippincott Company.

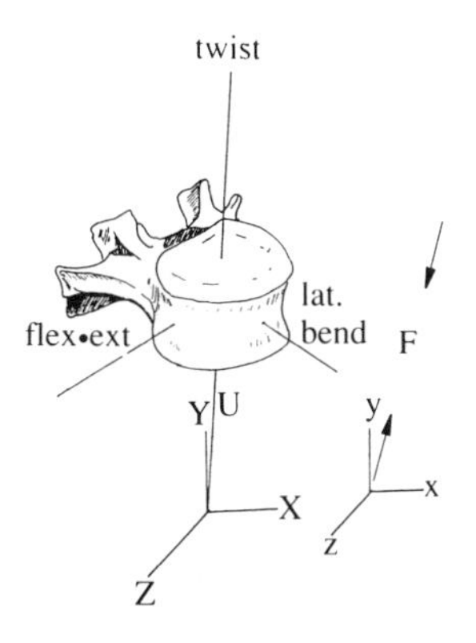

Figure 4.12 The convention of the orthopaedic axes of flexion-extension, lateral bend, and axial twist. Muscle forces contribute torque about each of these axes. *Note.* From ''Kinetic Potential of the Lumbar Trunk Musculature About Three Orthogonal Orthopaedic Axes in Extreme Postures'' by S.M. McGill, 1991, *SPINE*, **16**(7), pp. 809-815. Reprinted by permission of J.B. Lippincott Company.

In this section, we discuss some recent research findings as they relate to the issue of formulating reasonable and workable lifting instructions.

Flexing the Spine Versus Maintaining Lordosis

Very few lifting tasks in industry can be accomplished by bending the knees and not the back. Furthermore, most workers rarely adhere to this technique when repetitive lifts are required—a fact that is quite probably due to the increased physiological cost of squatting compared with stooping (Garg & Herrin, 1979). However, a case can be made for preserving lordosis—a posture of the lumbar spine that can be maintained independently of thigh and trunk angles—while lifting. There has been much confusion in the literature about trunk angle, or inclination, and the amount of lordosis or flexion specifically in the lumbar spine. Bending over is accomplished by either hip flexion, spine flexion, or both. Lumbar spine flexion is of concern here. Normal lordosis can be considered to be the curvature of the lumbar spine associated with the upright standing posture.

The many studies of disc loading reported by Adams and co-workers were summarized by Adams and Hutton (1988). They tested hundreds of lumbar motion units in compression and concluded that when compressive loads were applied to a joint in a neutral posture (neither flexed nor extended), the mode of failure was fracture of the end plate. Joints flexed 4° to 8° supported larger compressive loads, and generally the mechanism of failure involved bony components of the vertebrae. However, when the joint was fully flexed (as in the stooped lifting posture), the resultant failure was a posterior herniation of nuclear material through the annulus. Failure in the fully flexed posture was also achieved with lower compressive load.

Gracovetsky and colleagues have advocated conscious efforts to fully flex the lumbar spine while lifting to allow the posterior ligament system, including the lumbodorsal fascia, to support the extensor torque (Gracovetsky, Farfan, & Lamy, 1977; Gracovetsky, 1986). They believe the posterior ligaments have a larger moment arm than the extensor musculature, resulting in a mechanical advantage for the ligaments. They hypothesized that tensile forces in the lumbodorsal fascia produced by the transverse abdominis and internal oblique muscles could, when combined with lumbar flexion, support the extensor torque with a much smaller compressive penalty imposed on the lumbar joint. However, the assumption that muscular extensors are at a mechanical disadvantage appears to be erroneous, given the muscular moment arms reported for live, relatively young men by Nemeth and Ohlsen (1986) and McGill et al. (1988) and the moment arms of the individual posterior ligaments measured by McGill (1988). The grouped action of the posterior ligaments was used to estimate an equivalent moment arm of 5.8 cm (McGill, 1988), and the grouped action of the musculature appeared to be best represented with an equivalent moment arm of 6 cm.

The proposal by Gracovetsky, Farfan, & Lamy (1977) that muscular activation of the lumbodorsal fascia provides the major source of extensor torque was examined in cadaveric models by Tesh, Dunn, and Evans (1987) and Macintosh, Bogduk, and Gracovetsky (1987) and with analytical modeling techniques by McGill and Norman (1988). The extensor moment arm of the lumbodorsal fascia is 9 to 10 cm in the average adult male (McGill & Norman, 1988). All studies, including the one coauthored by Gracovetsky (Macintosh, Bogduk & Gracovetsky, 1987) concluded that the lumbodorsal fascia cannot produce significant contributions to lumbar extension. Some lifting theories have considered this hypothesis; therefore, it is covered in more detail in a subsequent section. Because ligaments are not recruited when lordosis is preserved, nor is the disc bent, it appears that the annulus is at low risk of failure. Hickey and Hukins (1980) demonstrated with an analytical model that the stress within the annual fibers is equalized when the neutral wedge shape of the disc is preserved, increasing its ability to withstand applied loads.

The issue of ligament recruitment upon spine flexion is more involved than simply a case of extensor mechanical advantage. When the interspinous ligaments are recruited and are generating large forces at full flexion, their line of action is such that additional shearing loads are placed on the joint. Unfortunately, unlike the shearing contributions of the pars lumborum muscle fascicles that help to offset the tendency of the superior vertebra to shear anteriorly on the inferior vertebra during lifting, the interspinous shearing forces add to the anterior load shear, increasing the total shear force to be supported by the facets, musculature, and disc. Preservation of lordosis insures that the moment-generating responsibility is provided by the muscles that minimize joint shear loads by offsetting the anterior shearing tendency caused by the load in the hands (Potvin, McGill & Norman, 1991).

Contributions of individual ligaments over the full range of joint flexion has never been experimentally determined. However, mathematical spine models have predicted that their primary role is to limit flexion and that they do not significantly impede flexion during moderate lumbar motion (McGill, 1988; Shirazi-Adl, Ahmed, & Shrivastava, 1986a). For these reasons, it is essential to monitor the level of lordosis, during analysis of lumbar loads, to determine the relative contribution of passive and active moment generators. For example, subjects lifting in our laboratory using a squat posture invariably elected to preserve enough lordosis so as to not signficantly recruit the ligaments, leaving the musculature with the responsibility to maintain equilibrium. Under these conditions, the ligaments and the disc in bending rarely contributed more than 4% of the restorative moment. In an effort to encourage ligament recruitment, work was conducted whereby subjects were requested to stoop rather than squat (Potvin et al., 1991). Dropout of the EMG signal confirmed that the load was supported by the ligaments, a phenomenon known as flexion-relaxation (Floyd & Silver, 1955). Workers are often observed in a stooped posture for lengthy periods of time but without a large load sustained in the hands. However, even in the stooped posture, subjects exhibit significant extensor EMG activity as soon as weight is taken in the hands, an observation also made by Gracovetsky et al. (1977). This evidence suggests that some mechanism is invoked that will not allow the ligaments to support a strenuous moment without auxiliary muscular contribution. We can only speculate that this is an attempt to provide a system of neuromuscular control to protect the supporting tissues. Thus, in most strenuous lifts it appears that the major burden of extensor moment generation during movement is relegated to the muscles.

A discussion such as the one presented in the previous paragraph is possible only because motion in the lumbar spine was extracted from total trunk motion. Often degrees of trunk flexion are reported in the literature without any regard as to whether the rotation occurred about the hips or from lumbar flexion. Obviously, normal lordosis can be preserved when rotating the trunk about the hips to minimize stress in the annulus. Separating trunk rotation about the hip from that about vertebral discs and reporting trunk kinematics in this way are critical for future evaluations of lumbar biomechanics.

The scientific evidence appears to point to the following positive effects if one lifts while maintaining a lordotic curvature: Conscious control of lumbar musculature is retained, reducing shear load on the facet joints and providing neurological protection not present if one depends on passive tissue to support the load; the disc is loaded uniformly permitting all of the annular fibers to share the stress rather than disabling some.

Lifting Shortly After Rising From Bed

The diurnal variation in spine length, together with the ability to flex forward, has been well documented. Losses in sitting height over a day have been measured up to 19 mm by Reilly, Tynell, and Troup (1984), who also noted that approximately 54% of this loss occurred in the first 30 min after rising. Over the course of a day, hydrostatic pressures cause a net fluid outflow from the disc, resulting in narrowing of the space between the vertebrae, which in turn reduces tension in the ligaments. When laying down at night, osmotic pressures exceed the hydrostatic pressure, causing the disc to expand. Adams, Dolan, and Hutton (1987) noted that the range of lumbar flexion increased by 5° throughout the day. The increased fluid content caused the lumbar spine to be more resistant to bending, whereas the musculature did not appear to compensate by restricting the bending range. They estimated that disc bending stresses increased by 300% and ligaments by 80% and concluded there is an increased risk of injury to these tissues when bending forward early in the morning.

Lifting Immediately Following Prolonged Flexion

For several years, it has been proposed that the nucleus within the annulus migrates anteriorly during spinal extension and posteriorly during flexion (McKenzie, 1981). McKenzie's program of passive extension of the lumbar spine (which is presently popular in physical therapy) was based on the supposition that an anterior movement of the nucleus would decrease pressure on the posterior portions of the annulus, which is the most potentially problematic site of herniation. Due to viscous properties of the nuclear material, such repositioning of the nucleus is not immediate upon a postural change but takes time. Whereas this hypothesis was conjecture for a period of time, several recent experiments have been reported verifying a repositioning of nuclear material upon forced extension of the lumbar spine. Krag, Seroussi, Wilder, and Pope (1987) demonstrated anterior movement, albeit quite minute, from an elaborate experiment that placed radio-opaque markers in the nucleus of cadaveric lumbar motion segments.

Whether this observation was just due to a redistribution of the centroid of the wedge-shaped nuclear cavity moving forward with flexion, or whether the whole nucleus migrates, remains to be seen. Nonetheless, hydraulic theory would suggest lower bulging forces on the posterior annulus if the nuclear centroid moved anteriorly during extension. If compressive forces were applied to a disc where the nuclear material was still posterior (as in lifting immediately after a prolonged period of flexion), then a concentration of stress would occur on the posterior annulus.

Although this specific area of research is in early stages of development, there does appear to be a time constant associated with this redistribution of nuclear material. If this is so, it would be unwise to lift an object immediately following prolonged flexion such as sitting or stooping (e.g., it would be unwise for a stooped gardener to stand erect and lift a heavy object). Furthermore, Adams and Hutton (1988) suggested that prolonged full flexion may cause the posterior ligaments to creep, which may allow damaging flexion postures to go unchecked if lordosis is not controlled during subsequent lifts. The data of McGill and Brown (1992) showed that even after 3 mintues following a bout of full flexion, subjects only regained half of their joint stiffness. Before lifting following a stooped posture, a case could be made for standing or even consciously extending the spine for a short period. This will allow ligaments to regain some of their stiffness and allow the nuclear material to equilibrate, or move anteriorly to a position associated with normal lordosis, decreasing forces on the posterior nucleus. However, further research is required to assess this potentially important possibility.

Role of Intra-Abdominal Pressure

It has been claimed for several decades that intra-abdominal pressure (IAP) plays an important role in support of the lumbar spine, especially during strenuous lifting. This issue has been considered in lifting mechanics for years and has formed a cornerstone for prescription of abdominal belts to industrial workers; it also has motivated various abdominal strengthening programs. Many have advocated the use of intra-abdominal pressure as a mechanism to reduce lumbar spine compression (Cyron et al., 1979; Thomson, 1988; Troup et al., 1983).

However, some believe the role of IAP in reducing spinal loads has been overemphasized (e.g., Bearn, 1961; Ekholm, Arborelius, & Nemeth, 1982; Grew, 1980). In fact, recent experimental evidence suggests that somehow, in the process of building up IAP, the net compressive load on the spine is increased! Increased low back EMG activity with higher IAP was noted by Krag, Byrne, Gilbertson, and Haugh (1986) during voluntary Valsalva's maneuvers. Nachemson and Morris (1964), and more recently Nachemson, Andersson, and Schultz (1986), showed an increase in intradiscal pressure during a Valsalva's maneuver, indicating a net increase in spine compression with an increase in IAP that was presumably a result of abdominal wall musculature activity.

Individual muscle force output from the model described in this chapter was used to evaluate the net benefit or penalty of the buildup of IAP and concomitant abdominal activity. It is important to review the pertinent components of the model, for they have an important effect on force estimates. During formulation of the anatomic model, a diaphragm was incorporated into the rib cage that was scaled in accordance with the dimensions of a 50th-percentile male. Its area and shape were confirmed from measures in the anatomy lab and from CT scans. Its surface area was 243 cm^2, and the centroid of this area was 3.8 cm anterior to the center of the T12 disc. It was difficult to conceive how the surface areas used in other models in the literature could fit into an average skeleton (compare with a 511-cm^2 pelvic floor and 465-cm^2 diaphragm). Moreover, the moment arms assumed for diaphragm forces on the spine were much larger (up to 11.4 cm). The size of the cross-sectional area of the diaphragm and the moment arm used to estimate force and moment produced by IAP have a major effect on conclusions reached about the role of IAP (see McGill & Norman, 1987).

It is usual to observe some level of abdominal activity during the lifting of loads. EMG records from 2 subjects are shown in Figure 4.13. Note the relatively low activity in the abdominals during the period of maximum extensor moment generation while internal oblique activity increased at the end of the lift to apparently help balance the load in an upright standing posture. To evaluate the proposed effects of IAP, the force-time histories of the individual abdominal muscles may be directly compared with the forces generated by IAP. During squat lifts, it appears that the net effect of the involvement of the abdominal musculature and IAP is to increase compression rather than alleviate joint load. (For a detailed description and analysis of the forces, see McGill & Norman, 1987.) IAP was not measured directly in this first study but was predicted from various regressive strategies. A later study directly measured pressures in conjunction with abdominal EMG, which confirmed the magnitudes of the predicted pressure, and observed pressure-time histories during activities such as sit-ups, running, lifting, and jumping (McGill & Sharratt, 1990). During the lifts, the forces and moments created from IAP did not overcome the compression and flexor moment created by the abdominal activation necessary to create the pressure. The compression of abdominal activity and compression relief from IAP are shown in Figure 4.14. This finding, predicted from our own modeled output, agrees with experimental evidence of Krag et al. (1986), who used EMG, and with Nachemson et al. (1986), who documented increased intradiscal pressure with an increase in IAP.

Grillner, Nilsson, and Thorstensson (1978) suggested that appreciable pressure could not be generated in the absence of abdominal muscle activity. More recently, however, Cresswell and Thorstensson (1989) showed that increases in IAP can be generated with no increase in abdominal activity during isometric extension efforts of subjects laying on their sides. Therefore, it may be possible to generate IAP in the absence of a compression penalty from abdominal activation in some types of movement. Some level of abdominal activity, albeit low, is usually observed during lifting. This notwithstanding, the size of the extensor moment and direct disc compression relief from IAP is not large if realistic surface areas and moment arm length of the diaphragm are modeled.

The generation of appreciable IAP during load-handling tasks is well documented. The role of IAP is not. Farfan (1973) has suggested that IAP creates a pressurized visceral cavity to maintain the hooplike geometry of the abdominals. However, the compression penalty of abdominal activity cannot be discounted. In fact, the presence of abdominal activity is proof that the mechanism of the lumbar spine does not work to minimize compression. Rather it appears that the spine prefers to sustain increased compression loads if intrinsic stability is increased. An unstabilized spine buckles under extremely low compressive load (e.g., approximately 20 N) (Lucas & Bresler, 1961). The geometry of the musculature suggests that individual components exert lateral and anterior-posterior forces on the spine, which perhaps can be compared to guy wires on a mast, to prevent bending and compressive buckling. Also, activated abdominals create a rigid cylinder of the trunk, resulting in a stiffer structure.

Increased IAP is commonly observed during many activities, including lifting, as well as in people experiencing back pain. However, the experimental evidence of others and the explanation of forces presented here suggest that IAP does not have a direct role in reducing spinal compression but rather is an agent used to stiffen the trunk and prevent tissue strain or failure from buckling.

Use of Abdominal Belts

The most popular research-based reason proposed for wearing a lifting belt is that the belt may assist in generating intra-abdominal pressure (IAP) without increasing abdominal muscle activity (Morris et al., 1961). While weight lifters and, quite recently, manual material handlers in industry have reported some perceived benefit from wearing an abdominal belt, there are documented contradictions.

We examined the effects of belt wearing on IAP and trunk EMG in 6 subjects performing eight lifts each on a lifting machine (McGill, Norman, & Sharratt, 1990). Two trials were performed while breath holding and another two performed while continuously expiring. These four trials were repeated both with and without wearing a belt for a total of eight trials per subject. The load was constrained to travel in a vertical track, and an outline for feet placement was marked on the floor in an attempt to standardize lifts.

The study findings demonstrated that the competition lifting belt slightly increased IAP (99 mmHg on average in the no-belt condition versus 120 mmHg when wearing a belt—a 21% increase, $p < 0.0001$), but there were no appreciable differences in rectus abdominis or erector spinae activity. Reduction in erector spinae activity has been used to assess whether IAP could contribute to the development of extensor moment and alleviate the erectors of a portion of their supporting responsibility (Krag et al., 1986). Krag et al. observed no significant reduction in erector spinae activity, observations that were corroborated by our work. Belts do not appear to contribute to support of the loaded lumbar spine in this way.

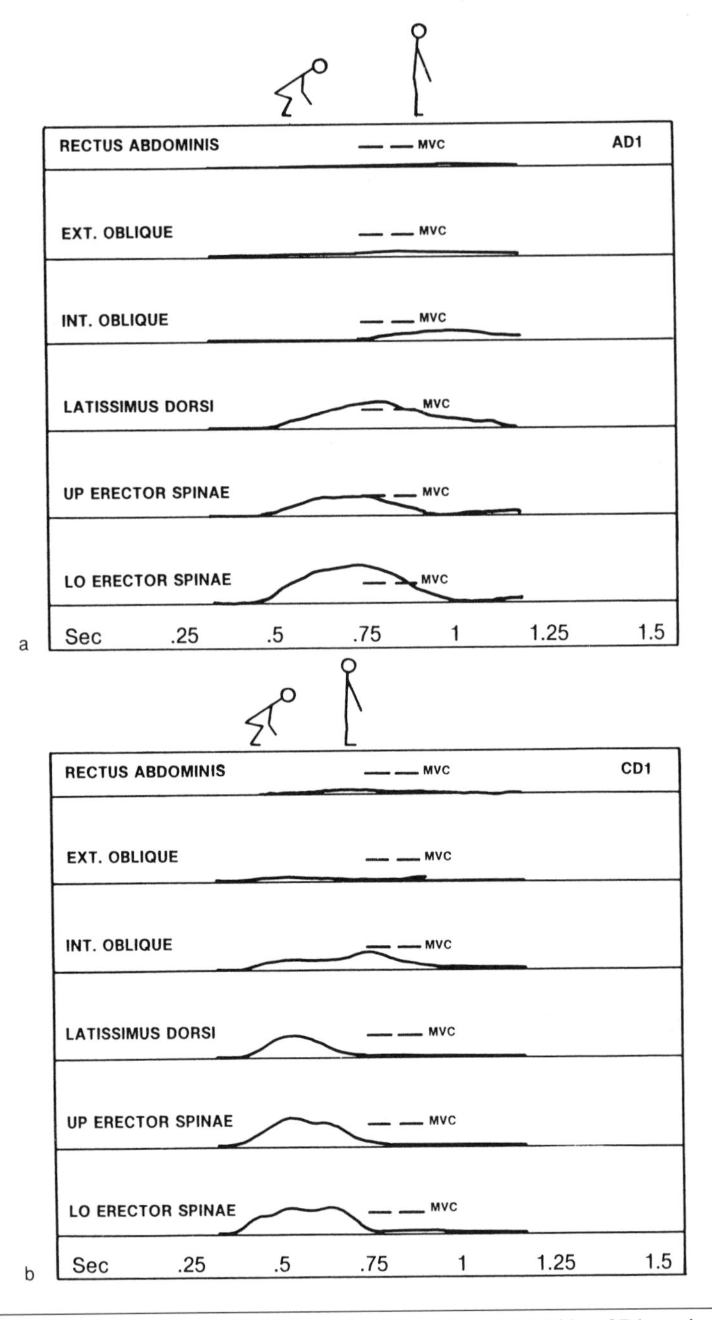

Figure 4.13 Linear envelope EMG from 2 subjects (a and b) lifting 27 kg using a squat style. EMG records show the level of abdominal activity for the rectus abdominis, external oblique, internal oblique, and the activity of the dorsal musculature: latissimus dorsi, upper erector spinae, and lower erector spinae. The maximum voluntary contraction (MVC) is shown for comparison. (Reprinted by permission from McGill, 1990.)

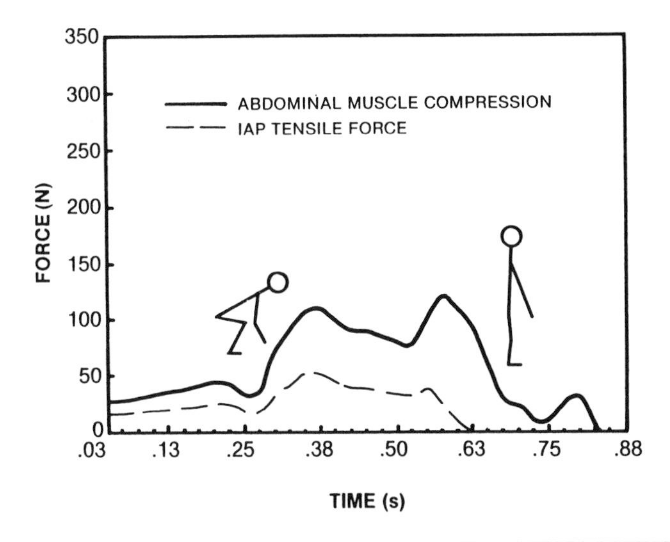

Figure 4.14 Compressive forces from activation of the abdominal wall and tensile forces produced by intra-abdominal pressure (IAP) exerted on the diaphragm for a typical sagittal plane lift. (Reprinted by permission from McGill & Norman, 1987.)

However, an important finding emerged from analysis of the effects of breath holding compared to exhaling. Holding the breath increases IAP and tends to reduce erector spinae activity whether a belt is worn or not, suggesting reduced lumbar compressive load (101 mmHg in the exhale condition vs. 118 mmHg during breath holding, $p < 0.017$—averages were obtained while both wearing and not wearing a belt). Histograms of the IAP and erector spinae activity are shown in Figure 4.15, a and b. Casual observation of all types of lifting indicated that people hold their breath during exertion. The measurements obtained in this study suggest a mechanism for such qualitative observations.

Our results differ in several ways from those reported by Kumar and Godfrey (1986), who studied the effects on IAP of six commonly prescribed spinal supports for 11 males and 9 females who lifted light loads of 9 and 7 kg, respectively. Kumar and Godfrey observed peak IAP in sagittal plane lifting of these loads with no bracing that was about 45 mmHg for males and 20 mmHg for females. Wearing supports did not result in appreciably higher IAP, contrary to the findings in our study. A major difference, however, was that our subjects handled loads that were 10 times heavier and produced IAP that was 2 to 3 times higher.

It is difficult to interpret this very limited information. In an industrial environment, lifting corsets and belts would hinder any axial twisting of the trunk, forcing a worker to pivot by moving the feet. The avoidance of twist while lifting has been advocated for some time by industrial safety associations to reduce overexertion injury. In addition, the belt might support anterior-posterior shear loads as the upper body tends to shear anteriorly on the pelvis due to the forward inclination of the trunk. Whereas the rib cage and pelvis tend to be more rigid

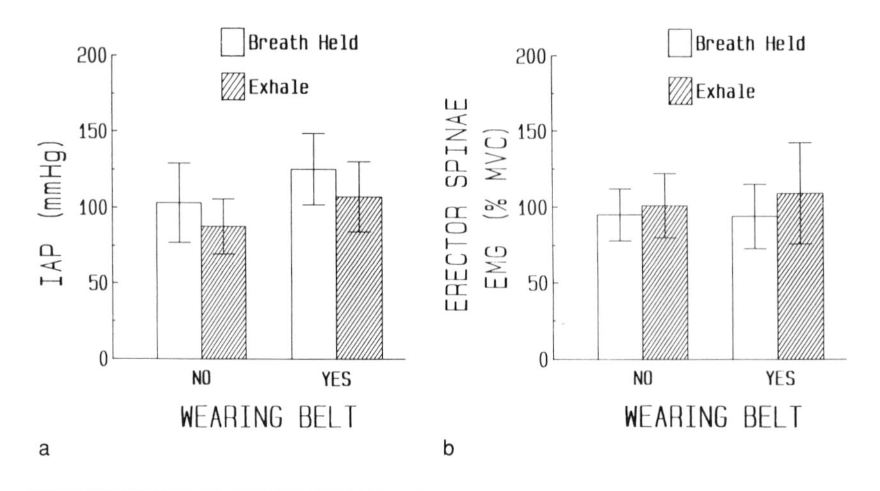

Figure 4.15 The effect of wearing an abdominal belt and breath holding on intra-abdominal pressure (IAP) in 6 lifters (a). The effect of belt wearing and breath holding on erector spinae activity in the same subjects (b). EMG measurements show percent maximum voluntary contraction (MVC). (Adapted by permission from McGill, Norman, & Sharratt, 1990.)

structures, hence better able to support shear, the abdomen might benefit from additional external contstraints to prevent shear. It appears likely that belts, causing an increase in IAP, act to stiffen the trunk. It does not appear that the extensor moment is appreciably reduced during lifting because no change in extensor muscle activity was observed when wearing a belt. Holding the breath appears to unload the lumbar spine somewhat, but this effect is not significantly altered by wearing a belt. Significant concerns are the increased blood pressure associated with belt wearing as reported by Hunter, McGuirk, Mitrano, Pearman, Thomas, and Arrington (1989) and the implications of this for those individuals who may have a compromised cardiovascular system. Therefore, it is difficult to justify the prescription of abdomninal belts to workers based on the muscle activity and IAP data. However, other considerations, including effects of long-term wearing, must be examined before a conclusive statement can be made about the efficacy of belt usage to reduce the incidence of occupational low back injury.

Recruiting the Lumbodorsal Fascia

Recent studies have attributed various mechanical roles to the lumbodorsal fascia (LDF). In fact, there have been some attempts to recommend lifting postures based on LDF hypthoses. Suggestions were originally made (Gracovetsky et al., 1981) that lateral forces generated by internal oblique and transverse abdominis are transmitted to the LDF via their attachments to the lateral border. This lateral tension was hypothesized to increase longitudinal tension, from Poisson's effect, pulling in the direction of the posterior midline of the lumbar spine, causing the posterior spinous processes to move together, resulting in lumbar extension. This proposed sequence of events formed an attractive proposition because the LDF has the largest moment arm of all extensor tissues. As a result, any extensor forces within the LDF would impose the smallest compressive penalty to vertebral components of the spine.

However, this hypothesis was examined by three studies, all published about the same time, and these collectively questioned its viability. Tesh et al. (1987) performed mechanical tests on cadaveric material, Macintosh et al. (1987) recognized the anatomic inconsistencies with the abdominal activation, and McGill and Norman (1988) tested the viability of LDF involvement with latissimus dorsi as well as with the abdominals. Regardless of the choice of LDF activation strategy, the LDF contribution to the restorative extension moment was negligible compared with the much larger low back reaction moment required to support the load in the hands. The moment contribution of the LDF for 2 subjects is shown in Figure 4.16. The LDF never produced more than 5% of the extensor moment required to complete the lift.

Although the LDF does not appear to be a significant active extensor of the spine, it is a strong tissue with a well-developed lattice of collagen fibers. Its function may be that of an extensor muscle retinaculum (Bogduk & Macintosh, 1984). The tendons of longissimus thoracis and iliocostalis lumborum pass under the LDF to their sacral and ilium attachments. Perhaps the LDF provides a form of strapping for the low back musculature. Hukins, Aspden, and Hickey (1990) have proposed, on theoretical grounds only at this time, that the LDF acts to increase by up to 30% the force-per-unit cross-sectional area that muscle can produce. They suggest that it does this by constraining bulging of the muscles when the muscles shorten. This contention remains to be proved. Tesh et al. (1987), suggested that the LDF may be more important for supporting lateral bending. No doubt, this notion will be pursued in the future. Given the confused state of knowledge about the role, if any, of the LDF, the promotion of lifting strategies based on intentional LDF involvement cannot be justified at this time.

Trunk Musculature Co-Contraction as a Means of Stabilizing the Spine

The ability of the joints of the lumbar spine to bend in any direction is accomplished with large amounts of muscle coactivation. Performing a simple lateral bend or axial twist recruits the trunk musculature on both sides of the body (see Figures 4.9 and 4.10; see McGill, in press). Such coactivation patterns are counterproductive to generating the torque necessary to support the applied load in a way that minimizes the load penalty imposed on the spine from muscle contraction. Several ideas have been postulated to explain muscular coactivation: The abdominals are involved in the generation of intra-abdominal pressure (Davis, 1959) or in providing support forces to the lumbar spine via the lumbodorsal

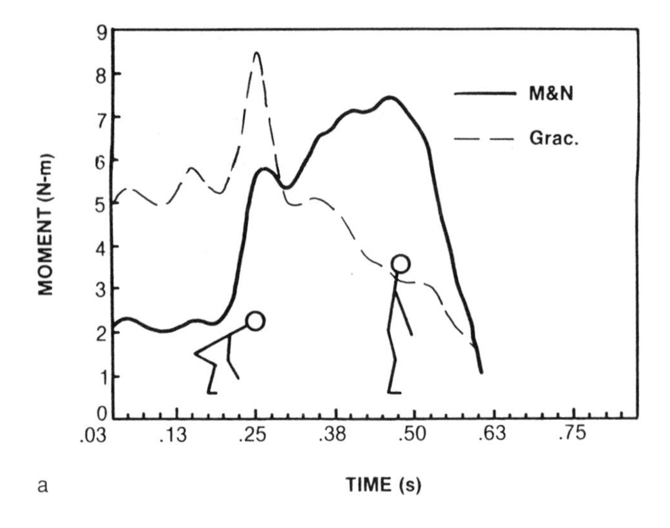

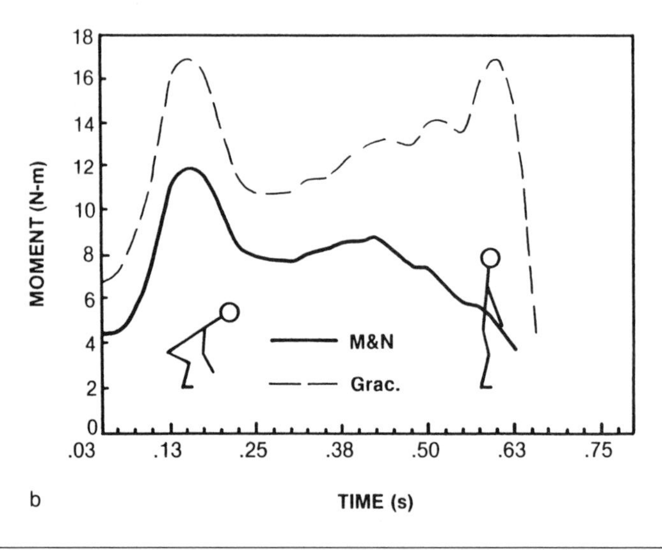

Figure 4.16 The lumbodorsal fascia (LDF) extensor moment contribution calculated from the McGill and Norman (M & N) method (latissimus dorsi activation) and Gracovetsky (Grac.) method (abdominal activation) for 2 subjects (a and b). The contribution of the LDF is small, regardless of the choice of activation strategy. The peak extensor moments generated during these lifts were 450 and 306 N·m for subjects a and b, respectively. (Reprinted by permission from McGill & Norman, 1988.)

fascia (e.g., Gracovetsky et al., 1981); however, some have opposed these ideas (see previous sections).

Another explanation for muscular coactivation seems tenable. A ligamentous spine will fail under compressive loading in a buckling mode at about 20 N (Lucas & Bresler, 1961). The spine can be likened to a flexible rod—under compressive loading, it will buckle. However, if the rod has guy wires connected to it, like the rigging on a ship's mast, the rod ultimately experiences more compression but is able to bear a much higher compressive load as it stiffens and becomes more resistant to buckling. The co-contracting musculature of the lumbar spine can perform the role of stablizing guy wires to each lumbar vertebra bracing against buckling. Recent work by Crisco and Panjabi (1990) has begun to quantify the influence of muscle architecture and the necessary coactivation on stability of the lumbar spine. The architecture of the lumbar erector spinae is especially suited for this role (see Macintosh & Bogduk, 1987; McGill & Norman, 1987). In order to invoke this antibuckling and stabilizing mechanism when lifting, one could justify lightly co-contracting the musculature to both minimize the potential of buckling and remove the possibility of any tissue having to bear a surprise load.

Twisting Lifts

Twisting of the trunk has been identified as a factor in the incidence of occupational low back pain (Frymoyer, Pope, Clements, Wilder, MacPherson, & Ashikaga, 1983; Troup, Martin, & Lloyd, 1981), but the mechanisms of risk require some explanation. Some hypotheses have been based on an inertia argument in that twisting at speed will impose dangerous axial torques upon braking the axial rotation of the trunk at the end range of motion. Farfan, Cossette, Robertson, Wells, and Kraus (1970) proposed that twisting the disc is the only way to damage the collagenous fibers in the annulus leading to failure. They reported that distortions of the neural arch permitted such injurious rotations. More detailed analyses of the annulus under twist were conducted by Shirazi-Adl, Ahmed, and Shrivastava (1986b), who supported Farfan's contention that twisting indeed can damage the annulus but also noted that twisting is not the sole mechanism of the annulus failure. In contrast, some research has suggested that twisting in vivo is not dangerous to the disc because the facet in compression forms a mechanical stop to rotation well before the elastic limit of the disc is reached, and thus the facet is the first structure to sustain torsional failure (Adams & Hutton, 1981).

Ligament involvement during twisting was studied by Ueno and Liu (1987), who concluded that the ligaments were under only negligible strain during a full physiological twist. However, an analysis of the L4-L5 joint by McGill and Hoodless (1990) suggested that posterior ligaments may become involved if the joint is fully flexed prior to twisting. This prompted McGill (in press) to examine the influence of the amount of lordosis on muscle-ligament interplay during twisting. These findings suggested that even though the ligaments are recruited when the spine is flexed and twisted, the ability to generate axial twist torque throughout the twist range appears to be impaired (see Figure 4.17). The mechanism for this impairment remains to be identified. Furthermore, no muscle in the trunk is specifically oriented to produce axial torque. Because muscles must act to produce torques about other axis (for example, flexion-extension and lateral bend), their axial contributions are limited and thus may not be a dependable source of torque when required to protect the spine (McGill, 1991a).

The mechanisms of injury from torsional loads applied under twisting conditions remain inconclusive. However, it is clear that the increase in compressive load on the spine is dramatic if a comparatively small amount of axial twist torque is required in addition to the dominant extensor torque. Data from a combination of our previous studies indicate that to support 50 N·m in extension imposes about 800 N of compression, but 50 N·m in axial twist would impose 2,500 N and 50 N·m of lateral bend, 1,400 N of compression. These differences result from the difference in coactivation of the trunk musculature, combined with small moment arms in many cases to generate the moments of force required. It appears that the joint pays dearly in order to support even small axial torques when extending during the lifting of a load.

Lifting Smoothly Versus Jerking

The often-repeated recommendation that a load should be lifted smoothly and not jerked was most likely rationalized on the basis that accelerating a load upward

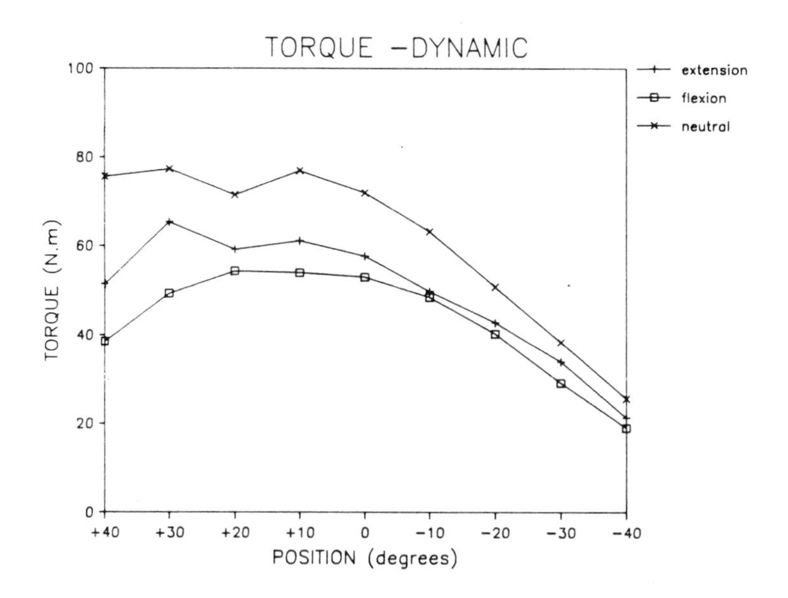

Figure 4.17 Axial torque (averages of 11 male subjects) generated throughout the twisting range with the lumbar spine in neutral lordosis, fully extended, and fully flexed.

increases its effective mass by virtue of an additional inertia force acting downward together with the gravitational vector. However, this may not always be the case. It is possible to lift a load by transferring momentum from an already moving segment. The concept of momentum transfer during lifting has been referred to by Troup and Chapman (1969) and by Grieve (1975); smaller low back moments associated with certain cases of dynamic lifts were recorded by McGill and Norman (1985).

For example, if a load is awkwardly placed, perhaps placed on a work table at a distance of 75 cm from the worker, a slow, smooth lift would necessitate the generation of a large lumbar extensor torque for a lengthy duration of time—a situation that is most strenuous on the back. However, this load could be lifted with a very low lumbar extensor moment or quite possibly with no moment at all. If the worker leaned forward and placed his hands on the load, with bent elbows, the elbow extensors and shoulder musculature could thrust upward, initiating upward motion of the trunk to create both linear and angular momentum. As the arms straighten, coupling takes place between the load and the large trunk mass, transferring some or all of the body momentum to the load, causing it to be lifted with a jerk. This highly skilled inertial technique is observed quite frequently throughout industry, but it must be stressed that such lifts are conducted by highly practiced and skilled individuals. In most cases, acceleration of loads to decrease low back stress in the manner described is not suitable for the lay individual when conducting the lifting chores of daily living. It is emphasized that for proper energy transfer, momentum must be built up in the torso before it is transfered to the load. Therefore, this type of lift can only be performed with objects at waist level, as momentum cannot be generated prior to object movement when the objects are placed near floor level or overhead.

The momentum-transfer technique is a skilled movement that requires practice, is only feasible for awkwardly placed lighter loads, and could not be justified for heavy lifts. However, there may be another mechanical variable to be integrated into the analysis of a dynamic technique. The tissue property of viscoelasticity enables tissues to sustain higher loads when loaded at rate (Burstein & Frankel, 1968). Troup (1977) suggested that the margin of safety for spine injury may be increased during a higher strain rate but cautioned that incorporation of this principle into lifting technique depends on the rate of increase in spinal stress, the magnitude of peak stress, and its duration. The lifting instruction to always lift a load smoothly may not invariably result in the least risk of injury. Indeed, it is possible to skillfully transfer momentum to an awkwardly placed object to position the load close to the body quickly and minimize the extensor torque required to support the load. Clearly, reduction of the extensor moment required to support the hand load is paramount in reducing the risk of injury; this is best accomplished by keeping the load as close to the body as possible.

Recommendations for Safer Lifting

The following lifting recommendations have been summarized from the biomechanical rationale developed in the previous section. Some are consistent with what has been advocated for years, and others contravene long-standing notions that were based on flawed biomechanical understanding. These guidelines are more versatile and move widely applicable than the commonly used instruction of "Bend the knees—not the back." However, they have not been subjected to rigorous scientific challenge. Due to the lack of conclusive evidence to support these recommendations, and because definitive studies will probably not be conducted in the near future, the guidelines are listed in the following form to provoke thought, stimulate discussion, and generate research interest. Perhaps it is overly optimistic to expect that these recommendations will withstand time and remain intact. However, these recommendations may have the potential to reduce tissue loads during the performance of a wide range of industrial lifting tasks, and they are able to accommodate all tasks including out-of-plane lifts. Furthermore, the exact instructions issued to a specific worker should not be taken verbatim from the following list, but rather the biomechanical principle should be explained in language that is familiar to the worker. In addition, often successful job incumbents have developed personal strategies for lifting that assist them in avoiding fatigue and injury. Their insights, gleaned from thousands of hours of performing the task, can be very perceptive—attempts should be made to accommodate them.

Tentative Lifting Guidelines

1. Maintain normal lordosis and rotate trunk using hips.
    - Strain in the annular fibers is equalized (Hickey & Hukins, 1980; Shirazi-Adl et al., 1986a).
    - Posterior ligaments are not strained and cannot be injured (McGill, 1988).
    - Facets are in contact and can bear some load (e.g., Nachemson, 1960).
    - The anterior shearing effect from ligament involvement is minimized, and the posterior supporting shear of the musculature is maximized (McGill, 1989).
    - Compressive testing of lumbar motion units has shown increases in tolerance with partial flexion but decreased ability to withstand compressive load at full flexion (Adams & Hutton, 1988).
2. Allow time for the disc nucleus to equilibrate and the ligaments to regain stiffness after prolonged flexion.
    - After prolonged sitting or stooping, spend time standing to allow the nuclear material within the disc to equilibrate and equalize the stress on the annulus.
    - Time is also required to allow the ligaments to regain their rest length and protect the disc from hyperflexion injury (McGill & Brown, 1992).
3. Avoid lifting shortly after rising from bed.

- Forward-bending stresses on the disc and ligaments are higher in the early morning than they are later in the day (Adams et al., 1987).

4. Follow this prestress system: Lightly co-contract the stabilizing musculature to remove the slack from the system and stiffen the spine.
   - Co-contraction and the corresponding increase in stability increase the margin of safety of material failure of the column under axial load (Crisco & Panjabi, 1990).
   - In this way, no tissue will have to bear a surprise load.
5. Choose a posture to minimize the reaction moment on the low back, but do not compromise Recommendation 1.
   - Lordosis is still maintained, but sometimes the load can be brought closer to the spine with bent knees (squat lift) or with relatively straight knees (stoop lift).
6. Avoid twisting.
   - Twisting reduces the intrinsic strength of the annulus by disabling some of its supporting fibers while increasing the stress in the remaining fibers under load (Shirazi-Adl et al., 1986b).
   - The facet in contact may shift the neutral axis of intervertebral stress away from the midline causing asymmetrical loading.
   - No muscle is designed to produce only axial torque, so the collective ability of the muscles to resist axial torque is limited and may not be able to protect the spine in certain postures (McGill, in press).
   - The additional compressive burden on the spine is substantial for even a low amount of axial torque production (McGill, in press).
7. Exploit the acceleration profile of the load.
   - This technique is only for highly skilled individuals performing repetitive tasks in which the load is placed at waist level.
   - This technique is dangerous for heavy loads and should not be attempted for such loads.
   - It is possible that a transfer of momentum from the upper trunk to the load can start moving an awkwardly placed load without undue low back load (McGill & Norman, 1985). Possibly, the viscoelastic property of biological material will safely absorb a momentary high load required to bring the load close to the trunk, which reduces the reaction moment (Grieve, 1975; Troup, 1977).

Conclusion

Recent advances in low back research and model sophistication have increased general understanding of lumbar mechanics. However, many issues pertaining to reducing the risk of industrial low back injury and to the recommendation of lifting guidelines remain unsolved. Future spinal biomechanists must address issues that will demand the utmost effort in creating models that capture biological fidelity. Continued effort must be directed toward obtaining more sophisticated anatomy for model components, for it is the fine details that unlock the secrets of force generation, transmission, and sharing strategies among tissues. Tissue properties such as strength, viscoelasticity, and fatigability must be better understood. The knowledge base of biomaterials is still in the developmental stage. With improvements in technology such as magnetic resonance imaging, CT scanning, and gamma mass techniques coupled with relatively simple materials testing apparatus, researchers can now proceed to obtain this anatomic information.

Static behavior has been quite well documented for some tissues, although not for all. However, with the recent development of quite involved dynamic models, dynamic tissue behavior is desperately required. The property of viscoelasticity is paramount in the determination of dynamic tissue load due to its time and loading-rate dependency. Examination of movement, particularly rapid movements of the trunk and limbs, is hindered by inadequate dynamic tissue information.

Whereas mechanical loading studies to understand the behavior of single ligaments, muscles, discs, and bones continue, additional work is required to describe the intact joint. Only very low bending moments in a flexion mode, for example, have been applied and reported. The passive response of a vertebral unit submitted to 300 to 400 Nm is unknown, yet moments of this magnitude are observed during quite routine tasks. Studies of coupled loads of compression, shear, and bending in other planes using physiological magnitudes are rare.

Few tasks of daily living can be defined in terms of one plane. Most models in the literature, however, were designed to quantify planar tasks even though some incorporated three-dimensional anatomical representations. Recent research has begun to address three-dimensional loading in modes of torsion, lateral bending, and complex movement. It is quite clear that only a model with sophisticated anatomy will have the necessary degree of biological integrity to satisfy such requirements. Muscle co-contraction is far more prevalent under complex motion conditions, which enforces the need to measure the activity of individual muscles. Both biological-EMG techniques and optimization strategies require basic research to improve predictions of the co-contracting muscle forces. Perhaps major developments in artificial intelligence will provide the required interface with motor control in the future. However, at present only the mind is capable of such sophisticated processing, and research efforts would appear to be best directed toward the improvement of biological techniques and the appropriate processing to obtain muscle force predictions. Work must continue to determine which muscles to monitor and where to place electrodes and to improve processing techniques to increase EMG reliability. Patterns of synergistic muscles must be cataloged to assess activation levels of those deep muscles that are inaccessible to surface electrodes. Some international effort is ongoing, with the continual reporting of some quite impressive predictions of individual muscle force measures in animal preparations, from processed EMG.

Analysis of the single task, with no provision for repeated movements, has dominated the literature. However, most tasks in industry, and those that are part of mundane daily events, are repeated. The effect of fatigue on the body system, intervertebral joints, muscles, and ligaments demands investigation. For example, at present the frequency of task repetition can only be recommended on the psychophysical criterion of what the individual thinks is appropriate. Data on repeated tissue loads are extremely scarce, although such data are appearing in the literature with greater frequency. Tissue fatigue must also be considered in the context of static holds that may occur in activities such as stooping for long periods of time while gardening. Whereas ergonomic design of the workplace is of utmost importance in facilitating work postures that minimize joint loads, much basic research remains to test and refine low back hypotheses and related issues such as lifting instructions for workers.

References

Adams, M.A., Dolan, P., & Hutton, W.C. (1987). Diurnal variations in the stresses on the lumbar spine. *Spine,* **12**(2), 130-137.

Adams, M.A., & Hutton, W.C. (1981). The relevance of torsion to the mechanical derangement of the lumbar spine. *Spine,* **6**, 241-248.

Adams, M.A., & Hutton, W.C. (1988). Mechanics of the intervertebral disc. In P. Ghosh (Ed.), *The biology of the intervertebral disc* (pp. 39-71). Boca Raton, FL: CRC Press.

Anderson, C., Chaffin, D.B., Herrin, G.D., & Matthews, L.S. (1985). A biomechanical model of the lumbosacral joint during lifting activities. *Journal of Biomechanics,* **18**, 571-584.

Bartelink, D.L. (1957). The role of abdominal pressure on the lumbar intervertebral discs. *Journal of Bone and Joint Surgery,* **39B**, 718-725.

Bearn, J.G. (1961). The significance of the activity of the abdominal muscles in weight lifting. *Acta Anatomica,* **45**, 83.

Bogduk, N. (1980). A reappraisal of the anatomy of the human erector spinae. *Journal of Anatomy,* **131**, 525-540.

Bogduk, N., & Macintosh, J.E. (1984). The applied anatomy of the thoracolumbar fascia. *Spine,* **9**, 164-170.

Buchthal, F., & Schmalbruch, H. (1970). Contraction times and fibre types in intact human muscle. *Acta Physiologica Scandinavica,* **79**, 435-452.

Burstein, A.H., & Frankel, W.H. (1968). The viscoelastic properties of some biological material. *Annals of the New York Academy of Sciences,* **146**, 158-165.

Chaffin, D.B. (1969). Computerized biomechanical models: Development of and use in studying gross body actions. *Journal of Biomechanics,* **2**, 429-441.

Cholewicki, J., & McGill, S.M. (1992). Lumbar posterior ligament involvement during extremely heavy lifts estimated (from flouroscopic measurements. *Journal of Biomechanics,* **25**(1), 1728.

Cresswell, A.G., & Thorstensson, A. (1989). Intra-abdominal pressure and patterns of abdominal muscle activation in isometric trunk flexion and extension. In R. Gregor, R. Zernicke, & W. Whiting (Eds.), *Proceedings of the XII International Congress of Biomechanics* (p. 252). Los Angeles.

Crisco, J.J., & Panjabi, M.M. (1990). Muscle architecture and the co-activation in the lateral postural stability of the human lumbar spine. *Proceedings of the 36th Annual Meeting of the Orthopaedic Research Society.*

Cyron, B.M., Hutton, W.C., & Stott, J.R. (1979). The mechanical properties of the lumbar spine. Mechanical Engineering, **8**(2), 63-68.

Davis, P.R. (1959). The causation of herniae by weight-lifting. *Lancet,* **2**, 155-157.

Dreyfuss, H. (1966). Anthropometric data: Standing 50 percentile American adult male.

Ekholm, J., Arborelius, U.P., & Nemeth, G. (1982). The load on the lumbosacral joint and trunk muscle activity during lifting. *Ergonomics,* **25**(2), 145-161.

Eycleshymer, A.C., & Shoemaker, P.M. (1911). *A cross-section anatomy.* New York: Appleton.

Farfan, H.F. (1973). *Mechanical disorders of the low back.* Philadelphia: Lea & Febiger.

Farfan, H.F., Cossette, J.W., Robertson, G.H., Wells, R.V., & Kraus, H. (1970). The effects of torsion on the lumbar intervertebral joints: The role of torsion in the production of disc degeneration. *Journal of Bone and Joint Surgery,* **52A**(3), 469-497.

Floyd, W.F., & Silver, P.H.S. (1955). The function of the erectores spinae muscles in certain movements and postures in man. *Journal of Physiology,* **129**, 184-203.

Frymoyer, J.W., Pope, M.H., Clements, J.H., Wilder, D.G., MacPherson, B., & Ashikaga, T. (1983). Risk factors in low back pain. *Journal of Bone and Joint Surgery,* **65A**, 213-218.

Frymoyer, J.W., Pope, M.H., Costanza, M.C., Rosen, S.C., Goggin, J.E., & Wilder, D.G. (1980). Epidemiologic studies of low back pain. *Spine,* **5**(5), 419-423.

Garg, A., & Herrin, G. (1979). Stoop or squat: A biomechanical and metabolic evaluation. *American Institute of Industrial Engineers Transactions,* **11**, 293-302.

Gracovetsky, S. (1986). Determination of safe load. *British Journal of Industrial Medicine,* **43**, 120-133.

Gracovetsky, S., Farfan, H.F., & Lamy, C. (1977). A mathematical model of the lumbar spine using an optimization system to control muscles and ligaments. *Orthopedic Clinics of North America,* **8**(1), 135-153.

Gracovetsky, S., Farfan, H.F., & Lamy, C. (1981). Mechanism of the lumbar spine. *Spine,* **6**(1), 249-262.

Granhed, H., Jonsson, R., & Hansson, T. (1987). The loads on the lumbar spine during extreme weight lifting. *Spine,* **12**(2), 114-149.

Grew, N.D. (1980). Intra-abdominal pressure response to loads applied to the torso in normal subjects. *Spine,* **5**(2), 149-154.

Grieve, D.W. (1975). Dynamic characteristics of man during crouch and stoop lifting. In R.C. Nelson & C.A. Morehouse (Eds.), *Biomechanics IV* (pp. 19-29). Baltimore: University Park Press.

Grillner, S.J., Nilsson, J., & Thorstensson, A. (1978). Intra-abdominal pressure changes during natural movements in man. *Acta Physiologica Scandinavica,* **103**, 275-283.

Heylings, D.J.A. (1978). Supraspinous and interspinous ligaments of the human lumbar spine. *Journal of Anatomy,* **123**, 127-131.

Hickey, D.S., & Hukins, D.W.L. (1980). Relation between the structure of the annulus fibrosus and the function and failure of the intervertebral disc. *Spine,* **5**(20), 106-116.

Hukins, D.W.L., Aspden, R.M., & Hickey, D.S. (1990). Thoracolumbar fascia can increase the efficiency of the erector spinae muscles. *Clinical Biomechanics,* **5**(1), 30-34.

Hunter, G.R., McGuirk, J., Mitrano, N., Pearman, P., Thomas, B., & Arrington, R. (1989). The effects of a weight training belt on blood pressure during exercise. *Journal of Applied Sport Science Research,* **3**(1), 13-18.

Keith, A. (1923). Man's posture: Its evolution and disorders [Lecture IV. The adaptations of the abdomen and its viscera to the orthograde posture]. *British Medical Journal,* **1**, 587-590.

Krag, M.H., Byrne, K.B., Gilbertson, L.G., & Haugh, L.D. (1986). Failure of intraabdominal pressurization to reduce erector spinae loads during lifting tasks. In P. Allard & M. Gagnon (Eds.), *Proceedings of the North American Congress on Biomechanics* (pp. 87-88).

Krag, M.H., Seroussi, R.E., Wilder, D.G., & Pope, M.H. (1987). Internal displacement distribution from in vitro loading of human thoracic and lumbar spinal motion segments: Experimental results and theoretical predictions. *Spine,* **12**(10), 1001-1007.

Kumar, S., & Godfrey, C.M. (1986). Spinal braces and abdominal support. In W. Karwowski (Ed.), *Trends in Ergonomics/Human Factors III* (pp. 717-726). New York: Elsevier Science.

Langenberg, W. (1970). Morphologic, physiologischer querschnitt and kraft des m. erector spinae in lumbalbereich des menshen. *Zeitschrift Für Anatomie Und Entwicklungsgeschicht,* **132**, 158-190.

Lucas, D., & Bresler, B. (1961). *Stability of the ligamentous spine* (Tech. Rep. No. 40). San Francisco: University of California, Biomechanics Laboratory.

MacGibbon, B., & Farfan, H.F. (1979). A radiologic survey of various configurations of the lumbar spine. *Spine,* **4**, 258-266.

McGill, S.M. (1988). Estimation of force and extensor moment contributions of the disc and ligaments at L4/L5. *Spine,* **13**, 1395-1402.

McGill, S.M. (1989). Loads of the lumbar spine and associated tissues. In V.K. Goel & J.N. Weinstein (Eds.), *Biomechanics of the spine: Clinical and surgical perspective* (pp. 65-95). Boca Raton, FL: CRC press.

McGill, S.M. (1991a). Electromyographic activity of the abdominal and low back musculature during the generation of isometric and dynamic axial trunk torque: Implications for lumbar mechanics. *Journal of Orthopaedic Research,* **9**, 91-103.

McGill, S.M. (1991b). Kinetic potential of the lumbar trunk musculature about three orthogonal orthopaedic axes in extreme postures. *Spine.*

McGill, S.M. (in press). The influence of lordosis on axial trunk torque and trunk muscle myoelectric activity. *Spine*, **16**, 809-815.

McGill, S.M., & Brown, S. (1992). Creep response of the lumbar spine to prolonged full flexion. *Clinical Biomechanics,* **7**, 43-46.

McGill, S.M., & Hoodless, K. (1990). Measured and modelled static and dynamic axial trunk torsion during twisting in males and females. *Journal of Biomedical Engineering,* **12**, 403-409.

McGill, S.M., & Norman, R.W. (1985). Dynamically and statically determined low back moments during lifting. *Journal of Biomechanics,* **18**(12), 877-885.

McGill, S.M., & Norman, R.W. (1986). Partitioning of the L4/L5 dynamic moment into disc, ligamentous and muscular components during lifting. *Spine*, **11**(7), 666-678.

McGill, S.M., & Norman, R.W. (1987). Reassessment of the role of intraabdominal pressure in spinal compression. *Ergonomics,* **30**(11), 1565-1588.

McGill, S.M., & Norman, R.W. (1988). The potential of lumbodorsal fascia forces to generate back extension moments during squat lifts. *Journal of Biomedical Engineering,* **10**, 312-318.

McGill, S.M., Norman, R.W., & Sharratt, M.T. (1990). The effect of an abdominal belt on trunk muscle activity and intra-abdominal pressure during squat lifts. *Ergonomics,* **33**(2), 147-160.

McGill, S.M., Patt, N., & Norman, R.W. (1988). Measurement of the trunk musculature of active males using CT Scan radiography: Implications for force and moment generating capacity about the L4/L5 joint. *Journal of Biomechanics,* **21**(4), 329-341.

McGill, S.M., & Sharratt, M.T. (1990). The relationship between intra-abdominal pressure and trunk EMG. *Clinical Biomechanics,* **5**, 59-67.

Macintosh, J.E., & Bogduk, N. (1987). The morphology of the lumbar erector spinae. *Spine,* **12**(7), 658-668.

Macintosh, J.E., Bogduk, N., & Gracovetsky, S. (1987). The biomechanics of the thoracolumbar fascia. *Clinical Biomechanics,* **2**, 78-93.

McKenzie, R.A. (1981). The Lumbar Spine: mechanical diagnosis and therapy, Spinal Publications, Lower Hutt, New Zealand.

Markolf, K.J., & Morris, J.M. (1974). The structural components of the intervertebral disc. *Journal of Bone and Joint Surgery,* **56A**, 675-687.

Morris, J.M., Lucas, D.B., & Bresler, B. (1961). Role of the trunk in stability of the spine. *Journal of Bone and Joint Surgery,* **43A**(3), 327-351.

Nachemson, A.L. (1960). Lumbar intradiscal pressure. *Acta Orthopaedica Scandinavica,* (Suppl. 43).

Nachemson, A. (1983). Work for all: For those with low back pain as well. *Clinical Orthopaedics and Related Research,* **179**, 77-99.

Nachemson, A., Andersson, G.B.J., & Schultz, A.B. (1986). Valsalva manoeuvre biomechanics: Effects on lumbar trunk loads of elevated intra-abdominal pressure. *Spine,* **11**(5), 476-479.

Nachemson, A.L., & Morris, J.M. (1964). In vivo measurements of intradiscal pressure. *Journal of Bone and Joint Surgery,* **46A**, 1077-1092.

National Institute for Occupational Safety and Health. (1981). *A work practices guide for manual lifting*. Cincinnati: Taft Industries.

Nemeth, G., & Ohlsen, H. (1986). Moment arm lengths of trunk muscles to the lumbosacral joint obtained in vivo with computer tomography. *Spine,* **11**(2), 158-160.

Park, K.S., & Chaffin, D.B. (1974). A biomechanical evaluation of two methods of manual load lifting. *American Institute of Industrial Engineers Transactions,* **6**, 105-113.

Potvin, J.R., McGill, S.M., & Norman, R.W. (1991). Trunk muscle and lumbar ligament contributions to dynamic lifts with varying degrees of trunk flexion, *Spine*, **16**(9), 1099-1107.

Reid, J.G., & Costigan, P.A. (1985). Geometry of adult rectus abdominis and erector spinae muscles. *Journal of Orthopaedic and Sports Physical Therapy,* **6**, 278-280.

Reilly, T., Tynell, A., & Troup, J.D.G. (1984). Circadian variation in human stature. *Chronobiology International,* **1**, 121-126.

Schultz, A.B., Andersson, G.B.J., Ortengren, R., Haderspeck, K., & Nachemson, A. (1982). Loads on the lumbar spine. *Journal of Bone and Joint Surgery,* **64A**(5), 713-720.

Schultz, A.B., Haderspeck, K., Warwick, D., & Portillo, D. (1983). Use of lumbar trunk muscles in isometric performance of mechanically complex standing tasks. *Journal of Orthopaedic Research,* **1**, 77-91.

Schultz, A.B., Warwick, D.N., Berkson, M.H., & Nachemson, A. (1979). Mechanical properties of the human lumbar spine motion segments: Part 1. Responses to flexion, extension, lateral bending and torsion. *Journal of Biomechanical Engineering,* **101**, 46-52.

Shirazi-Adl, A., Ahmed, A.M., & Shrivastava, S.C. (1986a). A finite element study of a lumbar motion segment subjected to pure sagittal plane moments. *Journal of Biomechanics,* **19**(4), 331-350.

Shirazi-Adl, A., Ahmed, A.M., & Shrivastava, S.C. (1986b). Mechanical response of a lumbar motion segment in axial torque alone and combined with compression. *Spine,* **11**(9), 914-927.

Shirazi-Adl, A., & Drouin, G. (1987). Load bearing role of the facets in the lumbar segment under sagittal plane loadings. *Journal of Biomechanics,* **20**(6), 601-603.

Tesh, K.M., Dunn, J., & Evans, J.H. (1987). The abdominal muscles and vertebral stability. *Spine,* **12**(5), 501-508.

Thomson, K.D. (1988). On the bending moment capability of the pressurized abdominal cavity during human lifting activity. *Ergonomics,* **31**(5), 817-828.

Thorstensson, A., Andersson, E., & Cresswell, A. (1989). Lumbar spine and psoas muscle geometry revised with magnetic resonance imaging. In R. Gregor, R. Zernicke, & W. Whiting (Eds.), *Proceedings of the XII International Congress of Biomechanics*, Los Angeles. (Abstract No. 251)

Troup, J.D.G. (1977). Dynamic factors in the analysis of stoop and crouch lifting methods: A methodological approach to the development of safe materials handling standards. *Orthopedic Clinics of North America,* **8**(1), 201-209.

Troup, J.D.G., & Chapman, A.E. (1969). The strength of the flexor and extensor muscles of the trunk. *Journal of Biomechanics,* **2**, 49-62.

Troup, J.D.G., Leskinen, T.P.J., Stalhammear, H.R., & Kuorinka, I.A. (1983). A comparison of intra-abdominal pressure increases, hip torque, and lumbar vertebral compression in different lifting techniques. *Human Factors,* **25**(5), 517-525.

Troup, J.D.G., Martin, J.W., & Lloyd, D.C. (1981). Back pain in industry: A prospective survey. *Spine,* **6**, 61-69.

Ueno, K., & Liu, Y.K. (1987). A three-dimensional nonlinear finite element model of lumbar intervertebral joint in torsion. *Journal of Biomechanical Engineering,* **109**, 200-209.

Warwick, R., & Williams, P.L. (Eds.) (1980). *Gray's anatomy: Descriptive and applied* (37th ed.). London: Longman.

Wood, G.A., & Hayes, K.C. (1974). A kinetic model of intervertebral stress during lifting. *British Journal of Sports Medicine,* **8**, 74-79.

Acknowledgment: Dr. Norman's financial support for our laboratory work was provided from the Defence and Civil Institute of Environmental Medicine from 1979 until 1989. Dr. McGill has been funded by the Natural Sciences and Engineering Research Council, Canada, since 1988.

Part II

Basic Tissue Biomechanics

One fundamental assumption underlying the mechanical approach (using film or video technology) to sport analysis is that bones act as rigid members and rotate about frictionless joints that are often assumed to be uniaxial. The bones and joints, then, are charged with rather simple and passive roles, but this belies their elegant structures and complex physiologies.

Part II addresses cellular and tissue biomechanics and looks in-depth at the complex and active natures of bone and ligament. The quantification of the loads placed on these tissues relative to normal tissue function, injury related to trauma, and the effects of exercise and of aging all are topics in the domain of the sports medicine and sport science practitioner.

Chapters 5 and 6 provide a broad view of the plasticity of bone and ligament, including tissue responses to exercise and aging. The authors offer insight into the mechanics of these tissues and discuss their interaction with the mechanisms of sport-related injuries and subsequent healing and rehabilitation. Chapter 7 focuses on skeletal muscle and integrates classical skeletal muscle contraction mechanics with very recent mechanical and physiological findings.

Chapter 5

Ligament Biomechanics

David Hawkins

David Hawkins begins Part II with a bioengineer's look at ligaments. Ligaments play an important role in stabilizing joints, and they are often injured during vocational and avocational activities. Because the knee is a common site of such injuries, Hawkins uses knee ligaments as a model for discussion. Chapter 5 examines the relationship between ligament structure and function and the mechanical testing techniques used to determine these relationships. This introduction leads into topics more directly related to sport such as the effects of exercise, aging, injury, and disuse on ligament structure and function. The chapter also addresses clinical considerations such as diagnosis, treatment, and rehabilitation.

From an engineering perspective, the human body is one of the most interesting and challenging systems to study and describe. Most structural components of the human body elicit nonhomogeneous (nonuniform structure or composition) and nonisotropic (exhibiting different properties along different axes) material characteristics. These characteristics make it difficult to derive accurate constitutive equations that describe the properties of these structures. Components of the human system also have inherent adaptive capabilities that allow them to hypertrophy, atrophy, and change their material constituents and mechanical properties to meet the physical demands placed upon them (see chapter 6). The complexity of the human body is further compounded by the fact that individual components act in a coordinated fashion with other components to achieve specific responses.

Presented in this chapter are current ideas pertaining to the structure and function of ligaments. Ligaments play a significant role in stabilizing joints and in specifying joint motion. They are commonly injured in sports, and therefore an understanding of their form and function is relevant for exercise and sport scientists. Such an understanding is fundamental both for designing training programs that minimize the risk of ligament damage and for recognizing early signs of ligamentous injuries.

Ligament Anatomy

General Structure and Function

Ligaments, lying internal or external to the joint capsule, bind bone to bone and supply passive support and guidance to joints. They function to supplement active stabilizers (i.e., muscles) and bony geometry (Akeson, S.L.Y. Woo, Amiel, & Frank, 1984). Ligaments generally are named according to their position in the body (e.g., collateral) or according to their bony attachments (e.g., coracoclavicular). Well suited for their functional roles, ligaments offer early and increasing resistance to tensile loading over a narrow range of joint motion. This allows joints to move easily within normal limits while causing increased resistance to movement outside this normal range.

Illustrated in Figure 5.1 is the structural arrangement of tendon, which like ligament is a collagenous tissue. The primary building unit of collagenous tissues is the tropocollagen molecule (Viidik, 1973). Tropocollagen molecules are organized into long, cross-striated fibrils that are arranged into bundles to form fibers.

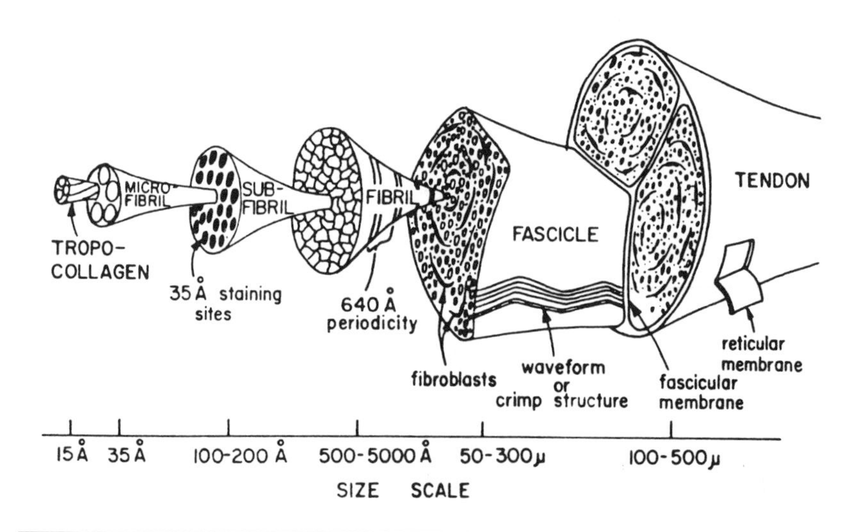

Figure 5.1 A schematic diagram of the structural hierarchy of tendon. Ligament and tendon are similar in structure; both are composed of smaller and smaller fiber bundles. The basic structural element is the tropocollagen molecule. (Reprinted by permission from Kastelic, Galeski, & Baer, 1978.)

Fibers are further grouped into bundles called fascicles, which group together to form the gross ligament or tendon.

Collagen fiber bundles are arranged in the direction of functional need and act in conjunction with elastic and reticular fibers along with ground substance (a composition of glycosaminoglycans and tissue fluid) to give ligaments their mechanical characteristics. In unstressed ligaments, collagen fibers take on a sinusoidal pattern. This pattern, referred to as a *crimp* pattern, is believed to be created by the cross-linking or binding of collagen fibers with elastic and reticular fibers. The crimp pattern is an important factor contributing to the mechanical behavior of ligaments, as discussed later in this chapter.

A large number of human ligamentous injuries occur to ligaments of the knee. Because of their clinical significance, knee ligaments are referred to throughout this chapter to illustrate various concepts. Four major ligaments assist in stabilizing the knee. (Basic knee anatomy is illustrated in Figure 5.2.) Two of these ligaments, the anterior cruciate (ACL) and posterior cruciate (PCL), are located

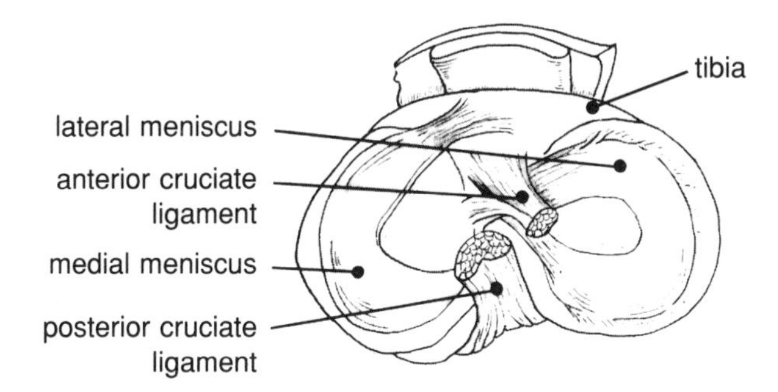

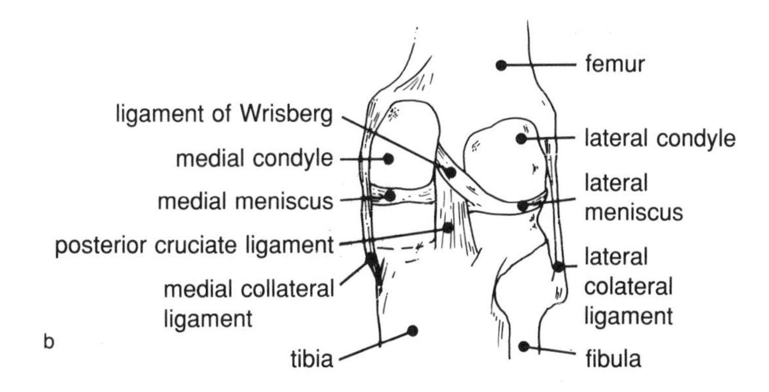

Figure 5.2 Schematic representations of the structures of the human knee viewed from both a superior (a) and posterior (b) position. The three bony structures are the femur, tibia, and fibula. The four major ligaments supporting the knee are the anterior cruciate, posterior cruciate, medial collateral (or tibial collateral), and the lateral collateral (or fibular collateral). (Adapted by permission from Donnelly, 1990.)

within the joint capsule. The ACL is attached to the posterior side of the lateral femoral condyle and to the anterior intercondylar fossa of the tibia. The PCL is attached to the anterior portion of the intercondylar notch on the femur and to the posterior intercondylar fossa of the tibia. The ACL and PCL touch as they span the joint with the ACL passing anterior and lateral to the PCL. The other two ligaments, the medial collateral (MCL), also called the tibial collateral, and the lateral collateral (LCL), also called the fibular collateral, are located external to the joint capsule lying medially and laterally to the joint respectively.

For each plane of knee mobility, there are primary and secondary ligament stabilizers. The primary and secondary knee stabilizers that resist valgus stresses (opening of the medial side of the knee) are the superficial MCL and the ACL respectively. Varus knee stresses (opening of the lateral side of the knee) are resisted by a combination of knee structures, depending on the angle of knee flexion. The LCL is the primary stabilizer at the midrange of knee flexion and the PCL, at 90° of flexion. At full extension no single structure acts as the primary stabilizer. The ACL is the primary check against anterior displacement of the tibia relative to the femur, and the PCL is the primary stabilizer preventing posterior displacement. In knee flexion, the superficial MCL is the first defense against external rotation, and the ACL acts as a secondary restraint. In knee extension, the ACL and superficial MCL act together as primary stabilizers against external rotation. With the knee flexed, internal rotation is prevented first by the cruciate ligaments and then by the LCL. In extension, the ACL is the primary stabilizer and the LCL, the secondary (Marshall & Rubin, 1977).

Biochemical Constituents

The major constituents of ligaments are collagen, elastin, glycoproteins, protein polysaccharides, glycolipids, water, and cells (mostly fibrocytes) (Akeson et al., 1984). Of these constituents, collagen and water exist in the greatest quantities. Collagen constitutes 70% of the dry weight of ligaments, the majority being Type I collagen, which is also found in tendon, skin, and bone. Water makes up about 60% to 80% of the wet weight of ligaments. A significant amount of this water is associated with the ground substance.

For practical purposes, the physical behavior of ligaments can be predicted based on the content and organization of collagen and ground substance (Akeson et al., 1984). Collagen has a relatively long turnover rate, its average half-life being between 300 and 500 days, which is slightly longer than that of bone. Therefore, several months may be required for a ligament to alter its structure to meet changes in physical loading conditions or to repair itself after injury. Although on a dry weight basis the ground substance comprises only about 1% of the total tissue mass, it provides ligaments with important properties. The ground substance probably provides lubrication and spacing that aid in the sliding of fibers. The presence of ground substance is also a source of ligamentous viscoelasticity.

Blood Supply

The extent of the vascular supply varies among ligaments and may depend on the ligament's location relative to the joint capsule. Blood is usually supplied to ligaments through periarticular arterial plexuses (Akeson et al., 1984; Butler, Grood, Noyes, & Zernicke, 1978; Dye & Cannon, 1988). Intraligament vasculature is rather limited, indicating that some degree of diffusion is necessary to supply the inner ligament fibers with needed nutrients. The blood supply to a ligament is important for the synthesis of new collagen (Butler et al.). As such, the extent of vascular damage that occurs during ligament trauma is a significant factor affecting the ability of the ligament to heal itself.

Ligament surgical and rehabilitative procedures should be designed to maintain or enhance the tissue's vascular and blood supplies. Many rehabilitation programs incorporate these ideas by using continuous passive motion, which moves the joint continuously and passively over a limited range of movement for a designated time period. Continuous passive motion is a means to facilitate blood flow to damaged tissues and promote tissue synthesis.

Neural Structures

In addition to a vascular network, ligaments also contain a variety of neural elements (Akeson et al., 1984; Dye & Cannon, 1988; Halata, Badalamente, Dee, & Propper, 1984; Kennedy, Alexander, & Hayes, 1982; Schutte, Dabezies, Zimny, & Happel, 1987; Schultz, Miller, Kerr, & Micheli, 1984; Zimny, Schutte, & Dabezies, 1986). Nerves present within ligaments originate from nerves innervating muscles. Reflex pathways appear to exist that may allow ligament strain to be communicated to the central nervous system (CNS). The CNS can respond by stimulating specific muscles, causing them to contract and prevent further joint displacement. The free nerve endings present in ligaments are believed to detect joint position, speed, and movement direction (Akeson et al., 1984).

Biomechanics

Structural and Mechanical Properties

Ligaments are composite, anisotropic structures exhibiting nonlinear time and history-dependent viscoelastic properties. Described in this section are the structural and mechanical properties of ligamentous tissue, the physiological origin of these properties, and their implications for ligament function during normal joint motion.

The structural properties of isolated tendons, ligaments, and bone-ligament-bone preparations are normally determined via tensile tests. In such a test, a ligament, tendon, or bone-ligament-bone complex is subjected to a tensile load applied at constant rate. A typical force-elongation curve obtained from a tensile test of a rhesus monkey ACL is shown in Figure 5.3. The force-elongation curve is initially upwardly concave, but the slope becomes nearly linear in the prefailure phase of tensile loading. The force-elongation curve represents structural properties of the ligament; that is, the shape of the curve depends on the geometry of the specimen tested (e.g., tissue length and cross-sectional area).

The mechanical, or material, properties of the ligament are expressed in terms of a stress-strain relationship. *Stress* is defined here as the force divided by the original specimen cross-sectional area. *Strain* is defined as the change in length of the specimen relative to its initial length, divided by its initial length. Hence, a tissue's mechanical properties may be obtained from force-elongation data by dividing the recorded force by the original cross-sectional area to give stress, and by dividing the difference between the specimen length and its original length by its original length to give strain. The advantage of constructing a stress-strain diagram is that to a first approximation the stress-strain behavior is independent of the tissue dimensions.

Stress-strain or force-elongation curves are typically described in terms of four regions. These four regions are illustrated in Figure 5.3. Region 1 is referred to as the *toe region*. The nonlinear response observed in this region is due to the straightening of the crimp pattern, resulting in successive recruitment of ligament fibers as they reach their straightened condition (Abrahams, 1967; Diamant, Keller, Baer, Litt, & Arridge, 1972). As the strain increases, the crimp pattern is

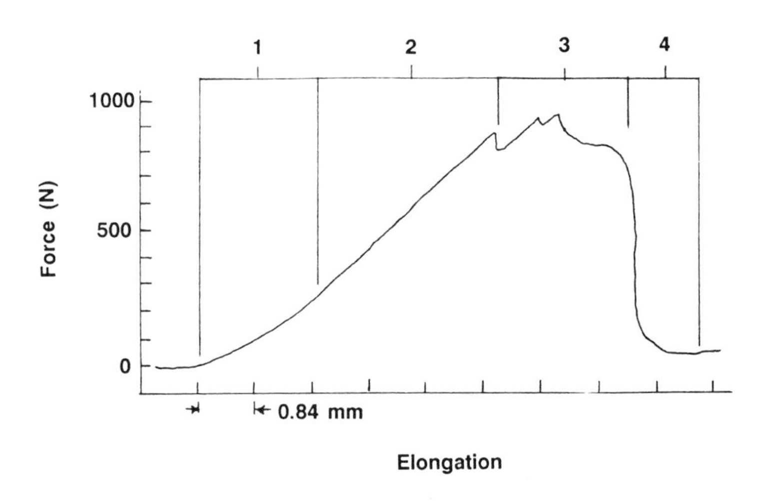

Figure 5.3 A force-elongation curve obtained from a tensile test to failure of a rhesus monkey femur-ACL-tibia preparation is shown. Four regions are commonly used to describe a force-elongation or stress-strain curve. Region 1, the toe region, elicits a nonlinear increase in load as the tissue elongates. Region 2 represents the linear region of the curve. In Region 3, isolated collagen fibers are disrupted and begin to fail. In Region 4, the ligament completely ruptures. (Reprinted by permission from Noyes, Keller, Grood, & Butler, 1984.)

lost and further deformation stretches the collagen fibers themselves (Region 2). As the strain is further increased, microstructural damage occurs (Region 3). Further stretching causes progressive fiber disruption and ultimately complete ligament rupture or bony avulsion at an insertion site (Region 4). A hypothetical curve relating the crimp pattern and collagen fiber stretch to various portions of a stress-strain curve is illustrated in Figure 5.4.

Clinically, it is important to know the normal operating conditions for ligaments acting within the body. Such information is needed to relate isolated ligament-bone test data to that of ligaments acting in vivo. This issue is considered again as more factors related to ligament function are presented throughout this chapter. Presently attention is given to the normal force and deformation levels experienced by ligaments in vivo.

Figure 5.5 illustrates a hypothetical force-elongation curve for the human ACL as postulated by Noyes, Butler, Grood, Zernicke, and Hefzy (1984). Levels of daily activity are shown along the right vertical axis, and hypothetical loading levels for the ACL are shown on the left vertical axis. This curve suggests that during daily activities (such as walking or light jogging) the ACL operates along the toe region of the force-elongation curve. The early part of Region 2 is

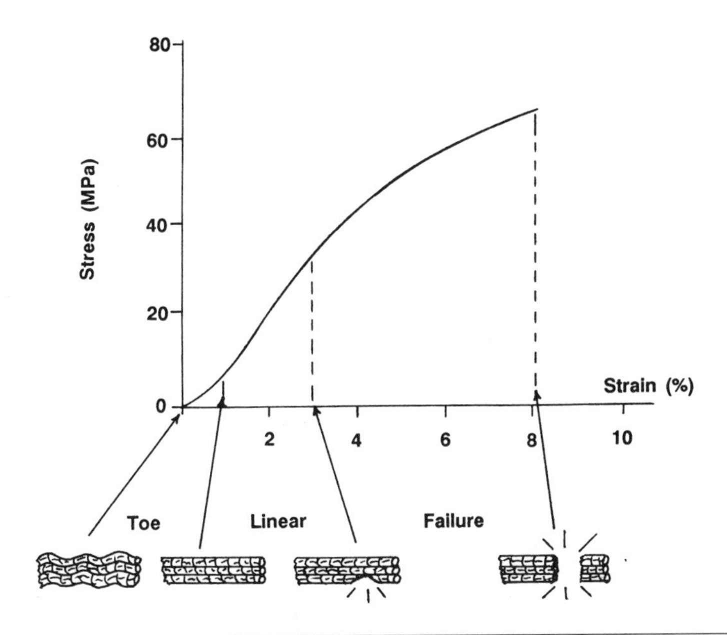

Figure 5.4 A stress-strain curve illustrating the relationship between changes in the collagen crimp pattern, or stretch, and ligament mechanical properties. Increases in ligament strain in the toe region of the curve result in straightening of the crimp pattern. In the linear portion of the curve, the collagen fibers are stretched. As the ligament is further strained, isolated ligament fibers begin to rupture, and if deformation continues, complete ligament failure occurs. (Reprinted by permission from Noyes, Keller, Grood, & Butler, 1984.)

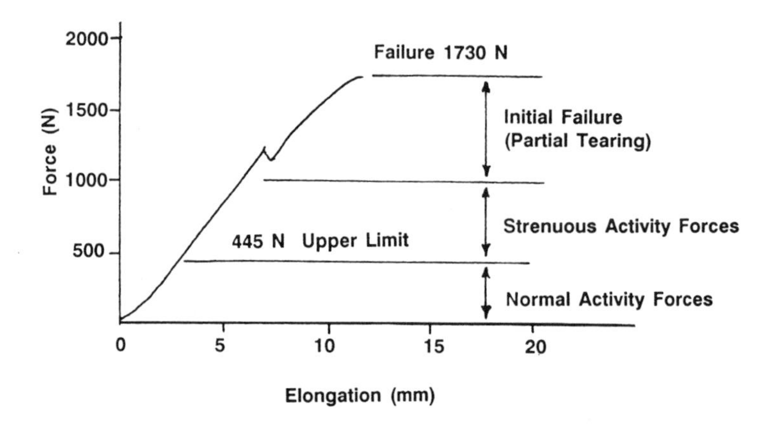

Figure 5.5 A hypothetical force-elongation curve for a human ACL-bone complex for daily activities that correspond to specific loading levels. During routine daily activities such as walking and standing, ligaments are loaded to less than one-fourth their ultimate tensile load. During strenuous activities such as fast cutting during intense running, loading levels may enter into Region 3, where isolated fiber damage takes place. (Reprinted by permission from Noyes, Butler, Grood, Zernicke, & Hefzy, 1984.)

considered the upper operating range of the ACL during strenuous activities such as fast cutting or pivoting while running. Loading of the ACL beyond Region 2 results in ligament damage and may be incurred during events like clipping in football, a ski accident, or an incorrect landing during a gymnastics floor exercise.

Viscoelastic Properties

Ligaments exhibit significant time- and history-dependent viscoelastic properties. A viscoelastic material possesses characteristics of strain rate sensitivity, stress relaxation, creep, and hysteresis.

Hysteresis is defined as the difference in the calculated area under the loading and unloading curves obtained from a force-displacement test (see Figure 5.6). Hysteresis represents the energy lost due to internal friction in the material as it is deformed.

Force relaxation (or stress relaxation) is a phenomenon that occurs in soft tissues stretched and held at a fixed length (see Figure 5.7a). Subjected to a fixed displacement, the force in a ligament will decrease with time. Force relaxation is strain rate sensitive. In general, the higher the strain rate, the larger the peak force and subsequently the greater the magnitude of the force relaxation.

In contrast to a stress-relaxation test in which the length is held fixed and the force monitored over time, during a creep test a constant force is applied and the displacement is recorded as a function of time. Under constant tensile load a ligament elongates with time (see Figure 5.7b). The general shape of the displacement-time curve depends on the past loading history (e.g., peak force, loading rate).

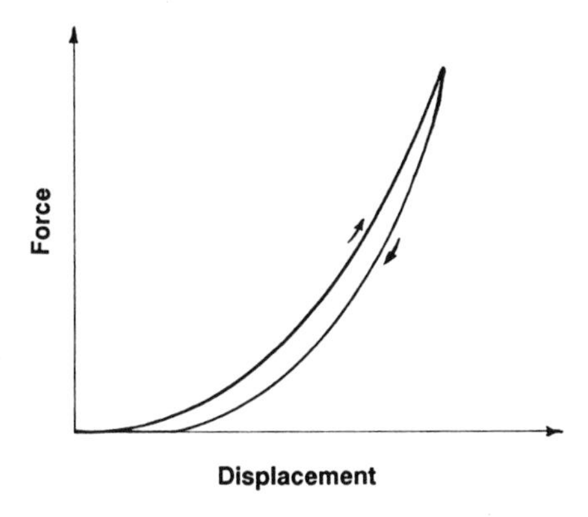

Figure 5.6 An example of an idealized force-displacement curve for cyclic loading and unloading of a ligament. Hysteresis is the difference in the area under the loading and unloading curves and represents the energy lost due to the ligament's internal friction.

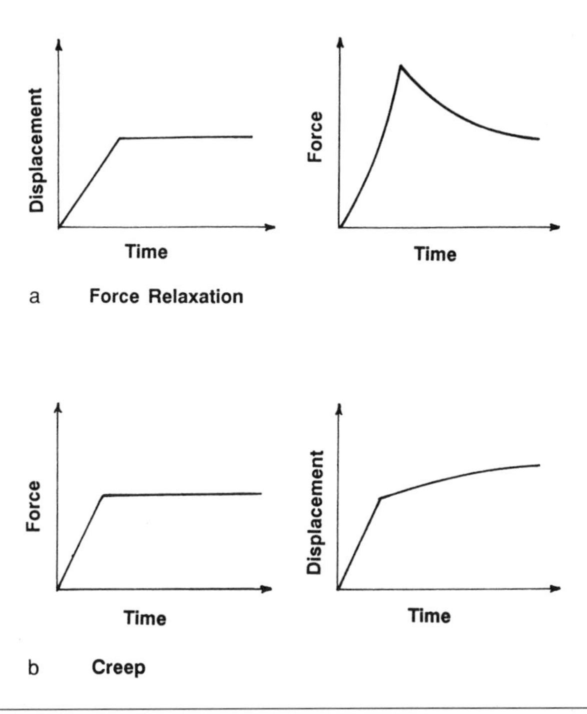

Figure 5.7 Idealized curves representing results obtained from force relaxation (a) and creep tests (b). These curves demonstrate the viscoelastic behavior of ligament. In a force- or stress-relaxation test, a ligament preparation is stretched to a given length and held fixed. With time, the force in the ligament decreases as shown on the right side (a). In a creep test, a force is applied across the ligament and maintained constant. With time, the ligament elongates as shown on the right side (b).

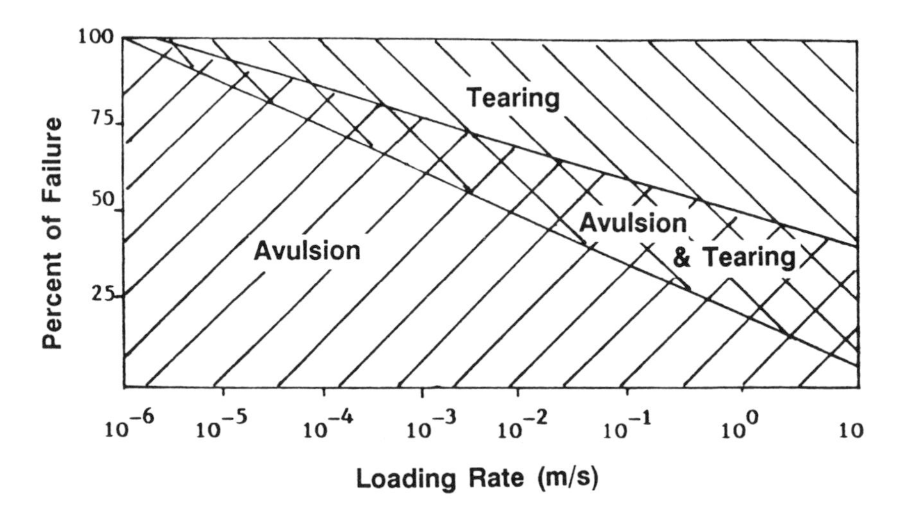

Figure 5.8 Experimental results indicate the probability of specific failure modes of bone-ligament-bone units as a function of loading rate. At slow loading rates, bony avulsion failures have the greatest probability of occurring. At fast loading rates, midsubstance failures are expected. (Adapted by permission from Crowninshield & Pope, 1976.)

Another time-dependent property is strain rate sensitivity. Very little difference has been reported in the stress-strain behavior of isolated ligaments subjected to tensile tests varying in strain rate over 3 decades (Akeson et al., 1984). However, when failure modes are considered for bone-ligament-bone complexes, strain rate becomes a significant factor due to the greater strain rate sensitivity of bone. A hypothetical diagram illustrating the relationship between the probability of failure mode of bone-ligament-bone units and the rate of loading is shown in Figure 5.8. During slow loading rates, bony avulsion failure is common. As the loading rate increases, the bone becomes stronger than the ligament substance, and the ligament substance fails first.

The time-dependent behavior of ligaments may be important during a variety of daily cyclic activities such as walking, running, cycling, and other forms of exercise. During these activities, ligament softening (peak ligament force decreases with successive loading cycles) may occur. Figure 5.9 illustrates ligament softening as observed during in vitro cyclic testing of a ligament to constant strain and at constant strain rate (S.L.Y. Woo, Gomez, & Akeson, 1981). Similar results have been observed for intact joints.

A commercial knee laxity testing device was used to quantify anterior and posterior laxity in basketball players after 90 min of practice and in runners after a 10-km race. Anterior-posterior laxity increased by more than 18% in both cases (Steiner, Grana, Chillag, & Schelberg-Karnes, 1986). In this same study it was demonstrated that muscle relaxation was not a factor in the laxity measurements

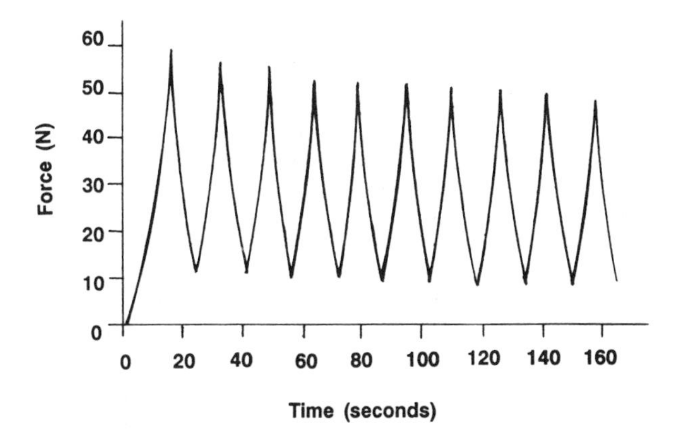

Figure 5.9 A typical ligament response to cyclic tensile loading and unloading. Peak loads decrease with each cycle, indicating ligament softening. (Reprinted by permission from S.L.Y. Woo, Gomez, & Akeson, 1981.)

because knee laxity did not vary appreciably for normal knees before and during general anesthesia. The implications of ligament softening to athletic performance are not yet known.

Temperature Sensitivity

Ligaments are temperature-sensitive; peak stresses increase with decreased temperature. On average, bone-ligament-bone preparations tested cyclically at 21 °C show 30% greater peak loads than the same preparation tested at 37 °C (Akeson et al., 1984). In vivo, peripheral joints are probably maintained near core temperature of 37 °C. However, temperature effects may become important under pathological conditions such as trauma, infection, or inflammation. One should consider this factor when conducting soft tissue experiments or comparing data from tests performed at different temperatures.

Correlation Between Structure and Function

The crimp pattern and the interaction and cross-linking of elastic, reticular, and collagen fibers are critical for normal joint mobility. These features allow ligaments a limited range of strains over which they produce minimal resistance to movement. As a result, joints may easily be moved in certain directions and over certain ranges. Additionally, if a joint is displaced toward the outer limit of some normal range of motion, the strains created in specific ligaments of that joint cause recruitment of collagen fibers from their crimp state to a straightened condition. Fiber recruitment causes these ligaments to quickly increase their resistance to further elongation, hence stabilizing the joint.

A second feature of ligaments that may be important for maintaining joint integrity is their neural network. Ligaments contain a variety of sensory receptors that may detect joint position and velocity. Sensory feedback from these receptors may contribute to the maintenance of joint integrity by initiating the recruitment (or derecruitment) of dynamic stabilizers such as muscles. More work is needed to determine the role of these neural components.

Ligament insertion sites are well suited for dissipating force. As the ligament passes through the insertion site, it is transformed from ligament to fibrocartilage and then to bone. The transition areas are less susceptible to disruption than are the extremes on either side (bone or peri-insertional ligament substance).

Individual ligaments act synergistically with other ligaments, bones, and muscles to maintain joint integrity. To understand the functional role of ligaments, all structures of the joint must be considered. Ligaments will alter their structure and mechanical properties in response to environmental changes. All these features should be considered prior to the development of training, injury treatment, or rehabilitation programs.

Failure Mechanisms

Many factors have been found to influence the failure properties of ligamentous tissue and ligament-bone units. The mechanisms of ligamentous injury may be quite different depending on the conditions during which the injury was incurred. For example, a skier quickly twisting his knee during a fall may rupture the midsubstance of both the MCL and ACL, whereas a football player slowly having his knee bent inward under the weight of several tacklers may experience a bony avulsion of the MCL.

Three principal failure modes have been observed in ligament-bone preparations (Butler et al., 1978). The first type is a ligamentous failure, characteristic of fast loading rate failures. This failure mode results in a "mop end" appearance of the disrupted ligament ends. The bundles of fibers fail at different locations due to shear and tensile mechanisms between fibers. The second mode of failure is bony avulsion fracture, which occurs during slow loading rate failures. Failure occurs through cancellous bone beneath the insertion site (Butler et al.). The third failure mode is cleavage (or pullout) at the ligament-bone interface. This mode of failure is less common than the first two due to the efficient force dissipation that occurs through the insertional zone. When a failure of this type does occur, it generally occurs through the mineralized fibrocartilage (Butler et al.).

When evaluating the extent of ligament injury, one should consider microstructural damage. Scanning electron microscopy has been used to evaluate the microstructural damage created in bone-ACL-bone preparations loaded to one-half their normal failure force (Grood, Noyes, Butler, & Suntay, 1981). Electron micrographs of these ligaments revealed that some fibers were disrupted by submaximal loading. In addition, some collagen fibers lost their wavy appearance, suggesting that permanent deformation had occurred. Results from this study suggest that ligaments may continually experience microstructural damage during strenuous activities. This may be an important factor contributing to overuse injuries. Injuries of this type occur when an individual does not allow enough reduced-activity time between exercise bouts for tissues to repair themselves. As a consequence, the tissue continues to be damaged and weakened, resulting in debilitating pain.

Effects of Aging

Numerous studies have examined the relationship between age-related processes and the structural and mechanical properties of collagenous tissues (Benedict, Walker, & Harris, 1968; Diamant et al., 1972; Elliott, 1965; Hall, 1976; Noyes & Grood, 1976; Tipton, Matthes, & Martin, 1978; Vogel, 1974; S.L.Y. Woo, Orlando, Gomez, Frank, & Akeson, 1986). However, very few studies have specifically considered ligaments (Noyes & Grood; Tipton et al., 1978; S.L.Y. Woo et al., 1986) and even fewer with bone-ligament complexes (S.L.Y. Woo et al., 1986). Two age-related processes, maturation and aging, affect bone-ligament properties.

During maturation the structure and mechanical properties of collagenous tissues change. Increases in collagen cross-linking, collagen glycosaminoglycan, and collagen-water ratios have been observed (Butler et al., 1978; Menard & Stanish, 1989). In addition, an increase in mechanical properties occurs during maturation. In studies of young and mature rabbits, S.L.Y. Woo et al. (1986) showed that the mechanical properties of bone-ligament units of younger rabbits are inferior to those of mature rabbits. During the early growth period, the strength of the bone-ligament complex rises quickly. Thereafter, strength changes occur at a slower rate. The changes in mechanical properties that occur during maturation are certainly related to the changes in cross-linking and water content that occur during this same period. The stabilization of collagen with maturity enhances tissue strength while the loss of water and elastin reduces tissue plasticity (Menard & Stanish).

Maturation affects bone and ligament properties differently. The bone-ligament junction of young animals is consistently weaker than that of the ligament substance. The reverse is true for mature animals. This suggests asynchronous rates of maturation between the bone-ligament junction and the ligament substance. A hypothetical curve illustrating the asynchronous rates of maturation for both the bone-ligament junction and the MCL substance is shown in Figure 5.10. Ligament substance matures earlier than the ligament-bone junction (S.L.Y. Woo et al., 1986). However, with maturity the bone-ligament junction strength increases to a value above that reached by the ligament substance.

After maturity is reached, ligament properties continue to change with increased age. A study of human femur-ACL-tibia complexes revealed that specimens from older donors showed lower stiffness and smaller force and deformation at failure than other specimens (Noyes & Grood, 1976). A higher incidence of bony avulsion

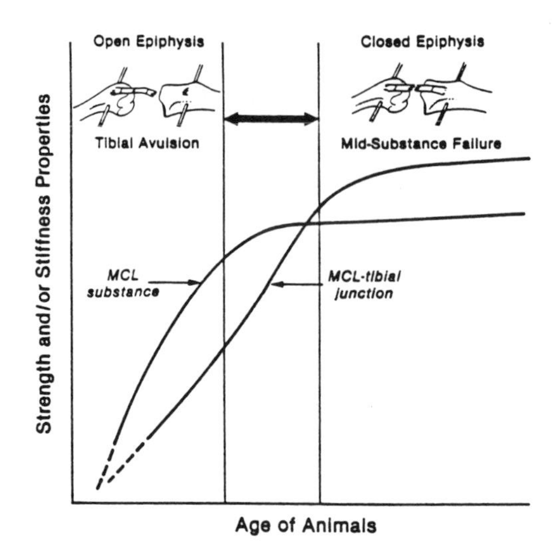

Figure 5.10 A hypothetical diagram relating the asynchronous rates of maturation for both the ligament substance and the ligament-tibia junction in terms of strength and stiffness characteristics. The strength of the ligament substance initially increases faster than that of the bone-ligament junction. The strength of the bone-ligament junction eventually surpasses that of the ligament substance. (Reprinted by permission from S.L.Y. Woo, Orlando, Gomez, Frank, & Akeson, 1986.)

failures occurred in specimens from donors ages 50 years and older than in specimens from younger donors. Aging connective tissue undergoes a generalized decrease in water content, which results in a reduction in tissue compliance. The elastic elements become coarser and more easily fractured. Thus, the alterations observed in connective tissue structure manifest themselves as changes in tissue mechanical properties (Menard & Stanish, 1989). It is difficult to say with certainty that the observed structural and mechanical changes are strictly the result of aging, and not the result of other factors such as either disease or changes in the level of a person's physical activity, which often accompany aging.

Both maturation and aging affect ligament properties. During maturation, rates of strength increase between ligament-bone substance and ligament substance appear to be asynchronous, with the ligament having the faster rate. The slower rate of strength increase of the bone-ligament unit may be due to the more complex structure of the bone-ligament interface, which requires more time to develop. It is difficult to speculate about the exact effects that aging has on bone-ligament units once maturity has been reached. Studies suggest a decrease in stiffness and smaller force and deformation at failure with increased age. However, these results may reflect changes in factors other than age. The effects of aging are highly individual and depend on factors such as genetics, past disease, and lifestyle. The observed alterations in the bone-ligament complex that occur in the later years of life are probably due to a combination of aging factors, disease, and decreased physical activity.

Similar to bone-ligament properties, gross joint properties are affected by maturation and aging. For a developing youth, maturation translates into less compliant joint structures. The ligaments stabilizing joints become stiffer, resulting in a reduction in overall joint compliance. The increased bone-ligament junction strength that occurs with maturation tends to equalize the strength between this region and the ligament substance. For mature individuals, reductions in tissue compliance and hence joint compliance can be expected with increased age. However, it has been estimated that regular exercise may retard the physiological decline associated with aging by as much as 50% (Menard & Stanish, 1989).

Exercise and Disuse

Both professional and recreational athletes experience periods of increased and decreased physical activity. These cycles are, in part, dictated by competition schedules and injuries. Alterations in activity levels can profoundly affect the structural and mechanical properties of ligaments.

Several studies have been conducted to determine the effects that joint immobilization and exercise have on the structure and mechanical characteristics of bone-ligament complexes (Akeson, Amiel, Mechanic, S.L.Y. Woo, Harwood, & Hamer, 1977; Akeson et al., 1984; Laros, Tipton, & Cooper, 1971; Noyes, 1977; Noyes, Torvik, Hyde, & DeLucas, 1974; Tipton, James, Mergner, & Tcheng, 1970; Tipton, Matthes, Maynard, & Carney, 1975; Wilson & Dahners, 1988; S.L.Y. Woo et al., 1982; S.L.Y. Woo, Gomez, Sites, Newton, Orlando, & Akeson, 1987). The effects of stress deprivation, induced by bed rest or joint immobilization, on intra-articular and extra-articular ligaments are profound. On a gross scale, ligaments appear less glistening and more grainy on dissection. Histologically, there is an increased randomness of fibers, cells, and matrix organization (Akeson et al., 1984). Some general changes that occur in bone-ligament units due to immobilization include a reduction in the failure strength and energy absorption at failure (Akeson et al., 1977, 1984; Laros et al.; Noyes; Noyes et al., 1974; Tipton et al., 1970, 1975; S.L.Y. Woo et al., 1982, 1987); bone resorption at the insertion sites (Akeson et al., 1977, 1984; Laros et al.; Noyes; Noyes et al., 1974; Tipton et al., 1975; S.L.Y Woo et al., 1982, 1987); and increased intramolecular cross-links (Akeson et al., 1977; Akeson, Amiel, & S.L.Y. Woo, 1980). In general these changes translate into increased joint stiffness and increased susceptibility to ligament or bone-ligament damage.

Joint immobilization, using casting or bracing, is often prescribed for an individual who has a joint injury. However, joint stiffness may result from this treatment as a result of ligament contracture (shortening of the structure), periarticular tissue adhesions, and decreased joint lubricity. Joint stiffness is of clinical interest for obvious reasons. As discussed previously, ligaments become less stiff with immobilization, which is contrary to what might be postulated from observed increases in joint stiffness after joint immobilization. However, if ligament contracture takes place while the joint is immobilized, then for a given joint angle that ligament will function on a stiffer portion of the force-elongation curve. So, though the general stiffness of the ligament has decreased, its contribution to joint stiffness may actually have increased. Wilson and Dahners (1988) demonstrated that ligaments do in fact contract during stress deprivation. It is difficult to determine exactly where on the force-elongation curve a ligament operates in vivo, and without such information it is difficult to assess how changes in ligament properties affect joint stiffness.

Of clinical relevance is the rate at which the properties of a bone-ligament complex return to normal after mobilizing a previously immobilized joint. Several studies have shown that considerably more time is needed to regain original strength than to deteriorate it (Noyes, 1977; Noyes et al., 1974; Tipton et al., 1975). Even after 5 months of recovery following 8 weeks of immobilization in which the joint was pinned, rhesus monkey ACLs showed only partial recovery toward normal failure strengths (the maximum load at failure being 79% that of controls) (Noyes; Noyes et al., 1974). The rates of recovery for ligament substances and bone-ligament junctions vary, with the insertion zones having slower recovery rates. As stated in the section on aging, these different rates may be due to the more complex structure of the insertion zones.

The effects of exercise on ligaments are less well defined due to inconsistent exercise protocols and methodological problems associated with quantifying physiological proof of exercise intensity. Increases in ligament mass, fiber diameter, and cross-sectional area have been reported from long-term exercise programs and the absolute separation force of bone-ligament-bone complexes is usually increased by exercise (Tipton et al., 1970, 1975). Most researchers have attributed this improvement to changes in the insertion zones rather than to changes in the ligament substance.

A hypothetical curve suggesting the relationship between duration of loading and the mechanical properties of ligaments is shown in Figure 5.11. Bone-ligament alterations due to exercise are less dramatic than those caused by stress deprivation. However, due to the slow collagen turnover rate, the resulting effects are probably long-lasting in an animal that maintains normal activity. As with stress deprivation, the extent to which exercise effects a bone-ligament complex depends on several factors including age, sex, species, nutrition, hormones, and mode of exercise.

Trauma

Ligament trauma occurs when loading of the collagenous tissue causes microscopic or macroscopic structural damage. Microscopic damage results when a few isolated fibers are torn or stretched. Macroscopic damage occurs when major portions of the tissue are disrupted or completely torn. Injury to collagenous tissue invokes an inflammatory and healing response. The initial responses of hemorrhage and inflammation usually dominate for the first 1 to 7 days following trauma. Following the initial response, a proliferation of connective tissue cells

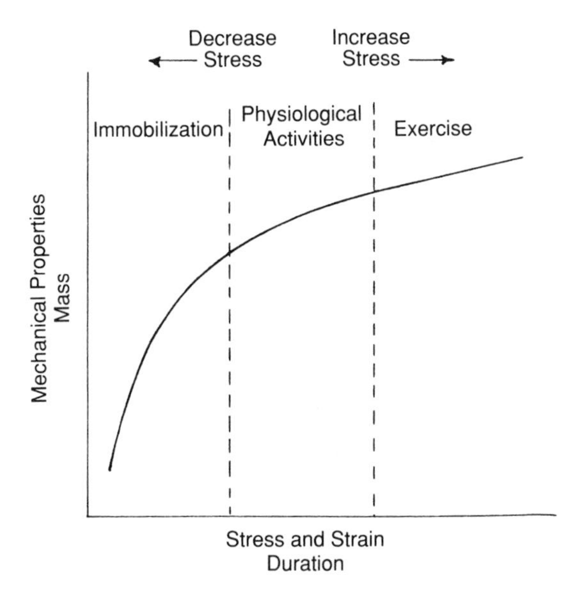

Figure 5.11 A hypothetical curve demonstrating the nonlinear relationship between soft tissue mechanical properties (and mass) and the level (and duration) of tissue stress and strain is shown. During stress deprivation or immobilization, the mechanical properties can decrease rapidly. Exercise levels increased above that of daily activities cause the mechanical properties to increase, but less dramatically than the decrease observed during stress deprivation. (Reprinted by permission from S.L.Y. Woo, Gomez, Y.K. Woo, & Akeson, 1982.)

takes place, with peak levels reached 2 to 3 weeks after injury (Akeson et al., 1984). Three to 6 weeks after injury, the number and size of fibroblasts decrease, and their nuclei begin to align with the long axis of the ligament. Remodeling and cell alignment continue thereafter (Frank, S.L.Y. Woo, Amiel, Harwood, Gomez, & Akeson, 1983). Progressive changes in cell number, size, distribution, and orientation occur throughout stages of inflammation, proliferation, and remodeling. Variability in the extent and duration of these changes depends on a number of systematic factors (age, nutrition, hormones) and local factors (extent of trauma, blood supply, infection, mechanical stress, temperature, chemical environment).

From a clinical perspective, these processes are indicative of the amount of time required for a ligament to heal. Remodeling and cell alignment do not begin until 3 to 6 weeks after trauma. It may be several more months before the remodeling process is complete. Thus, ligament healing is a slow process and should be recognized and treated as such.

Soft tissue trauma may or may not be perceived by an individual and may or may not affect joint function. Microstructural damage and remodeling often take place without being recognized by the individual. However, repeated minor traumas may compound if sufficient healing time is not given between injuries. Under these conditions inflammation and pain may become noticeable and hinder joint function. Clinically, these conditions are manifested as lateral epicondylitis (tennis elbow), Achilles tendonitis, and other collagenous overload injuries.

Drugs

Frequently, corticosteroids are prescribed to treat inflammation. Various studies have investigated the effects of steroid injections on the mechanical properties of ligaments (Hollander, 1972, 1974; Mankin, 1974; Noyes, 1977). In the study by Noyes, multiple intra-articular injections of methylprednisolone acetate, a corticosteroid, were administered into the knees of rhesus monkeys. Animals were assigned to a control group, a sham group in which saline was substituted for the steroid, a 20-mg (large-dosage) group studied at 6 and 15 weeks, or a 4-mg (small-dosage) group studied at 15 weeks following injection. Femur-ACL-tibia preparations were tested to failure at a rapid rate. Statistically significant decreases in load at failure occurred in both the large-dosage (11% at 6 weeks, 20% at 15 weeks) and small-dosage (9% at 15 weeks) groups compared with controls. An 11% decrease in energy to failure was present at 15 weeks in the large-dosage group, and an 8% decline occurred at 8 weeks in the small-dosage group. Further work in this area reported by Noyes, Keller, Grood, and Butler (1984) indicate that single intraligamentous injections of methylprednisolone acetate may have greater deleterious effects than multiple intra-articular injections. Results from these studies suggest that single intraligamentous or multiple intra-articular steroid injections have the potential to cause deleterious effects on the mechanical properties of ligaments. Corticosteroids are known to have inhibitory effects on the synthesis of glycosaminoglycans, proteins, and collagen. However, the specific mechanisms responsible for the observed deleterious effects are not known.

Clinical Considerations

In today's health-conscious society, with more and more people participating in fitness programs and strenuous sports, ligament injuries are on the rise and the number of clinical repairs and replacements is considerable. Accompanying the rise in ligament injuries has been a rise in biomechanical research. Clinically, this research has contributed to the treatment of ligament injuries in four major areas: injury diagnosis, treatment, rehabilitation, and joint bracing.

Diagnosis

Determination of the nature and degree of ligament injury is one of the most important factors in successful treatment of the damaged structure. Diagnosis of a damaged ligament should begin with a history of the trauma. The activity during which the injury occurred, the sensation of a "pop," swelling, and joint instability are all indicators of the structures involved and the extent of injury. Beyond this step, physical examination of the injured joint is required.

Numerous test protocols have been developed for diagnosing knee ligament injuries (anterior drawer, Lachman, MacIntosh pivot shift, Losse, Slocum, "jerk," and recurvatum). For detailed discussions of these tests and others, refer to the articles by Hughston, Andrews, Cross, and Moschi (1976); Galway and MacIntosh (1980); Losse (1985); and Torg, Conrad, and Kalen (1976). During these tests a clinician attempts to produce a displacement of the joint in a given direction by applying a prescribed force. The resulting displacement indicates the joint structures damaged. For example, during an anterior drawer test, the patient's knee is flexed to 90°, the tibia is held in neutral rotation, and the clinician attempts to produce an anterior displacement of the tibia relative to the femur. The amount of displacement indicates whether the ACL has been damaged.

To understand the significance of these clinical procedures, one must understand the role that various structures play to stabilize the joint. Studies have been conducted in which intact cadaver knees were loaded or displaced a specific amount in specially designed fixtures. Individual ligaments were then cut, and the change in force or joint position was recorded (Nielson, Ovesen, & Rasmussen, 1984; Noyes, Keller, Grood, & Butler, 1984). Resulting changes in the load or joint position suggest the role that the cut structure played in resisting the applied load or displacement. For example, when an initial anterior displacement is applied to the tibia with the femur held fixed and all ligaments of the knee intact, a restraining force can be measured. If the ACL is subsequently cut, then the restraining force decreases. The amount this force decreases is indicative of the contribution the ACL makes to stabilize the joint in the configuration tested.

Several interesting findings have been reported from work that involved similar tests. The knee has primary and secondary restraints (Noyes, Keller, Grood, & Butler, 1984). The ACL is the primary restraint of anterior drawer when the knee is flexed between 30° and 90°. The PCL is the primary restraint of posterior drawer at similar knee angles. Nielson et al. (1984) proposed that injuries caused by external rotation of the tibia, as might occur in skiing when the inside of a ski edge hits a bump and causes the tibia to rotate externally, damage the MCL first and then the ACL, whereas internal rotation, as might occur when a football player attempts to pivot on one foot while that foot is held fixed to the ground, injures the ACL first and then the MCL. This information indicates the diagnostic usefulness of obtaining the history of the trauma; specific movements are likely to damage specific structures. These data also suggest that protocols or test devices that examine the amount of anterior tibial displacement as a function of knee angle can be useful in determining ACL insufficiency.

Devices to quantify knee laxity have been devised to assist in the diagnosis of ligament injuries. These devices generally apply some force or torque to the joint. The resulting linear or angular displacement is recorded; compared to *normals*, or results from testing of the contralateral joint; and used to evaluate the extent of specific ligament damage. Such knee devices include the Stryker knee laxity tester (KLT) and the Medmetric KT-1000 or KT-2000. Support for clinical use of the Stryker KLT has been given by Boniface, Fu, and Ilkhanipour (1986) and for both the Stryker KLT and Medmetric KT-2000 by Daniel, Malcom, Losse, Stone, Sachs, and Burks (1985). This support is based on the successful use of these devices in predicting ACL damage as indicated by corresponding arthroscopic studies. In most cases where an ACL tear was present, the laxity measurements performed in the clinic were suggestive of a pathologic anterior laxity. Despite this support, the usefulness of these devices may be limited by the fact that under the low loads (90 N) generally applied when using these devices, secondary stabilizers may be sufficient to prevent joint displacement and hence may mask the injury present in the primary stabilizer.

In addition to passive knee laxity tests, some researchers advocate dynamic tests (Tegner, J. Lysholm, M. Lysholm, & Gillquist, 1986; Tibone, Antich, Fanton, Moynes, & Perry, 1986). During dynamic tests such as walking, stair climbing, one-legged hopping, running, and cutting, quantities such as muscle activity (recorded using electrodes) and joint forces (determined using a force plate, cinematography, and inverse dynamics) may indicate abnormal responses and hence be useful for diagnosing injuries. Information from dynamic tests may also be beneficial for determining when an injured knee is sufficiently rehabilitated to withstand strenuous activities.

Treatment

Damaged ligaments may be treated conservatively or surgically by primary suture, augmentation, and synthetic or allograft replacement. For a review of the history of ACL repair techniques, refer to Burnett and Fowler (1985).

The selection of treatment procedure depends on the needs of the individual. A conservative approach may be adequate for mildly active individuals who can compensate for the damaged tissue with other ligaments, muscles, or braces. However, as shown in a study by McDaniel and Dameron (1980), untreated ruptures of the ACL may result in increased anterior laxity, rotatory instability, and meniscal tears. In addition, osteoarthritis may result from long-term joint laxity.

Some form of ligament repair may be necessary for individuals with severe joint instability or for those desiring to participate in activities placing the injured ligaments and joint at further risk. In these cases, ligament repair is performed with the goal of returning the joint to near-normal functional integrity. To do this requires an understanding of the mechanical properties of the normal ligament, the repair process and its effect on ligament mechanical properties, immune responses to implant materials, joint structure interactions, and the effect of attachment geometry and ligament tension on joint motion and stability.

A ligament substitute should possess mechanical characteristics similar to that of the original tissue. Ligament mechanical properties obtained from testing bone-ligament-bone preparations have been used to assess the adequacy of possible ligament substitutes (allograft or synthetic). Noyes, Keller, Grood, and Butler (1984) compared the ultimate load of the ACL with various allografts that have been used to replace a ruptured ACL. The medial third of the bone-patellar-bone tendon has been found to have an ultimate load 159% that of the ACL. The retinaculum-patellar tendon, in comparison, has an ultimate load 21% that of the ACL. Both of these structures have been used as ACL replacements. However, from simple tensile test experiments, it is evident that the quadriceps-patellar-bone tendon is inadequate as an ACL replacement, whereas the medial third of the bone-patella-bone tendon has a greater likelihood of success.

Ultra-high-molecular-weight polyethylene (UHMWPE) was once used as an ACL replacement. However, biomechanical testing of UHMWPE showed it to possess poor creep characteristics compared with a normal ACL (Chen & Black, 1980). When the artificial ligament was repeatedly stressed, it lengthened and hence lost its joint-stabilizing effect. UHMWPE implants demonstrated a high failure rate clinically and were removed from the market. Unfortunately, adequate biomechanical testing was not performed until after the artificial ligament had been used clinically.

Carbon fiber is another material that has been used in the construction of synthetic ligaments. Although carbon fibers have high tensile strengths and low creep characteristics, studies of the use of these implants have shown that carbon fibers fragment within the body and subsequently migrate to other areas (Amis, Kempson, Campbell, & Miller, 1988; Zoltan, Reinecke, & Indelicato, 1988). The long-term effects of these fragments and increased carbon concentrations are not yet known.

A promising ligament substitute appears to be polyester. Polyester ligaments have been shown to have similar mechanical properties to the human ACL (Amis et al., 1988). ACLs replaced using this material have shown good success clinically, but long-term data is still being compiled.

These few examples illustrate the importance of properly testing and evaluating ligament substitutes before allowing their use clinically. Unfortunately, rigorous biomechanical testing of replacement ligaments has often taken place after their clinical use has begun.

Other important factors contributing to the successful replacement of a ligament are the attachment locations and preload. Errors in either locating the ligament attachment sites or preloading the ligament will cause joint stiffness or laxity. Arms, Pope, Johnson, Fischer, Arvidsson, and Eriksson (1984) showed that misplacement of ligament attachment sites causes the ligament to strain more during normal joint motion than if it were located properly. Incorrect ligament attachment may lead to excessive ligament strain and early failure. Likewise, even if a ligament substitute is attached in the proper location, if the preload is too large, then the ligament will limit the joint range of motion, causing excessive loading and possibly failure. If the preload is too small and the ligament is slack, then the ligament will not lend proper support to the joint.

Suture of the original ligament is the least-used surgical procedure. If the attachment sites are intact, then the major issues are uniting the ruptured ends and restoring proper tension. Unfortunately, these are not simple issues to resolve. Ligaments repair themselves by scar formation followed by cell differentiation and alignment. The scar material, having different material properties than ligament, may stretch during healing and result in a slack ligament. This may lead to joint laxity and conditions causing cartilage degeneration. Early, controlled joint mobilization appears to enhance the healing rate of sutured ligaments by increasing cellularity, collagen content, and tensile strength (Noyes, Butler, Grood, Zernicke, & Hefzy, 1984; Amis et al., 1988). Whether early mobilization reduces ligament stretching during healing needs further investigation.

Rehabilitation

Biomechanical study of knee ligaments has contributed significantly to the development of rehabilitation protocols. Arms et al. (1984) placed a strain gauge on the anteromedial aspect of the ACL of cadavers and measured the change in strain during knee flexion. During passive knee flexion, the ACL is minimally strained when the knee is flexed between 30° and 45° (0° being full extension). Simulated quadriceps muscle forces increased the strain in the ACL above that of passive knee flexion for angles less than 60° of flexion and decreased the strain for larger flexion angles. Therefore, according to these data, to reduce ACL strain after reconstruction or replacement, knee motion should be allowed only within the range 30° to 40° of flexion during rehabilitation programs calling for no muscle activity. If isometric quadriceps muscle contractions are used during rehabilitation to prevent muscle atrophy, then motion should be limited to angles greater than 45° of flexion.

Bracing

There are three primary brace categories:

- Prophylactic
- Rehabilitative
- Functional

Prophylactic braces are intended as preventive devices and hence are used prior to any ligament or joint injury. Rehabilitative braces are designed to allow a limited range of joint motion during recovery from a joint injury or a surgical procedure. Functional braces are intended to assist in stabilizing joints that may be prone to injury due to joint laxity or ligament weakness.

Prophylactic knee braces consist of either a lateral bar or both medial and lateral bars suspended by either straps or taping, or both. They are intended to prevent or reduce the severity of injuries by reducing medial joint opening. Yet clinical studies of knee injuries sustained by football players wearing and not wearing prophylactic braces indicate that prophylactic braces do not reduce the incidence of knee injury and may, in fact, increase the incidence (Garrick &

Requa, 1987; Hewson, Mendini, & Wang, 1986; Rovere, Haupt, & Yates, 1987). Some braces have lateral bars that may act as fulcrums, causing greater displacement of the tibia relative to the femur during certain loading conditions. The mechanics of the brace may provide only a partial explanation for increased injury. Braces may also give athletes a false sense of security, causing them to act more aggressively than they would without the braces.

In addition to the clinical investigations of braces, biomechanical studies have been conducted. Paulos, France, Jayaraman, Abbot, and Rosenberg (1987); Baker, VanHanswyk, Orthotist, Bogosian, Werner, and Murphy (1987); and Baker, VanHanswyk, Orthotist, Bogosian, and Werner (1989) looked at static and dynamic response characteristics of prophylactic braces on cadaver knees. They reported that prophylactic braces provide little if any protective effect. In general, there is no evidence to support the use of prophylactic braces at this time (Millet & Drez, 1988).

Rehabilitative knee braces are designed to allow protected motion of injured knees treated operatively or nonoperatively by allowing controlled joint motion that may be beneficial during ligament healing. Rehabilitative knee braces consist of hinges, brace arms, and thigh and calf enclosures. Eight commercially available rehabilitative knee braces were compared using a mechanical surrogate leg (Cawley, France, & Paulos, 1989). Most of the braces tested reduced both lower limb translations and rotations compared with the nonbraced limb. Although these braces appear to be effective in various treatment programs, these braces provide little anterior-posterior stability and more knee joint motion may occur than prescribed by the device (due to motion of the knee relative to the brace).

Functional braces are designed to assist or provide stability for unstable joints. Two basic construction types are available. Both types use hinges and posts. Differences exist in whether limb enclosures or suspension straps are employed. Several studies have been performed to determine the effectiveness of these braces (Beck, Drez, Young, Cannon, & Stone, 1986; Colville, Lee, & Ciullo, 1986; Knutzen, Bates, & Schot, 1987; Wojtys, Goldstein, Redfern, Trier, & Matthews, 1987). Beck et al. tested several functional braces on ACL-deficient knees with commercially available knee laxity testers and concluded that braces constructed with enclosures rather than with straps perform better in preventing anterior tibial displacement. They also noted that as the force increased, the effectiveness of the braces in controlling anterior tibial displacement decreased. Similar results were obtained by other investigators (Colville et al.; Wojtys et al.). During athletic competition, when conditions of high loading exist, the ability of functional braces to control pathologic anterior laxity is minimal (Millet & Drez, 1988).

Knee bracing continues to be a complex and controversial topic. Further research is needed to develop brace designs that will provide the support and flexibility required by normal functional joints.

Conclusion

Ligaments are important structures for maintaining joint function and integrity. They act as passive joint stabilizers, but they are dynamic in their ability to adapt to changing environmental conditions. Because ligaments are commonly injured during sports, an understanding of their structural and mechanical properties along with the various factors that affect these properties is relevant for exercise and sport scientists. Although significant advances have been made in understanding the biology, biochemistry, and biomechanics of soft tissues, much work remains to be done. Limited information is available pertaining to in vivo tissue mechanical characteristics and behavior. Without accurate values of such in vivo information, extrapolations from animal and human in situ bone-ligament-bone testing to the function of intact human ligaments cannot be made confidently.

References

Abrahams, M. (1967). Mechanical behavior of tendon *in-vitro*: A preliminary report. *Medical and Biological Engineering*. **5**(5), 433-443.

Akeson, W.H., Amiel, D., Mechanic, G.L., Woo, S.L.Y., Harwood, F.L., & Hamer, M.L. (1977). Collagen cross-linking alterations in joint contractions: Changes in the reducible cross-links in periarticular connective tissue collagen after nine weeks of immobilization. *Connective Tissue Research*, **5**(1), 15-19.

Akeson, W.H., Amiel, D., & Woo, S.L.Y. (1980). Immobility effects on synovial joints the pathomechanics of joint contracture. *Biorheology*, **17**(1/2), 95-110.

Akeson, W.H., Woo, S.L.Y., Amiel, D., & Frank, C.B. (1984). The chemical basis of tissue repair. In L.Y. Hunter and F.J. Funk (Eds.), *Rehabilitation of the injured knee* (pp. 93-104). St. Louis: Mosby.

Amis, A.A., Kempson, S.A., Campbell, J.R., & Miller, J.H. (1988). Anterior cruciate ligament replacement: Biocompatibility and biomechanics of polyester and carbon fibre in rabbits. *Journal of Bone and Joint Surgery*. **70B**(4), 628-634.

Anthony, C.P., & Thibodeau, G.A. (1983). *Textbook of anatomy and physiology* (p. 157). Toronto: Mosby.

Arms, S.W., Pope, M.H., Johnson, R.J., Fischer, R.A., Arvidsson, I., & Eriksson, E. (1984). The biomechanics of anterior cruciate ligament rehabilitation and reconstruction. *American Journal of Sports Medicine*, **12**(1), 8-18.

Baker, B.E., VanHanswyk, E., Orthotist, C., Bogosian, S., & Werner, F.W. (1989). The effect of knee braces on lateral impact loading of the knee. *American Journal of Sports Medicine*, **17**(2), 182-186.

Baker, B.E., VanHanswyk, E., Orthotist, C., Bogosian, S., Werner, F.W., & Murphy, D. (1987). A biomechanical study of the static stabilizing effect of knee braces on medial stability. *American Journal of Sports Medicine*, **15**(6), 566-570.

Beck, C., Drez, D., Young, J., Cannon, W.D., & Stone, M.L. (1986). Instrumented testing of functional knee braces. *American Journal of Sports Medicine*, **14**(4), 253-256.

Benedict, J.V., Walker, L.B., & Harris, E.H. (1968). Stress-strain characteristics and tensile strength of unembalmed human tendon. *Journal of Biomechanics*, **1**(1), 53-63.

Boniface, R.J., Fu, F.H., & Ilkhanipour, K. (1986). Objective anterior cruciate ligament testing. *Orthopedics*, **9**(3), 391-393.

Burnett, Q.M., & Fowler, P.J. (1985). Reconstruction of the anterior cruciate ligament: Historical overview. *Orthopedic Clinics of North America*, **16**(1), 143-157.

Butler, D.L., Grood, E.S., Noyes, F.R., & Zernicke, R.F. (1978). Biomechanics of ligaments and tendons. *Exercise and Sports Science Reviews*, **6**, 125-181.

Cawley, P.W., France, E.P., & Paulos, L.E. (1989). Comparison of rehabilitative knee braces: A biomechanical investigation. *American Journal of Sports Medicine*, **17**(2), 141-146.

Chen, E.H., & Black, J. (1980). Materials design analysis of the prosthetic anterior cruciate ligament. *Journal of Biomedical Materials Research*, **14**(5), 567-586.

Colville, M.R., Lee, C.L., & Ciullo, J.V. (1986). The Lenox Hill brace: An evaluation of effectiveness in treating knee instability. *American Journal of Sports Medicine*, **14**(4), 257-261.

Crowninshield, R.D., & Pope, M.H. (1976). The strength and failure characteristics of rat medial collateral ligaments. *Journal of Trauma*, **16**(2), 99-105.

Daniel, D.M., Malcom, L.L., Losse, G., Stone, M.L., Sachs, R., & Burks, R. (1985). Instrumented measurement of anterior laxity of the knee. *Journal of Bone and Joint Surgery*, **67A**(5), 720-726.

Diamant, J., Keller, A., Baer, E., Litt, M., & Arridge, R.G.C. (1972). Collagen: Ultrastructure and its relations to mechanical properties as a function of ageing. *Proceedings of the Royal Society of London: Series B: Biological Sciences*, **180**(1060), 293-315.

Dye, S.F., & Cannon, W.D. (1988). Anatomy and biomechanics of the anterior cruciate ligament. *Clinics in Sports Medicine*, **7**(4), 715-725.

Elliott, D.H. (1965). Structure and function of mammalian tendon. *Biological Reviews*, **40**(3), 392-421.

Frank, C., Woo, S.L.Y., Amiel, D., Harwood, F., Gomez, M., & Akeson, W. (1983). Medial collateral ligament healing: A multidisciplinary assessment in rabbits. *American Journal of Sports Medicine*, **11**(6), 379-389.

Galway, H.R., & MacIntosh, D.L. (1980). The lateral pivot shift: A symptom and sign of anterior cruciate ligament insufficiency. *Clinical Orthopaedics and Related Research*, **147**, 45-50.

Garrick, J.G., & Requa, R.K. (1987). Prophylactic knee bracing. *American Journal of Sports Medicine*, **15**(5), 471-476.

Grood, E.S., Noyes, F.R., Butler, D.L., & Suntay, W.J. (1981). Ligamentous and capsular restraints preventing straight medial lateral laxity in intact human cadaver knees. *Journal of Bone and Joint Surgery*, **63A**(8), 1257-1269.

Halata, Z., Badalamente, M.A., Dee, R., & Propper, M. (1984). Ultrastructure of sensory nerve endings in monkey (macaca fascicularis) knee joint capsule. *Journal of Orthopaedic Research*, **2**(2), 169-176.

Hall, D.A. (1976). *The aging of connective tissue*. London: London Academic Press.

Hewson, G.F., Mendini, R.A., & Wang, J.B. (1986). Prophylactic knee bracing in college football. *American Journal of Sports Medicine*, **14**(4), 262-266.

Hollander, J.L. (1972). *Intrasynovial corticosteroid therapy: Arthritis and allied conditions: A textbook of rheumatology*. Philadelphia: Lea & Febiger.

Hollander, J.L. (1974). Collagenase, cartilage and cortisol. *New England Journal of Medicine*, **290**(1), 50-51.

Hughston, J.C., Andrews, J.R., Cross, M.J., & Moschi, A. (1976). Classification of knee ligament instabilities: Parts I and II. *Journal of Bone and Joint Surgery*, **58A**(2), 159-179.

Kastelic, J., Galeski, A., & Baer, E. (1978). The multicomposite structure of tendon. *Connective Tissue Research*, **6**(1), 11-23.

Kennedy, J.C., Alexander, I.J., & Hayes, K.C. (1982). Nerve supply of the human knee and its functional importance. *American Journal of Sports Medicine*, **10**(6), 329-335.

Knutzen, K.M., Bates, B.T., & Schot, P. (1987). A biomechanical analysis of two functional knee braces. *Medicine and Science in Sports and Exercise*, **19**(3), 303-309.

Laros, G.S., Tipton, C.M., & Cooper, R.R. (1971). Influence of physical activity on ligament insertions in the knees of dogs. *Journal of Bone and Joint Surgery*, **53A**(2), 275-286.

Losse, R.E. (1985). Diagnosis of chronic injury to the anterior cruciate ligament. *Orthopedic Clinics of North America*, **16**(1), 83-97.

Mankin, J.J. (1974). The reaction of articular cartilage to injury and osteoarthritis. *New England Journal of Medicine*, **291**(24), 1285-1292.

Marshall, J.L., & Rubin, R.M. (1977). Knee ligament injuries: A diagnostic and therapeutic approach. *Orthopedic Clinics of North America*, **8**(3), 641-668.

McDaniel, W.J., & Dameron, T.B. (1980). Untreated ruptures of the anterior cruciate ligament. *Journal of Bone and Joint Surgery*, **62A**(5), 696-705.

Menard, D., & Stanish, W.D. (1989). The aging athlete. *American Journal of Sports Medicine*, **17**(2), 187-196.

Millet, C.W., & Drez, D.J. (1988). Principles of bracing for the anterior cruciate ligament-deficient knee. *Clinics in Sports Medicine*, **7**(4), 827-833.

Nielson, S., Ovesen, J., & Rasmussen, O. (1984). The anterior cruciate ligament of the knee: An Experimental Study of its importance in rotatory knee instability. *Archives of Orthopaedic and Traumatic Surgery*, **103**(3), 170-174.

Noyes, F.R. (1977). Functional properties of knee ligaments and alterations induced by immobilization: A correlative biomechanical and histological study in primates. *Clinical Orthopaedics and Related Research*, **123**, 210-242.

Noyes, F.R., Butler, D.L., Grood, E.S., Zernicke, R.F., & Hefzy, M.S. (1984). Biomechanical analysis of human ligament grafts used in knee ligament repairs and reconstruction. *Journal of Bone and Joint Surgery*, **66A**(3), 344-352.

Noyes, F.R., & Grood, E.S. (1976). The strength of the anterior cruciate ligament in humans and rhesus monkeys: Age related and species related changes. *Journal of Bone and Joint Surgery*, **58A**(8), 1074-1082.

Noyes, F.R., Keller, C.S., Grood, E.S., & Butler, D.L. (1984). Advances in the understanding of knee ligament injury, repair, and rehabilitation. *Medicine and Science in Sports and Exercise*, **16**(5), 427-443.

Noyes, F.R., Torvik, P.J., Hyde, W.B., & DeLucas, J.L. (1974). Biomechanics of ligament failure: 2. An analysis of immobilization, exercise, and reconditioning effects in primates. *Journal of Bone and Joint Surgery*, **56A**(7), 1406-1418.

Paulos, L., France, E.P., Jayaraman, G., Abbot, P., & Rosenberg, T. (1987). Biomechanics of lateral bracing. *American Journal of Sports Medicine*, **15**(5), 419-429.

Rovere, G.D., Haupt, H.A., & Yates, C.S. (1987). Prophylactic knee bracing in college football. *American Journal of Sports Medicine*, **15**(2), 111-116.

Schultz, R.A., Miller, D.C., Kerr, C.S., & Micheli, L. (1984). Mechanoreceptors in human cruciate ligaments. *Journal of Bone and Joint Surgery*, **66A**(7), 1072-1076.

Schutte, M.J., Dabezies, E.J., Zimny, M.L., & Happel, L.T. (1987). Neural anatomy of the human anterior cruciate ligament. *Journal of Bone and Joint Surgery*, **69A**(2), 243-247.

Steiner, M.E., Grana, W.A., Chillag, K., & Schelberg-Karnes, E. (1986). The effect of exercise on anterior-posterior knee laxity. *American Journal of Sports Medicine*, **14**(1), 24-29.

Tegner, Y., Lysholm, J., Lysholm, M., & Gillquist, J. (1986). A performance test to monitor rehabilitation and evaluate anterior cruciate ligament injuries. *American Journal of Sports Medicine*, **14**(2), 156-159.

Tibone, J.E., Antich, T.J., Fanton, G.S., Moynes, D.R., & Perry, J. (1986). Functional analysis of anterior cruciate ligament instability. *American Journal of Sports Medicine*, **14**(4), 276-284.

Tipton, C.M., James, S.L., Mergner, W., & Tcheng, T.K. (1970). Influence of exercise on strength of medial collateral knee ligaments of dogs. *American Journal of Physiology*, **218**(3), 894-902.

Tipton, C.M., Matthes, R.D., & Martin, R.K. (1978). Influence of age and sex on the strength of bone-ligament junctions in knee joints of rats. *Journal of Bone and Joint Surgery*, **60A**(2), 230-234.

Tipton, C.M., Matthes, R.D., Maynard, J.A., & Carney, R.A. (1975). The influence of physical activity on ligaments and tendons. *Medicine and Science in Sports*, **7**(3), 165-175.

Torg, J.S., Conrad, W., & Kalen, V. (1976). Clinical diagnosis of anterior cruciate ligament instability in the athlete. *American Journal of Sports Medicine*, **4**(2), 84-93.

Viidik, A. (1973). Functional properties of collagenous tissue. *International Review of Connective Tissue Research*, **6**, 127-215.

Vogel, H.G. (1974). Correlation between tensile strength and collagen content in rat skin: Effect of age and cortisol treatment. *Connective Tissue Research*, **2**(3), 177-182.

Wilson, C.J., & Dahners, L.E. (1988). An examination of the mechanisms of ligament contracture. *Clinical Orthopaedics and Related Research*, **227**, 286-291.

Wojtys, E.M., Goldstein, S.A., Redfern, M., Trier, E., & Matthews, L.S. (1987). A biomechanical evaluation of the Lenox Hill knee brace. *Clinical Orthopaedics and Related Research*, **220**, 179-183.

Woo, S.L.Y., Gomez, M.A., & Akeson, W.H. (1981). The time and history-dependent viscoelastic properties of the canine medial collateral ligament. *Journal of Biomechanical Engineering*, **103**, 293-298.

Woo, S.L.Y., Gomez, M.A., Sites, T.J., Newton, P.O., Orlando, C.A., & Akeson, W.H. (1987). The biomechanical and morphological changes in the medial collateral ligament of the rabbit after immobilization and remobilization. *Journal of Bone and Joint Surgery*, **69A**(8), 1200-1211.

Woo, S.L.Y., Gomez, M.A., Woo, Y.K., & Akeson, W.H. (1982). Mechanical properties of tendons and ligaments: 2. The relationships of immobilization and exercise on tissue remodeling. *Biorheology*, **19**(3), 397-408.

Woo, S.L.Y., Orlando, C.A., Gomez, M.A., Frank, C.B., & Akeson, W.H. (1986). Tensile properties of the medial collateral ligament as a function of age. *Journal of Orthopaedic Research*, **4**(2), 133-141.

Zimny, M.L., Schutte, M., & Dabezies, E. (1986). Mechanoreceptors in the human anterior cruciate ligament. *Anatomical Record*, **214**(2), 204-209.

Zoltan, D.J., Reinecke, C., & Indelicato, P.A. (1988). Synthetic and allograft anterior cruciate ligament reconstruction. *Clinics in Sports Medicine*, **7**(4), 773-784.

Chapter 6

Skeletal Adaptation to Functional Stimuli

Ted S. Gross and Steven D. Bain

Bone tissue is dynamic, constantly adapting to the demands of its environment. An understanding of how the skeleton adapts is important to those concerned with health and fitness and also to those interested in bone diseases such as osteoporosis. In chapter 6, Ted Gross and Steven Bain discuss how the cells of bone tissue respond to the functional loads placed upon them. Although the study of bone mechanics can be traced back about 150 years, much of contemporary thought on bone is exactly that—contemporary. Research of the past 10 years has contributed greatly to our understanding of bone mechanics. As with other scientific disciplines, however, much remains to be discovered.

Bones act as lever arms to translate the action of muscles into motion and, in addition, form protective shields for sensitive inner organs. As the mineral storehouse of the body, the skeleton is also an active participant in the systemic regulation of calcium, an element vital to the function of all cells. The skeleton successfully fulfills its dual structural and metabolic roles via complex cellular interactions that permit bone to express an attribute unusual among structural materials: the ability to adapt shape and mass to perceived functional demands. Although we recognize the intimate interaction of metabolic factors with the skeleton, we put these considerations aside as we focus on how the skeleton, as an organ, adapts to altered levels of activity.

The adaptation of the skeleton is elegant in its simplicity: Decreased activity precipitates a loss of bone mass, and increased activity stimulates skeletal hypertrophy. Although phenomenological evidence of skeletal adaptation is abundant, little is known of the mechanisms that govern this process. What cell or cells monitor the mechanical environment? Which mechanical parameter triggers the adaptive response? Once triggered, does the adaptive process aspire to a predetermined structural goal? An understanding of the mechanisms that govern skeletal adaptation not only is important to those interested in health and fitness but also holds the potential to profoundly influence the treatment of skeletal pathologies such as postmenopausal and age-induced osteoporosis.

The cellular milieu of bone is first introduced to promote an understanding of the entities ultimately responsible for adaptation. The second section provides a historical perspective of research and observations leading to Wolff's law, the widely recognized theory describing activity-induced skeletal adaptation. Subsequent sections address recent research in the field and are grouped by their relevance to the three tenets of Wolff's law. The final section briefly addresses proposed concepts describing the feedback system responsible for perceiving the functional stimuli. We intend for this chapter to foster an appreciation for current research on skeletal adaptation and encourage consideration of the skeleton as a dynamic, interactive biological organ.

Cellular Basis of Skeletal Adaptation

Skeletal adaptation demonstrates that bone is not a static scaffold but a dynamic tissue comprised of cell populations with the capacity to mold and sculpt bone into competent load-bearing structures. Bone, as a result, possesses an intrinsic ability to adapt mass and architecture to its functional milieu. In growing skeletons, this process establishes both the transverse and longitudinal bone geometries and is referred to as *modeling* (Enlow, 1963; Frost, 1973). The analogous events in the adult skeleton are termed *remodeling* and are operationally distinct from bone modeling (Frost). Whether considered in young or adult animals, adaptation in response to perceived functional stimuli is ultimately modulated by cellular interactions that can adjust, renew, and replace the structural elements of bone.

At the cellular level, the strategic placement and adjustment of bone mass are accomplished by the activity of two basic cell types: the *osteoblast* and the *osteoclast* (see Figure 6.1). The osteoblast produces the collagen and noncollagen proteins comprising the organic matrix and defining the three-dimensional architecture of the mineralized skeleton. The osteoclast, a large multinucleated cell, is specially designed to remove mineralized bone by a process termed resorption. As a result, the ability to modify the location, size, and orientation of skeletal components depends on the coordinated interplay between bone production and bone destruction.

In growing animals, skeletal architecture is modeled by moving the bone's outer periosteal and inner endosteal surfaces through tissue space via osteoblastic

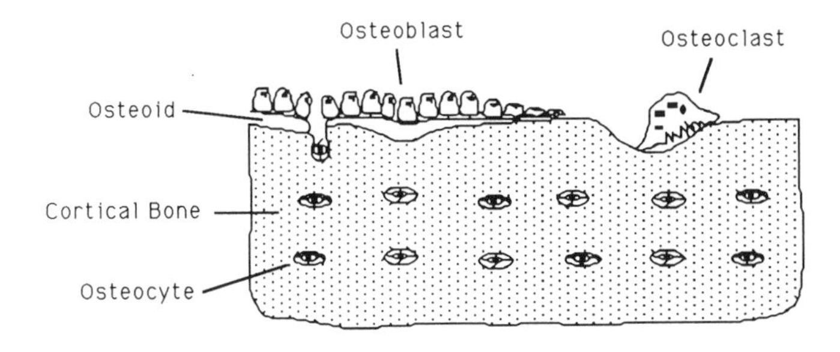

Figure 6.1 Schematic of the spatial relationship between an osteoclast resorbing bone on a free surface, osteoblasts forming organic matrix, or osteoid, and osteocytes located within mineralized cortical bone. (Adapted by permission from Marks & Popoff, 1988.)

formation drifts and osteoclastic resorption drifts, respectively (Frost, 1973). These cellular drifts act in concert to produce the characteristic cross-sectional and longitudinal geometries observed in the diaphysis of any bone (Frost, 1982). Therefore, at any point in time, formation and resorption can occur simultaneously on different bone surfaces.

In contrast, resorptive and formative phenomena in the mature skeleton occur in a tightly regulated sequence that can be divided into three distinct phases (Frost, 1964):

1. Activation
2. Resorption
3. Formation

Activation, the initial event in the remodeling cycle, begins with the transformation of a quiescent bone surface into a site of bone resorption. The key event in this phase is the recruitment of osteoclast precursors from progenitor cell populations harbored in the bone marrow. Once a bone surface is activated, the resorption stage proceeds rapidly as teams of osteoclasts excavate mineralized bone from well-defined unit areas. As osteoclastic activity subsides, osteoblasts arrive at the resorption site and, via formation of an organic matrix, fill in the area previously resorbed by the osteoclast. Differences between modeling and remodeling cellular activities are summarized in Table 6.1.

A unique feature of bone remodeling is that resorption of a bone surface is always followed by bone formation, linking the two activities both spatially and temporally. This phenomena is called *coupling*. One can conceive that the coordination of bone destruction and renewal could, over time, alter trabecular orientation and mass in response to functional stimuli. However, although it is possible to imagine how bone adapts functionally, little is known about the cellular mechanisms that perceive and ultimately transduce mechanical stimuli into the biochemical events illiciting adaptive change at the level of the bone cell.

Table 6.1 Summary of Modeling-Remodeling Differences

	Modeling	Remodeling
Timing	Continuous	Cyclical (ARF)[a]
Resorption and formation surfaces	Different	Same
Surfaces affected	100%	20%
Activation	Not required	Required
Balance	Net gain	Net loss
Coupling of formation and resorption	Systemic? (no ARF)[a]	Local

Note. Adapted from Parfitt, 1984 by permission.
[a]ARF = Activation-Resorption-Formation.

One potential effector in the perception of mechanical stimuli is the *osteocyte*, a cell that is actually osteoblastic in origin. During bone formation, osteoblasts become entrapped in the advancing mineralization front and are then termed osteocytes. Significantly, even though the osteocyte is imprisoned within the bone, it retains its viability and remains in contact with its neighbors via an intricate network of cell processes that travel through the mineralized matrix in tiny channels called *canaliculi*. From a teleologic perspective, this cell network appears ideally positioned for the reception of mechanically engendered signals. Indeed, recent evidence suggests that osteocytes alter their metabolic activity in response to physical stimuli (Skerry, Bitensky, Chayen, & Lanyon, 1989). Whether this represents a relevant event in the transduction process is not clear. What is clear, however, is that a rigorous understanding of how bone tissue adaptation occurs will depend on experiments that can isolate the specific components of the mechanical milieu affecting bone at the cellular level.

A Historical Perspective

The response of bone, as an organ, to altered mechanical stimuli is well defined. When bed rest, cast immobilization, or spaceflight decreases the level of stimulus, bone decreases its mass. Conversely, increased stimulus, such as would result from a rigorous exercise regimen, precipitates an increase in bone mass. This phenomenon has intrigued investigators for over 150 years, and although many of the experimental and analytical techniques used to study skeletal adaptation have been developed only recently, the focus of modern research is better understood within the context of early study in the area. This section provides a brief summary of that work, but those with further interest in the pioneering efforts to describe and understand skeletal adaptation are referred to D'Arcy-Thompson (1942) and Roesler (1981).

The most recognized theory describing skeletal adaptation is referred to as *Wolff's law*. The theory is comprised of three principal components:

1. Bone attempts to provide optimal strength with minimal mass.
2. Bone adapts to its functional environment.
3. Skeletal adaptation is ordered and can be described mathematically.

Although Wolff receives historical credit for these concepts, they were each originally proposed by others. It was Wolff, in the late 1800s, who synthesized the thoughts of his contemporaries with his own observations of normal and pathological skeletons to formulate broad hypotheses regarding skeletal adaptation.

In his extensive review, Roesler (1987) credits Bourgery (1832) with the origin of the first component of Wolff's law, that bone attempts to provide optimal strength with minimal mass. Bourgery suggested that "the condition for a successful and stable bone is slightness of the critical volume" (Roesler, 1987, p. 1026). As noted by Fung (1981), this thesis was restated in more formal terms by Roux (1895), who postulated that an inherent feature of tissue adaptation is the desire to promote maximal strength with minimum material. That bone can, by some mechanism, seek the required level of strength, while continually striving to minimize mass, is further confirmation that biological systems strive to streamline themselves.

Roux, in 1881, proposed what has become the second tenet of Wolff's law, that bone adapts to changes in mechanical stimuli in a self-regulated manner. He termed this process *functional adaptation*. Roux also hypothesized that the process was in some way mediated by cellular events and interactions. This was a particularly farsighted observation; current research suggests that cellular interactions both regulate and mediate skeletal adaptation (Lippiello, Kaye, Neumata, & Mankin, 1985; Rodan, Baurret, Harvey, & Mensi, 1975; Somjen, Binderman, Berger, & Harell, 1980).

The final component of Wolff's law, that skeletal adaptation can be described mathematically, was stimulated by the visual order displayed by femoral trabeculae. Wyman (1857) and Dwight (1886) reported that Bourgery described the structural order of cancellous bone in 1832. However, D'Arcy-Thompson (1942) quotes Sir Charles Bell remarking in 1827, "This minute lattice-work, or the cancelli which constitute the interior structure of bone, have still reference to the forces acting on the bone" (D'Arcy-Thompson, 1942, p. 976). In addition, Sir John Herschel, in 1830, reportedly described bone as a "framework of the most curious carpentry: in which occurs not a single straight line nor any known geometrical curve, yet all evidently systematic, and constructed by rules which defy our research" (D'Arcy-Thompson, 1942, p. 976).

In 1867, Meyer related the observation of the engineer Culmann, who suggested that organization of femoral trabeculae corresponded to the trajectories of principal stresses in the neck of a curved crane (Roesler, 1981). This observation has been subsequently labeled the trajectorial hypothesis. While Meyer cautiously drew conclusions based on his collaborative observations with Culmann, Wolff showed little of Meyer's caution and, in 1870, proposed that the trajectorial hypothesis represented a perfect correlation between the trabecular structure of the proximal femur and the lines of principal stress in the femoral head as described by Culmann. Interestingly, although the Culmann crane analogy is a less-than-rigorous analysis of femoral stress distributions (Roesler), recent analytical work with the human patella supports the relationship between trabecular orientation and lines of principal stress (Hayes & Snyder, 1981).

The work of Bourgery, Roux, Meyer, and Culmann, and others, were synthesized by Wolff (1892, 1988) in his monograph, *The Law of Bone Remodeling*. In this work, Wolff stated,

> The law of bone remodeling is the law according to which alterations of the internal architecture clearly observed, as well as secondary alterations of the external form of the bones following the same mathematical rules, occur as a consequence of the primary changes in the shape and stressing or in the stressing of the bones. (p. 91)

Wolff proposed that skeletal adaptation followed from the static stress state of the bone and that this adaptation followed strict mathematical laws. As a consequence, Wolff has been acknowledged as the father of orthopedic research. However, the details of Wolff's theories—such as that trabeculae of the femoral head always meet at right angles, as mathematical theory would predict for the principal stress trajectories—were not based on rigorous experimentation. Moreover, in view of current research, Wolff's hypothesis that the skeleton was in equilibrium with a static stress state is less attractive than the self-regulating adaptive process proposed by Roux. However, in the absence of contradictory experimental evidence, Wolff was able to defend his phenomenological theories, often by using the observations of others, and the resulting amalgamation of thought is now referred to as Wolff's law. It is within this framework that modern research continues to investigate how bone perceives and responds to mechanical stimuli.

Minimum Mass

The skeletal imperative to satisfy structural requirements with minimal mass is vividly demonstrated by the adaptation induced by the removal of activity. Rather than a pathological response, the loss of bone concomitant to reduced mechanical stimuli may be viewed as a complementary process to the skeletal hypertrophy precipitated by exercise. Research examining bone loss in response to disuse has addressed the extent and time course of this loss, whether bone loss after a period of disuse may be reversed by a return to activity, and whether bone loss in space is related to the microgravity environment encountered by astronauts.

Disuse in the adult human skeleton stimulates bone loss via endosteal resorption and increased intracortical porosity (Dequeker, 1975) (see Figure 6.2, a and b). The loss occurs in both cortical and trabecular bone, but the increased surface area of the trabeculae makes them particularly sensitive to the cellular events responsible for removing bone. Donaldson, Hulley, Vogel, Hattner, Bayers, and McMillan (1970), in a bed-rest study involving 3 healthy adult males, reported an average 34% loss of calcaneal bone mineral content, a correlate of bone mass, after 30 to 36 weeks of non-weight-bearing. Fortunately, there appears to be a genetic limit to the extent of the loss; Minaire, Meunier, Edouard, Bernard, Courpron, and Bourret (1974) found that trabecular bone loss in immobilized patients reached 33% after 25 weeks but stabilized thereafter. The maximum extent of cortical bone loss appears to be similar in magnitude. Jaworski and Uhthoff (1986) found that dogs experienced a 38% reduction in third metacarpal cortical bone and a 47% loss of metaphyseal trabecular bone after 32 weeks of forelimb cast immobilization. The loss was rapid for the first 6 to 12 weeks and appeared to equilibrate between 24 and 32 weeks of immobilization. Primates also demonstrate a rapid loss of cortical bone upon immobilization, a process that appears to slow as a new steady state is established (Young & Schneider, 1981). These studies emphasize the site-specific response of the skeleton to disuse. Immobilization of the leg does not engender bone loss in the arm. However, within a bone subjected to disuse, recent preliminary evidence suggests that the skeletal response is consistent in magnitude throughout the diaphyseal shaft (Gross & Rubin, 1989).

If a healthy person is immobilized, can the lost bone be recovered upon return to ambulation? Using the rat hindlimb suspension model, which was developed

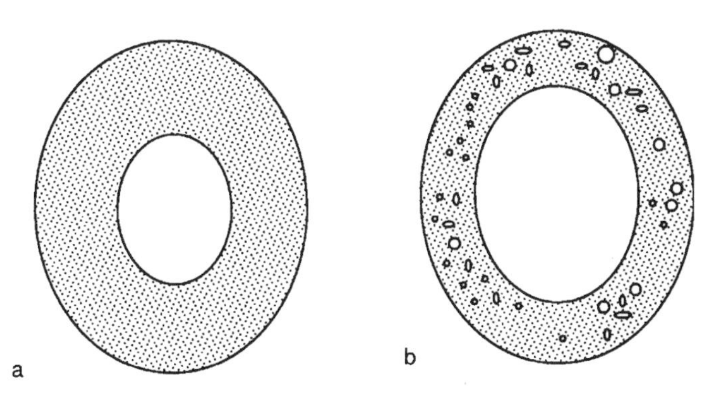

Figure 6.2 Bone loss in response to a reduction in mechanical stimuli, as would be engendered by extended bed rest or casting in the adult skeleton, is characterized by expansion of the endosteal, or inner surface, and increased porosity within the cortex. Here, a section of healthy cortical bone (a) is contrasted to the same cross section after a hypothetical period of disuse (b).

to mimic the skeletal unloading that occurs during spaceflight, Sessions, Halloran, Bikle, Wronski, Cone, and Morey-Holton (1989) found that 2 weeks of reambulation following 2 weeks of suspension increased bone formation rates 30% to 34% above normal, but that was not sufficient to restore the tibia and vertebra. Difficulty in restoring bone mass after a period of disuse has also been observed in dogs. Jaworski and Uhthoff (1986) found that 28 weeks of reambulation following 32 weeks of immobilization restored only 40% of lost cortical bone and 60% of trabecular bone in the third metatarsal. Interestingly, after just 16 weeks of immobilization, the same 28-week period of ambulation restored approximately 50% of cortical bone and all trabecular bone. Similarly, Rubin, McLeod, Brand, and Lanyon (1987) found that 8 weeks of isolation from functional stimuli induced an average 13% loss of cortical bone in the turkey ulna, and that a subsequent 8-week application of an osteogenic loading protocol recovered only 65% of the lost bone. Despite species variability, these data indicate that bone mass can be at least partially restored after disuse atrophy, that the rate of restoration is slow compared to the rate of loss, and that an early return to weight bearing increases the percentage of bone that is restored.

With the advent of space exploration, a question of particular importance is whether disuse adaptation is responsible for bone loss concomitant to spaceflight. When exposed to microgravity environments, urinary calcium excretion in astronauts and cosmonauts increases 80% to 100% above preflight levels (Rambaut, Leach, & Whedon, 1979; Rambaut & Johnston, 1979), and the origin of this excess mineral appears correlated with the loss of bone from weight-bearing sites in the appendicular skeleton (Oganov, 1981; Rambaut & Johnston, 1979; Smith, Rambaut, Vogel, & Whittle, 1977). This response was further investigated by exposing rats to microgravity while the rats traveled on three separate Soviet biosatellites. Results indicated that periosteal bone formation in the normally weight-bearing tibia and humerus was arrested during the 18- to 19-day flights (Morey & Baylink, 1978; Morey-Holton, Turner, & Baylink, 1989; Wronski & Morey, 1983). The non-weight-bearing rib demonstrated no difference in formation rates between the flight and control rats, although possible differences may have been masked by low formation rates in the rib under normal conditions (Wronski & Morey).

Although these data suggest mechanically based microgravity effects on the skeleton, confounding variables such as vibrations during takeoff and reentry, cramped conditions in space, and endocrine response to spaceflight make interpretation of the data difficult. Earthbound models designed to isolate mechanical effects from these other factors have proven difficult to develop. The most commonly used simulation is rat hindlimb suspension, which induces a transient reduction in tibial bone formation rate similar to spaceflight but also results in decreased trabecular bone volume in both the tibia and humerus (Morey-Holton & Wronski, 1981). This suggests that the skeletal response to decreased loading may also be influenced by systemic variables such as corticosteroid excess (Wronski & Morey, 1983), although a recent report contradicts this conclusion (Halloran,

Bikle, Cone, & Morey-Holton, 1988). Moreover, these investigators have acknowledged the difficulty of extrapolating data from actively modeling skeletons, such as the young rat, to the relatively quiescent remodeling skeleton of a healthy adult. This is substantiated by recent studies demonstrating different disuse responses in the immature versus mature skeletons of rats (Parazynski, Morey-Holton, Cone, & Martin, 1989), dogs (Jaworski & Uhthoff, 1986), and turkeys (Bain & Rubin, 1990). A clear solution to this multifactorial problem will occur only with the development of an animal model that allows independent control of mechanical and systemic stimuli on the skeleton.

Adaptation to Mechanical Stimuli

Functional adaptation, as proposed by Roux and incorporated into Wolff's law, suggests that bone architecture adjusts according to perceived structural demands engendered by functional stimuli. Exercise stimulates a site-specific increase in bone mass (Dalen & Olsson, 1974; Jones, Priest, Hayes, Tichenor, & Nagel, 1977; Nilsson, Andersson, Havdrup, & Westlin, 1978) rather than an alteration of bone mineral composition (Woo et al., 1981); therefore, some component of this stimulus must be considered osteogenic (see Figure 6.3, a and b). Indeed, this feature may eventually allow the exploitation of exercise as a prophylaxis to treat skeletal pathologies characterized by excessive bone loss. However, the success of this strategy will depend on the isolation of osteogenic parameters from within the functional milieu.

General exercise studies, while providing valuable clinical insight, do not permit mechanistic inference due to a lack of rigorous knowledge of the mechanical stimulus responsible for the observed adaptation. To gain further insight into this process, several models have been developed, each with varying levels of

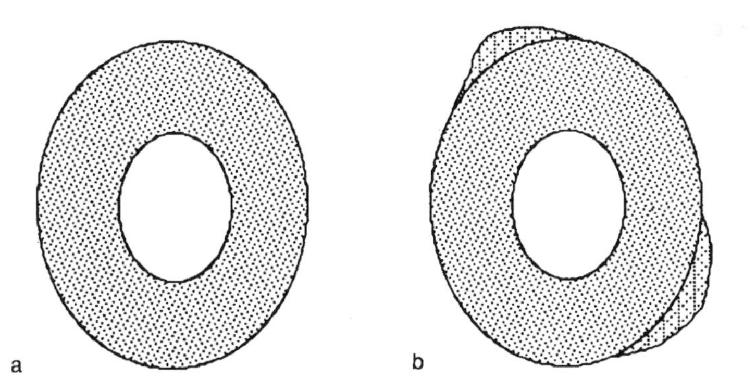

Figure 6.3 A cross section of healthy adult cortical bone (a) perceives an increased mechanical stimulus and responds by hypertrophy at specific sites upon the periosteal cortex (b). The relationship between the location of the newly formed bone and the applied stimulus remains unknown.

control over the mechanical stimulus applied to the bone. Osteotomy models have provided the opportunity to assess the adaptive response generated by a step increase in loading. After osteotomy of either the radius or ulna, and a return to ambulation, the overloaded bone increased mass adjacent to the osteotomy site, presumably to compensate for the removed cross-sectional area (Chamay & Tschantz, 1972; Goodship, Lanyon, & MacFie, 1979; Lanyon, Goodship, Pye, & MacFie, 1982). In vivo strain gage recordings suggested that a primary objective of the skeletal response was to restore the initial strain milieu, suggesting a site-specific feedback control mechanism. Attempts to correlate the adaptive response to the imposed stimulus were restricted by the complexity of the locomotion-generated stimulus, but O'Conner, Lanyon, and MacFie (1982) found that adaptation was most correlated to the experimentally imposed change in strain rate.

Further experimental control of the mechanical stimuli input into a bone has been attained by the development of models allowing externally applied loading regimens via surgically implanted transcortical pins. Using this approach, workers have found that static loading may trigger an adaptive response when superimposed upon locomotion (Meade, Cowin, Klawitter, Van Buskirk, & Skinner, 1984), but that response appears to be minimal compared to superimposed dynamic loading of similar magnitude (Chamay & Tchantz, 1972; Churches, Howlett, Waldron, & Ward, 1979; Hert, Liskova, & Landrgot, 1969; Liskova & Hert, 1971). However, further interpretation of the adaptive response in relation to specific external stimuli is potentially obscured by the animal's normal activity during nonloading periods.

Isolating the mechanical stimulus responsible for precipitating the adaptive process requires the capacity to selectively exclude or enhance the mechanical stimulus applied to a given bone. By isolating the avian ulna from functional stimuli with parallel epiphyseal osteotomies, and using transcortical pins to apply external loading regimens, Rubin and Lanyon (1984b) developed a model that has led to an improved understanding of the osteogenic potential of mechanical stimuli. In this model, static loading precipitates an osteopenic response similar to disuse, whereas physiological magnitude dynamic loading generates a 24% increase in cross-sectional area (Rubin & Lanyon, 1984b). Rubin and Lanyon (1985) found that the extent of the osteogenic response correlated with the peak strain engendered during daily loading regimens (*strain* is a measure of deformation in a body and is defined such that a 0.1% lengthening of a given initial length is equivalent to 1,000 με). Peak strains of 1,000 με maintained bone, and peak strains of 2,000 με, 3,000 με, and 4,000 με precipitated increasingly osteogenic responses. The maintenance of bone at 1,000 με, when peak strains during flapping reach well above 2,000 με, was of particular interest because decreases in engendered strain magnitude are normally associated with a reduction of bone mass. Because the distribution of the experimentally applied strain environment differed from the normal environment, the investigators suggested that the distribution of strain played an important role in regulating bone mass.

The extreme sensitivity of the skeleton in perceiving an altered mechanical stimulus was demonstrated by Pead, Skerry, and Lanyon (1988), who, using the ulna model, found that a single 300-cycle burst of loading generating peak strains of 3,000 $\mu\varepsilon$ was sufficient to stimulate activation of periosteal osteoblasts. Furthermore, McLeod, Bain, and Rubin (1990) found that high-frequency mechanical stimuli were more osteogenic than low-frequency stimuli of identical amplitude. These results suggest that, as with sensory systems associated with hearing and touch, the feedback system responsible for perceiving changes in the mechanical milieu is sensitive to the frequency content of the applied stimulus.

The Mathematical Order of Skeletal Adaptation

The ordered manner by which the skeleton adapts leads to the hypothesis that bone architecture can be predicted mathematically. The first formally stated thesis reflecting this concept was the trajectorial hypothesis. As originally proposed by Meyer and Culmann, and restated by Wolff, it suggested that the trabeculae of the head of the femur align themselves to the principal stress directions generated during load bearing. Koch (1917) attempted the first experimental assessment of the trajectorial hypothesis by rigorously analyzing the stress distribution of the femur with beam theory. He reported a strong relationship between trabecular organization and the orientation of principal stress generated by the single load distribution studied. However, the material, loading, and structural simplifications necessary to make this problem tractable by hand calculation cast doubt on the strength of the correlations.

Computer technology—specifically, the engineering technique of finite element modeling—has greatly facilitated analyses similar in intent to that of Koch. The finite element method involves breaking a very complicated structural problem, such as the stresses generated in the femur during locomotion, into a large number of smaller problems suitable for computer analysis (see Figure 6.4). Using this technique, Hayes and Snyder (1981) were able to relate trabecular orientation in the patella with principal stress directions estimated during slow stair-climbing. These data were the first to quantitatively establish a significant correlation between an engendered stress environment and trabecular organization.

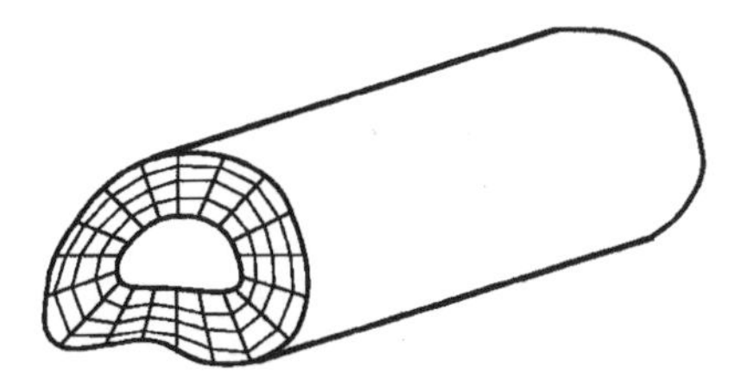

Figure 6.4 Finite element mesh superimposed on a long bone cross section. The complex geometry is divided into small, regular geometric shapes that are termed *elements*, and the computer's ability to perform repetitive calculations is used to define stresses and strains within each element.

Mathematical algorithms have also been used to correlate trabecular and cortical morphology with mechanical loading environments. Due to the complexity of analyzing stresses and strains in trabeculae, and a paucity of experimental trabecular adaptation data, work has focused on relating morphology to numerically derived mechanical stimuli. Fyhrie and Carter (1986) described trabecular bone as a self-optimizing material whose orientation was governed by the directions of principal stress and whose density was based on maximizing structural integrity with minimal bone mass. Their flexible approach to self-optimization used an effective stress parameter, based on either fracture or elastic energy criteria, as the signal driving the adaptive process.

Using this foundation, Carter, Orr, and Fyhrie (1989) implemented a multiple loading condition finite element analysis of femoral trabecular structure and found that morphology could be explained on the basis of calculated stresses only when multiple loading conditions were considered. Using a more sophisticated nonlinear three-dimensional finite element model, Fyhrie and Carter (1990) compared density predictions from self-optimizing criteria based on the imposed mechanical environment. Predictions were more successful when the optimization criteria encompassed all components of the mechanical environment, as opposed to a specialized stress variable based on shear contributions to material failure. In addition, Whalen, Carter, and Steele (1988) related calcaneal bone density to stress histories estimated from daily activities. Both the magnitude of engendered stress and the number of cycles influenced the prediction of skeletal mass, but stress magnitudes were the dominant correlate. Current efforts to relate trabecular adaptation to mechanical stimuli involve the use of even more sophisticated models to calculate stresses and strains within individual trabeculae, each of which may have varying material properties (Fyhrie & Hollister, 1990).

Cortical bone adaptation presents a unique modeling problem because the pattern of adaptation does not visually appear as ordered as trabecular alignment in the femoral head. Cowin and Hegedus (Cowin & Hegedus, 1976; Hegedus & Cowin, 1976) proposed a theory of adaptive elasticity to describe cortical adaptation. This theory differs from the trabecular self-optimization approach in that the model assumes a mechanical environment as a baseline, and alterations in the environment stimulate the skeleton to adapt its mass to reestablish the initial state of equilibrium. Cowin, Hart, Balser, and Kohn (1985) used beam theory to apply the theory to five published experimental studies of skeletal disuse and hypertrophy. Defining basal mechanical stimuli from in vivo strain data, the authors established remodeling rate coefficients based on the magnitude of surface adaptation occurring over the life of the experiment. Despite differing bones and initial strain environments, the coefficients were within an order of magnitude of each other across the experiments, suggesting an ordered adaptation was engendered by the altered strain environment. This theory has also been incorporated into a finite element model whose surface elements change shape and material properties across time according to predicted remodeling phenomenon (Hart, 1990; Hart, Davy, and Heiple, 1984).

Although current analytical theories may have direct applications, such as employing modified adaptive elasticity theory to improve prosthetic design (Huiskes, Weinans, Grootenboer, Dalstra, Fudala, & Slooff, 1987), experimental validation of predictive models has been limited. Cowin (1987) used adaptive elasticity to predict cortical remodeling in a state of pure torsion, but experimental data to validate the predictions have not yet been obtained. Using a combined experimental and analytical approach, Brown, Pedersen, Gray, Brand, and Rubin (1990) isolated components of an applied mechanical environment, such as principal tensile strains and strain energy density, that were correlated to cortical adaptation. However, the predictive value of these correlations has not yet been validated with an alternative mechanical stimulus. The success of mathematical algorithms in relating mechanical stimuli to skeletal architecture may indicate the existence of an underlying causal relationship. However, until these models are validated by their ability to predict adaptation within the context of an experimentally applied stimulus, only the elegant adaptation of bone as an organ can be appreciated.

Possible Control Mechanisms

Although ample evidence exists that trabecular and cortical bone adapt to mechanical stimuli, the means by which bone perceives variations in its physical environment remains elusive. Based on experimental and clinical observations, Frost (1964, 1987) has suggested that skeletal adaptation is governed by structural imperatives in which cortical remodeling would be triggered by an increase in strain above some acceptable level, and strains below this level would satisfy an equilibrium state. In addition, Frost proposed that the objective of cortical remodeling was to minimize deformation by bending, thereby accentuating axial load bearing. Another observation (Frost, 1980) suggested that the feedback mechanism governing intracortical remodeling was a desire to repair microdamage created during physiological loading. This theory is supported by the work of Carter, Caler, Spengler, and Frankel (1981), who demonstrated that fatigue properties of bone in vitro are strain related, and by the work of Burr, Martin, Schaffler, and Radin (1985), who reported microfractures in cortical bone exposed to external loads of physiological magnitude.

An alternative theory was proposed by Rubin and Lanyon (1984a), who, after observing a similarity in functionally engendered strains across a wide range of vertebrates and activities, suggested that adaptation was stimulated by a desire to maintain a predetermined strain environment, rather than as a response to a potentially deleterious structural challenge. Rubin and Lanyon (1987) and Lanyon (1987) further proposed that feedback control was genetically programmed and that deviations from the appropriate strain environment were sensed by cells, possibly osteocytes, that communicated the appropriate adaptive signal to osteoblasts and osteoclasts. The magnitude of the response would then be mediated by parameters such as strain magnitude and strain rate (O'Conner et al., 1982; Rubin & Lanyon, 1985).

Conclusion

Although this chapter focuses on the adaptation of bone to altered stimuli, it would be remiss to imply that only mechanical parameters are responsible for the regulation of skeletal adaptation. In fact, mechanical deformation of bone may be only a mechanism for creating the phenomena actually sensed, such as fluid flow or electric field shifts adjacent to cells. Regardless of its mechanism, skeletal adaptation in response to mechanical stimuli represents one of nature's most amazing manifestations. The ability of a structure to regulate its mass and architecture according to perceived structural obligations demands a complexity far beyond the capacity of any man-made material and, as such, is a source of continued research interest. An understanding of the mechanisms responsible for this process may potentially affect the treatment of skeletal pathologies such as osteoporosis and space-induced osteopenia. Just as one would aerobically exercise today to maintain cardiovascular fitness, the coming century may witness the introduction of exercise regimens designed to maximize osteogenic benefit with minimal time commitment as a means of prophylactically conditioning the skeleton.

References

Bain, S.D., & Rubin, C.T. (1990). Skeletal modeling objectives in growing bone: What role do mechanical factors play? [Abstract]. *Transactions of the 36th Orthopaedic Research Society,* **36**, 418.

Brown, T.D., Pedersen, D.R., Gray, M.L., Brand, R.A., & Rubin, C.T. (1990). Identification of mechanical parameters initiating periosteal remodeling: A complementary experimental and analytic approach. *Journal of Biomechanics,* **23**, 893-905.

Burr, D.B., Martin, R.B., Schaffler, M.B., & Radin, E.L. (1985). Bone remodeling in response to in vivo fatigue microdamage. *Journal of Biomechanics,* **18**, 189-200.

Carter, D.R., Caler, W.E., Spengler, D.M., & Frankel, V.H. (1981). Fatigue behaviour of adult cortical bone: The influence of mean strain and strain range. *Acta Orthopaedica Scandinavica,* **52**, 481-490.

Carter, D.R., Orr, T.E., & Fyhrie, D.P. (1989). Relationships between loading history and femoral cancellous bone architecture. *Journal of Biomechanics,* **22**, 231-244.

Chamay, A., & Tschantz, P. (1972). Mechanical influences in bone remodeling: Experimental research on Wolff's law. *Journal of Biomechanics,* **5**, 173-180.

Churches, A.E., Howlett, C.R., Waldron, K.J., & Ward, G.W. (1979). The response of living bone to controlled time-varying loading: Method and preliminary results. *Journal of Biomechanics,* **12**, 35-45.

Cowin, S.C. (1987). Bone remodeling of diaphyseal surfaces by torsional loads: Theoretical predictions. *Journal of Biomechanics,* **20**, 1111-1120.

Cowin, S.C., Hart, R.T., Balser, J.R., & Kohn, D.H. (1985). Functional adaptation in long bones: Establishing in vivo values for surface remodeling rate coefficients. *Journal of Biomechanics,* **18**(9), 665-684.

Cowin, S.C., & Hegedus, D.H. (1976). Bone remodeling: 1. Theory of adaptive elasticity. *Journal of Elasticity,* **6**, 313-326.

Dalen, N., & Olsson, K.E. (1974). Bone mineral content and physical activity. *Acta Orthopaedica Scandinavica,* **45**, 170-174.

D'Arcy-Thompson, W. (1942). *On growth and form.* Cambridge, UK: Cambridge University Press.

Dequeker, J. (1975). Bone and ageing. *Annals of the Rheumatic Diseases,* **34**, 100-115.

Donaldson, C.L., Hulley, S.B., Vogel, J.M., Hattner, R.S., Bayers, J.H., & McMillan, D.E. (1970). Effect of prolonged bed rest on bone mineral. *Metabolism,* **19**(12), 1071-1084.

Dwight, Y. (1886). The significance of bone-architecture. *Memoirs of the Boston Society of Natural History,* **4**, 1-15.

Enlow, D.H. (1963). *Principles of bone remodeling.* Springfield, IL: Charles C Thomas.

Frost, H.M. (1964). *The laws of bone structure.* Springfield, IL: Charles C Thomas.

Frost, H.M. (1973). *Bone modeling and skeletal modeling errors.* Springfield, IL: Charles C Thomas.

Frost, H.M. (1980). A lamellar bone modeling theory. *Proceedings of the Japanese Orthopaedic Research Society,* **7**, 571-628.

Frost, H.M. (1982). Mechanical determinants of bone modeling. *Metabolic Bone Disease and Related Research,* **4**, 217-229.

Frost, H.M. (1987). The mechanostat: A proposed pathogenic mechanism of osteoporoses and the bone mass effects of mechanical and nonmechanical agents. *Bone and Mineral,* **2**, 73-85.

Fung, Y.C. (1981). *Biomechanics: Mechanical properties of living tissues.* New York: Springer-Verlag.

Fyhrie, D.P., & Carter, D.R. (1986). A unifying principle relating stress to trabecular bone morphology. *Journal of Orthopaedic Research,* **4**, 304-317.

Fyhrie, D.P., & Carter, D.R. (1990). Femoral head apparent density distribution predicted from bone stresses. *Journal of Biomechanics,* **23**, 1-10.

Fyhrie, D.P., & Hollister, S.J. (1990). A tissue strain remodeling theory for trabecular bone using homogenization theory [Abstract]. *Transactions of the 36th Orthopaedic Research Society,* **36**, 76.

Goodship, A.E., Lanyon, L.E., & MacFie, J.H. (1979). Functional adaptation of bone to increased stress. *Journal of Bone and Joint Surgery,* **61A**, 539-545.

Gross, T.S., & Rubin, C.T. (1989). Disuse osteopenia at the organ level: Is the cortical response uniform? [Abstract]. *Proceedings of the Fifth American Society of Gravitational and Space Biology,* **5**, 115.

Halloran, B.P., Bikle, D.D., Cone, C.M., & Morey-Holton, E.R. (1988). Glucocorticoids and inhibition of bone formation induced by skeletal unloading. *American Journal of Physiology,* **255**, E875-E879.

Hart, R.T. (1990). A theoretical study of the influence of bone maturation rate on surface remodeling predictions: Idealized models. *Journal of Biomechanics,* **23**, 241-257.

Hart, R.T., Davy, D.T., & Heiple, K.G. (1984). A computational method for stress analysis of adaptive elastic materials with a view toward application in strain-induced bone remodeling. *Journal of Biomechanical Engineering,* **106**, 342-350.

Hayes, W.C., & Snyder, B. (1981). Toward a quantitative formulation of Wolff's law in trabecular bone. In S.C. Cowin (Ed.), *Mechanical properties of bone* (Applied Mechanics Division Vol. 45, pp. 43-68). New York: American Society of Mechanical Engineers.

Hegedus, D.H., & Cowin, S.C. (1976). Bone remodeling: 2. Small strain adaptive elasticity. *Journal of Elasticity,* **6**, 337-352.

Hert, J., Liskova, M., & Landrgot, B. (1969). Influence of the long-term, continuous bending on the bone. *Folia Morphologica,* **17**, 389-399.

Huiskes, R., Weinans, H., Grootenboer, H.J., Dalstra, M., Fudala, B., & Slooff, T.J. (1987). Adaptive bone-remodeling theory applied to prosthetic-design analysis. *Journal of Biomechanics,* **20**, 1135-1150.

Jaworski, Z.F.G., & Uhthoff, H.K. (1986). Reversibility of nontraumatic disuse osteoporosis during its active phase. *Bone,* **7**, 431-439.

Jones, H.H., Priest, J.D., Hayes, W.C., Tichenor, C.C., & Nagel, D.A. (1977). Humeral hypertrophy in response to exercise. *Journal of Bone and Joint Surgery,* **59A**(2), 204-208.

Koch, J.C. (1917). The laws of bone architecture. *American Journal of Anatomy,* **21**, 177-298.

Lanyon, L.E. (1987). Functional strain in bone tissue as an objective, and controlling stimulus for adaptive bone remodeling. *Journal of Biomechanics,* **20**, 1083-1093.

Lanyon, L.E., Goodship, A.E., Pye, C.J., & MacFie, J.H. (1982). Mechanically adaptive bone remodeling. *Journal of Biomechanics,* **15**, 141-154.

Lanyon, L.E., & Rubin, C.T. (1984). Static vs dynamic loads as an influence on bone remodeling. *Journal of Biomechanics,* **17**, 897-905.

Lippiello, L., Kaye, C., Neumata, T., & Mankin, H.H. (1985). In vitro metabolic response of articular cartilage segments to low levels of hydrostatic pressure. *Connective Tissue Research,* **13**, 99-107.

Liskova, M., & Hert, J. (1971). Reaction of bone to mechanical factors: 2. Periosteal and endosteal bone apposition in the rabbit tibia due to intermittent stressing. *Folia Morphologica,* **19**, 301-317.

Marks, S.C., & Popoff, S.N. (1988). Bone cell biology: The regulation of development, structure, and function in the skeleton. *American Journal of Anatomy,* **183**, 1-44.

McLeod, K.J., Bain, S.D., & Rubin, C.T. (1990). Dependence of bone adaptation on the frequency of induced dynamic strains [Abstract]. *Transactions of the 36th Orthopaedic Research Society,* **36**, 103.

Meade, J.B., Cowin, S.C., Klawitter, J.J., Van Buskirk, W.C., & Skinner, H.B. (1984). Bone remodeling due to continuously applied loads. *Calcified Tissue International,* **36**, S25-S30.

Minaire, P., Meunier, P., Edouard, C., Bernard, J., Courpron, P., & Bourret, J. (1974). Quantitative histological data on disuse osteoporosis: Comparison with biological data. *Calcified Tissue Research,* **17**, 57-73.

Morey, E.R., & Baylink, D.J. (1978). Inhibition of bone formation during space flight. *Science,* **201**, 1138-1141.

Morey-Holton, E.R., Turner, R.T., & Baylink, D.J. (1989). Does centrifugation at 1G during spaceflight protect bone strength? [Abstract]. *Calcified Tissue International,* **44**, S98.

Morey-Holton, E.R., & Wronski, T.J. (1981). Animal models for simulating weightlessness. *The Physiologist,* **24**, S45-S48.

Nilsson, B.E., Andersson, S.M., Havdrup, T., & Westlin, N.E. (1978). Ballet-dancing and weight-lifting: Effects on BMC. *American Journal of Roentgenology,* **131**, 541-542.

O'Conner, J.A., Lanyon, L.E., & MacFie, J.H. (1982). The influence of strain rate on adaptive bone remodeling. *Journal of Biomechanics,* **15**, 767-781.

Oganov, V.S. (1981). Results of biosatellite studies of gravity-dependent changes in the musculo-skeletal system of mammals. *The Physiologist,* **24**, S55-S58.

Parazynski, S.E., Morey-Holton, E.R., Cone, C.M., & Martin, B.R. (1989). Musculoskeletal unloading in skeletally mature female Sprague-Dawley rats [Abstract]. *Proceedings of the Fifth American Society of Gravitational and Space Biology,* **5**, 112.

Parfitt, A.M. (1984). The cellular basis of bone remodeling: The quantum concept reexamined in light of recent advances in the cell biology of bone. *Calcified Tissue International,* **36**, S37-S45.

Pead, M.J., Skerry, T.M., & Lanyon, L.E. (1988). Direct transformation from quiescence to bone formation in the adult periosteum following a single brief period of bone loading. *Journal of Bone and Mineral Research,* **3**, 647-656.

Rambaut, P.C., & Johnston, R.S. (1979). Prolonged weightlessness and calcium loss in man. *Acta Astronomica,* **6**, 1113-1122.

Rambaut, P.C., Leach, L.C., & Whedon, G.D. (1979). A study of metabolic balance in crew members of Skylab IV. *Acta Astronomica,* **6**, 1313-1322.

Rodan, G.A., Baurret, L.A., Harvey, A., & Mensi, T. (1975). Cyclic AMP and cyclic GMP: Mediators of the mechanical effects on bone remodeling. *Science,* **189**, 467-469.

Roesler, H. (1981). Some historical remarks on the theory of cancellous bone structure (Wolff's law). In S.C. Cowin (Ed.), *Mechanical Properties of Bone*

(Applied Mechanics Division Vol. 45, pp. 27-42). New York: American Society of Mechanical Engineers.

Roesler, H. (1987). The history of some fundamental concepts in bone biomechanics. *Journal of Biomechanics,* **20**(11/12), 1025-1034.

Rubin, C.T., & Lanyon, L.E. (1984a). Dynamic strain similarity in vertebrates: An alternative to allometric limb bone scaling. *Journal of Theoretical Biology,* **107**, 321-327.

Rubin, C.T., & Lanyon, L.E. (1984b). Regulation of bone formation by applied dynamic loads. *Journal of Bone and Joint Surgery,* **66A**, 397-402.

Rubin, C.T., & Lanyon, L.E. (1985). Regulation of bone mass by mechanical strain magnitude. *Calcified Tissue International,* **37**, 411-417.

Rubin, C.T., & Lanyon, L.E. (1987). Osteoregulatory nature of mechanical stimuli: Function as a determinant for adaptive remodeling in bone. *Journal of Orthopaedic Research,* **5**, 300-310.

Rubin, C.T., McLeod, K.J., Brand, R.A., & Lanyon, L.E. (1987). Reversal of disuse osteopenia by controlled dynamic loading [Abstract]. *Transactions of the 33rd Orthopaedic Research Society*, **33**, 351.

Sessions, N. De V., Halloran, B.P., Bikle, D.D., Wronski, T.J., Cone, C.M., & Morey-Holton, E.R. (1989). Bone response to normal weight bearing after a period of skeletal unloading. *American Journal of Physiology,* **257**, E606-E610.

Skerry, T.M., Bitensky, L., Chayen, J., & Lanyon, L.E. (1989). Early strain-related changes in enzyme activity in osteocytes following bone loading in vivo. *Journal of Bone and Mineral Research,* **4**, 783-788.

Smith, M.C., Rambaut, P.C., Vogel, J.M., & Whittle, M.W. (1977). Bone mineral measurements: Experiment M078. In R.S. Johnston and L.F. Dietlein (Eds.), *Biomedical results from skylab* (pp. 183-190), Washington, DC: National Aeronautics and Space Administration.

Somjen, D., Binderman, I., Berger, E., & Harell, A. (1980). Bone remodeling induced by physical stress is Prostaglandin E2 mediated. *Biochimica et Biophysica Acta,* **627**, 91-100.

Whalen, R.T., Carter, D.R., & Steele, C.R. (1988). Influence of physical activity on the regulation of bone density. *Journal of Biomechanics,* **21**, 1988.

Wolff, J. (1870). Uber die innere architektur der knochen und ihre bedeutung fur die frage vom knochenwachstum. *Virchows Archiv fur Pathologische Anatomie und Physiologie und Klinische Medizin*, **50**, 389-453.

Wolff, J. (1988). *The law of bone remodeling.* (P. Maquet & R. Furlong, Trans.). New York: Springer-Verlag. (Original work published 1892)

Woo, S.L.Y., Kuei, S.C., Amiel, D., Gomez, M.A., Hayes, W.C., White, F.C., & Akeson, W.H. (1981). The effect of prolonged physical training on the properties of long bone: A study of Wolff's law. *Journal of Bone and Joint Surgery,* **63A**, 780-787.

Wronski, T.J., & Morey, E.R. (1983). Effect of spaceflight on periosteal bone formation in rats. *American Journal of Physiology,* **244**, R305-R309.

Wyman, J. (1857). On the cancellous structure of some of the bones of the human body. *Memoirs of the Boston Society of Natural History,* **4**, 125-140.

Young, D.R., & Schneider, V.S. (1981). Radiographic evidence of disuse osteoporosis in the monkey (M. nemestrina). *Calcified Tissue International,* **33**, 631-639.

Chapter 7

Skeletal Muscle Mechanics and Movement

Robert J. Gregor

Robert Gregor's chapter discusses what is probably of greatest familiarity to the exercise and sport scientist—skeletal muscle mechanics. From the standpoint of the musculoskeletal system, simple modeling can demonstrate that, for some tasks, the dominant source of joint reaction forces is the muscle contraction about the joint. However, from a practical standpoint, the actual measurement of muscle force is one of today's more problematic areas and, therefore, tomorrow's opportunities. The chapter presents muscle mechanics from the cellular to the total joint level, from in situ to in vivo, using animal models as well as human models. Gregor presents some realistic challenges to areas long thought to be dogma, such as the stretch-shorten cycle, and also identifies relatively unexplored areas of research such as load sharing between elements of muscle groups.

Skeletal muscle is a transformer designed to convert signals from the nervous system and chemical energy derived from food into force for the control of our actions in the environment. Some general properties of muscle include

- irritability, or the ability to respond to a stimuli,
- conductivity, or the ability to propagate a wave of excitation,
- contractility, or the ability to shorten, and
- the limited ability to grow and potentially, in some situations, to regenerate.

Whereas muscle fibers possess certain physiological characteristics, they are joined to bone through tendon. Some properties of muscle are described with respect to the muscle fibers, but more realistically the performance of muscle must be considered in light of its interaction with tendon. It is this muscle-tendon unit that deals with the real world and possesses certain mechanical properties, modulated by impulses from sensory and motor portions of the nervous system, that perform selected functions in movement control.

This discussion is devoted to the mechanical properties of skeletal muscle and the application of our understanding of these properties to function in normal movements. The text begins with the fundamental structural unit of skeletal muscle, the sarcomere, and continues with the arrangement of sarcomeres into fibers, fibers into motor units, and motor units into whole muscle. Muscle architecture is then related to the mechanical properties of length-tension and force-velocity. The treatment of length changes includes a discussion on muscle stiffness and concepts related to the use of elastic strain energy during movement. Muscle strategies to increase force and procedures related to the quantification of the electrical output of the muscle (EMG) are briefly discussed. Finally, the mechanical properties of muscle measured directly during normal movement and the application of these data to our understanding of movement control are explored. Data related to both human and animal models are presented.

Muscle Architecture

The sarcomere is the basic structural unit of skeletal muscle fibers. This repeating unit represents the zone of a myofibril from one Z band to another and comprises an interdigitating set of thick and thin filaments (see Figure 7.1, a & b). At rest, the sarcomere's length is between 2.0µm and 2.25 µm. The exact length, however, may be species-dependent. Each myofibril is composed of bundles of myofilaments (thick and thin contractile proteins). The obvious striations in skeletal muscle are due to the differential refraction of light as it passes through the contractile proteins, resulting in the traditional A band (anisotrophic), I band (isotrophic), H band, M band, and Z bands. Six thin filaments surround a single thin filament.

The thick and thin filaments and the different proteins found within each filament are considered the fundamental elements controlling both force and contraction velocity. Although the thin filament structure is dominated by the actin molecule, it actually contains three proteins: actin, troponin, and tropomyosin, each described as having fairly well-established functions. Each thin filament contains approximately 350 actin monomers and 50 molecules each of troponin and tropomyosin. The major role of the actin filament is to interact with the myosin, whereas tropomyosin and troponin have selected control functions within the sarcomere. The actin filament, which contains both G- and F-actin is a polarized structure with the sense of the polarization being different on either side of the Z band. Since both thick and thin filaments are polarized, force generated by the union of these two filaments acts in a direction toward the center

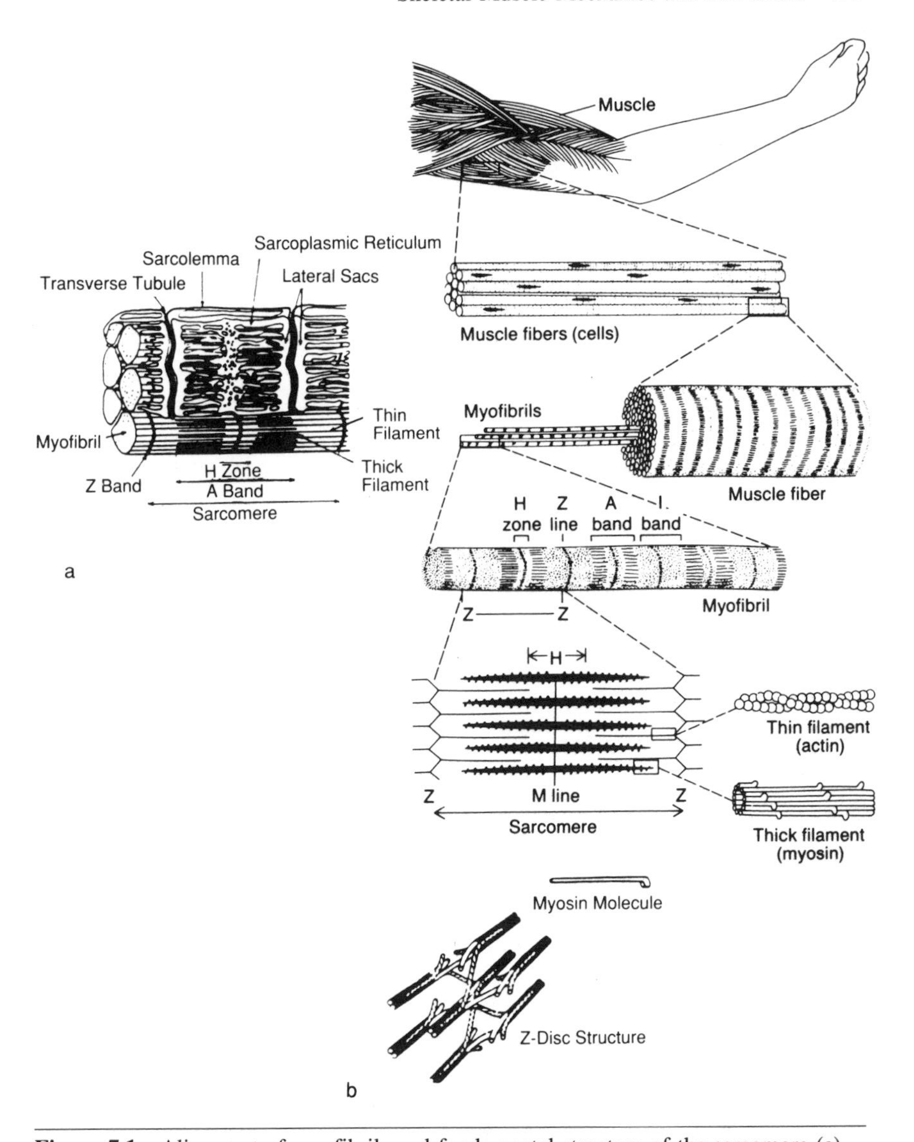

Figure 7.1 Alignment of myofibrils and fundamental structure of the sarcomere (a). (Reprinted by permission from Fawcett, 1968.) Organization of skeletal muscle from the whole muscle to the Z-disc structure (b). (Reprinted by permission from McMahon, 1984.)

of the sarcomere. Free, unattached muscle pulls its ends toward the center when stimulated.

Each tropomyosin filament is bound to an actin strand and lies in the grooves between the actin strands (see Figure 7.2). Each troponin complex consists of troponin-I, troponin-C, and troponin-T and binds to both tropomyosin and actin.

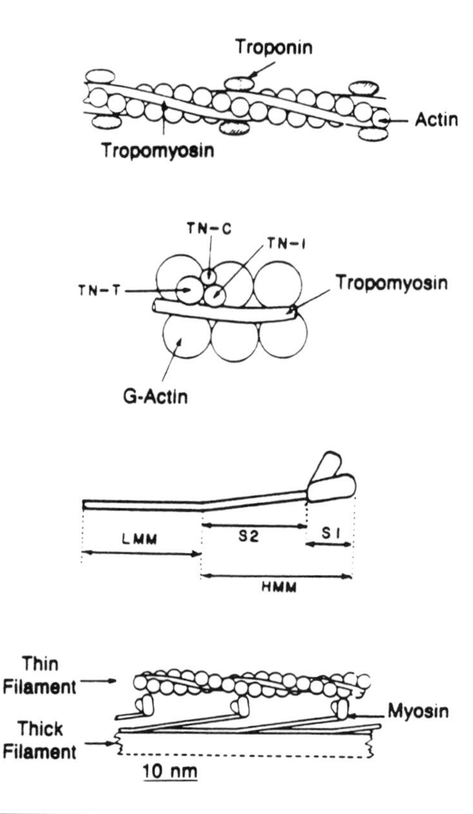

Figure 7.2 Schematic of the basic structure and orientation of selected protein molecules in the actin and myosin filaments. (Reprinted by permission from Enoka, 1988.)

Troponin-I together with tropomyosin inhibits the activation of myosin adenosine triphosphatase (ADP) by actin, possibly by physically blocking the myosin-combining site on the actin. Each tropomyosin filament inhibits seven actin molecules, probably by each of the actins interacting with one of seven rather similar regions on the tropomyosin molecule. Because of its affinity for calcium, troponin sensitizes the thin filament to calcium. When troponin-C combines with calcium, it removes the inhibitory effects of tropomyosin, and contraction takes place.

The main constituent of the thick filament is the protein myosin. Myosin is a large molecule composed of two identical heads connected to a long tail (see Figure 7.2). A flexible hinge region can be found in the tail approximately 43 nm from the heads. The two halves of the filaments on either side of the M line are of opposite polarity, and the tails of the myosin molecules point toward the center of the filament. The heads of the myosin molecules project from all areas of the thick filament except for a region approximately 0.15 μm long in the center where there are only overlapping myosin tails. The exact nature of the packing of the myosin tails in the thick filament remains uncertain. The myosin heads projecting from the thick filament form a helical array with a periodicity of approximately

43 nm and an axial interval of 14.3 nm. In addition to myosin, the thick filaments contain much smaller amounts of other proteins whose role may be to control and define the structure of the filaments themselves. The specific function of the cross-bridge region, commonly referred to as the *S2* or *S1 segment* of the myosin filament, lies in more detailed analyses of the sliding filament theory and the molecular basis for the dynamics of contraction. Suffice it to say that the tail of the myosin filament is composed of light meromyosin (LMM), whereas the S2 region and two globular heads in the S1 region are composed of heavy meromyosin (HMM). The mechanical rotation of these heads and the ability of the globular heads to interact with sites on the actin filament are critical to the generation of force and, at times, to the storage of strain energy in the muscle fiber. As a force-generating system, the development of force by myofibrils in their act of shortening requires some way of transmitting force from the thick to the thin filament, and the sliding filament theory, substantiated by light and electron microscopy as well as by x-ray diffraction, explains most aspects of this force-generating unit.

Muscle Fiber Length

Sarcomeres arranged in series form a skeletal muscle fiber, and fiber length exhibits a direct relationship to shortening velocity. Given similar angles of pinnation and biochemical properties, a muscle with more sarcomeres in series will record a faster shortening velocity at the tendon. The relationship between fiber length and maximum shortening velocity has been presented using a number of examples. Bodine et al. (1982) illustrated in the cat semitendinosus that the combined V_{max} of both proximal and distal segments, each having the same angle of pinnation and biochemical profiles, equaled the sum of each segment's V_{max} measured separately. Sacks and Roy (1982) presented data on the cat adductor femoris, which also has relatively homogeneous histochemistry and pinnation angle, indicating that although relative fiber lengths remained near 85%, absolute lengths in the distal fibers were three times longer than in the proximal fibers because to maintain a constant angular velocity of the femur, the distal fibers had to contract three times faster than the proximal fibers.

Absolute and relative fiber lengths (presented as a percentage of muscle length) have been reported to vary between muscles as well as in the same muscle between subjects. Wickiewicz, Roy, Powell, and Edgerton (1983) presented architectural data on 27 human lower extremity muscles, Friederich and Brand (1990) reported data on 38 muscles in the human lower extremity, and Sacks and Roy (1982) studied 24 muscles in the cat hindlimb. Although absolute values varied in these reports, some similar trends emerged. For example, Table 7.1 summarizes data on five muscles acting at the knee and ankle. In general, it appears that the flexors have higher fiber length to muscle length ratios (FL/ML) than the extensors, which these authors conclude is the result of longer fibers, that is, more sarcomeres in series. Although the cat soleus muscle is very different from the human soleus (i.e., the cat soleus contains 100% Type S

Table 7.1 Fiber Length to Muscle Length Ratios in Two Muscles Acting at the Knee and Three Muscles Acting at the Ankle

Muscle	Wickiewicz et al. (1983)	Friederich & Brand (1990)	Sacks & Roy (1982)
Semimembranosus	.24	.31	.70
Vastus lateralis	.17	.26	.30
Tibialis anterior	.26	.25	.58
Soleus	.06	.09	.49
Medial gastrocnemius	.14	.17	.24

Note. Data are taken from the human lower extremity and the cat hindlimb. References are specified in the text.

fibers), the trend between flexors and extensors appears to be the same for cats and humans. To further support the idea that extensors are designed for force and flexors for speed (i.e., longer fibers), Wickiewicz, Roy, Powell, Perrine, and Edgerton (1984) showed lower torques in knee and ankle flexors than in extensors, as measured on a Cybex device. Additionally, Ingen Schenau (1989) contended that single joint extensors, because they produce high maximal torques, are power producers, whereas biarticular flexors, with their longer fibers, are power distributors. The practical application of this information has relevance to the study of energy transfer between adjacent body segments via active biarticular muscle.

Physiological Cross-Sectional Area

A final feature of muscle relating design to function concerns its physiological cross-sectional area (PCSA). This variable is typically computed using the following formula:

$$\text{PCSA} = \text{muscle mass (g)} \times \cos \Theta/\text{fiber length (cm)} \times \text{muscle density (g/cm}^2)$$

where Θ is the average angle of pinnation.

The greater the cross-sectional area, the better suited a muscle is for force production. Because force per unit of cross-sectional area (specific tension) is considered to be relatively constant (22.5 N/cm^2), with some differences observed in different species, muscle force is directly related to the percent activation of the muscle's physiological cross-sectional area. Force modulation through excitation by the nervous system is considered in the section on electromyography. In an effort to understand the intricate relationships between muscle architecture, biochemistry, force-velocity characteristics, and function, Spector, Gardiner, Zernicke, Roy, and Edgerton (1980) studied the isometric and isotonic contractile properties and architecture of the medial gastrocnemius (MG) and soleus (SOL) muscles in 7 adult cats. They concluded that a combination of architectural and biochemical features accounted for the large differences observed in force production, maximum shortening velocity, and power output between these two hindlimb extensors. With the effects of muscle architecture eliminated, (a) the threefold difference in maximum shortening velocities (MG > SOL) was concluded to be a result of the intrinsic differences in biochemistry (i.e., the MG is approximately 85% FF [fast fatigable] and the SOL is 100% S [slow]) and (b) the fivefold difference in peak tetanic tension (MG > SOL) yielded the same specific tension in each muscle, namely, 22.5 N/cm^2. The most interesting aspect of this work lies in its application to motor control. It appears that the biochemical and structural properties of these two different muscles are very consistent with the way they are recruited, as evidenced by electromyogram (EMG) and muscle force records as well as the way they behave in response to varying environmental demands (Edgerton et al., 1986).

Since a complex relationship exists between muscle structure and function (Woittiez, Huijing, Boom, & Rosenthal, 1984), a final interesting note related to muscle architecture and function lies in recent findings that indicate muscle fibers taper and connect in a complicated way to tendon (Ounjian, Roy, Eldred, & Edgerton, 1987). Muscle fibers do not run the entire length of the muscle but end in the middle of a fascicle. Consequently, forces must be transmitted to adjacent fibers and to the network of connective tissue elements forming the interfiber matrix. Architecture of the tapered interface between the contractile protein molecules in the fiber and the connective tissue molecules (e.g., collagen) has been explored in an effort to understand the effective transmission of force generated by the fiber to the environment. Garrett and Tidball (1987) presented a schema of the structures involved in this force transmission and proposed that shear forces generated at this juncture, as well as those between adjacent active and passive fibers through the interfiber matrix, can have significant implications on the motor output of the muscle-tendon unit. For example, exercise involving predominantly lengthening action of the muscle showed marked disruption of myofibrillar material and tears along the myotendinous junction.

Two final questions to consider when studying muscle architecture are these:

- Fibers taper and yield a reduction in cross-sectional area. Is maximum tension related to the largest or smallest cross section?
- It is generally thought that exercises involving predominantly lengthening muscle actions display marked disruption of myofibrillar material and tears along the myotendinous junction. How are tissue strains distributed in the tapered region?

The Motor Unit

The functional unit within a single skeletal muscle is the motor unit. A single motor unit consists of the cell body and dendrites of an alpha motoneuron, the multiple branches of its axon, and the muscle fibers it innervates. It is thought that each muscle fiber is innervated by a single motoneuron, but that each motoneuron innervates more than one muscle fiber. The number of muscle fibers innervated by a single motoneuron, referred to as its *innervation ratio*, can vary from approximately 1:2,000 to 1:10. One motoneuron has a possible range of 10 to 2,000 innervated muscle fibers. The central nervous system activates the motoneuron and sends action potentials to all muscle fibers in a particular motor unit. This all-or-none phenomenon results in a single depolarization of an alpha motoneuron and subsequent depolarization of its muscle fibers. If there are only a few muscle fibers, the motor unit yields small increments in tension and demonstrates the capability for very precise control. Extraocular muscles, for example, are capable of such fine control because each motor unit has only three or four fibers. In contrast, in motor units of the large back and lower extremity muscles, a single alpha motoneuron may excite more than 2,000 fibers.

A major focus of current research on motor unit properties concerns the distribution of fibers within a single motor unit in a region of muscle or in the whole muscle. Are these fibers clustered, or are they randomly distributed in the muscle? What principles govern the distribution of fibers within the same motor unit or with respect to fibers in adjacent motor units with either similar or different histochemical profiles? Bodine, Garfinkel, Roy, and Edgerton (1988) isolated single motor units via glycogen depletion techniques in the cat tibialis anterior and soleus muscles. Measurement of the distances between all motor unit fibers showed a greater tendency for grouping than for dispersion, but this tendency was not statistically significant. This finding implies the existence of some potential mechanism responsible for governing the domain and absolute distribution of fibers of a single motor unit in the muscle. Given that each fiber has a certain cross-sectional area, the PCSA of a motor unit and subsequently of the entire muscle considers the summation of all fibers in each motor unit as well as their angles of pinnation. Knowing the domain of each motor unit, the spatial arrangement of the fibers within the unit both proximally-distally and throughout the cross section appears significant to our understanding of how the nervous system uses each motor unit in the control of movement.

Physiologically, motor units are compared on a number of properties including

- discharge characteristics,
- speed of contraction,
- force-production capabilities, and
- resistance to fatigue.

A direct evaluation of motor unit properties includes measurement of both electrical and mechanical discharge characteristics. These output measures usually are observed with reference to variable inputs in a controlled laboratory environment. In contrast, indirect measurements involve quantification of both histochemical and biochemical profiles of fibers within the same motor unit.

When one makes direct measurements of the physiological properties, two different types of input to the motor unit are usually employed. The first, a single action potential, generates what is referred to as a *twitch*, and the second, involving a series of action potentials, produces a fused *tetanus*. The ratio between twitch and tetanus can vary from 1:1.5 to 1:10 and is apparently related to the type of muscle studied. Measurements commonly made when evaluating the twitch response include

- the time from force onset to peak force (i.e., contraction time),
- the magnitude of the peak force, and
- the time from peak force to the time at which the force has declined to one half of its peak value (i.e., one-half relaxation time).

Essentially, the contraction time is used as a measure of the speed of the contractile machinery, and variations in contraction speed among motor units are considered to be a function of variations in enzyme activity (i.e., myosin ATPase). If the contraction time is long (e.g., 100 ms), the motor unit is referred to as a *slow-twitch* motor unit and if the contraction time is short (e.g., 45 ms), the motor unit is referred to as a *fast-twitch* unit.

Peak force is related to the number of fibers and the PCSA of the motor unit, as previously discussed. It is currently believed that the peak force exerted during a twitch response is related to and can increase with training. Additionally, force produced in a single tetanus declines over time if the motor unit is required to produce a series of tetanic contractions. The ability of a motor unit to prevent such a decline is taken as a measure of its resistance to fatigue. A standard fatigue test usually involves a 2- to 6-min period in which the motor unit is stimulated at a rate of 1 Hz, with each tetanus lasting approximately 330 ms. In fatigue-resistant fibers, this protocol will yield a relatively continuous output. For fatigable fibers, however, a decline in peak force is observed after 2 min of stimulation and may continue to decline for up to 6 min.

On the basis of contraction time and fatigue resistance, motor units can be classified into three categories (see Figure 7.3). The categories are

- slow-contracting fatigue-resistant (S),
- fast-contracting fatigue-resistant (FR), and
- fast-contracting fatigable (FF).

The fatigue-resistant units, namely, Types S and FR, have a fatigue index approximately three times that of the FF fiber, whereas Type S units typically produce the least amount of force, and FF fibers produce the greatest amount. These differences are due not only to variation in the number of muscle fibers within a motor unit (i.e., the innervation ratio for a slow motor unit is typically lower than that for a fast motor unit) but also to the size of the individual muscle fibers (i.e.,

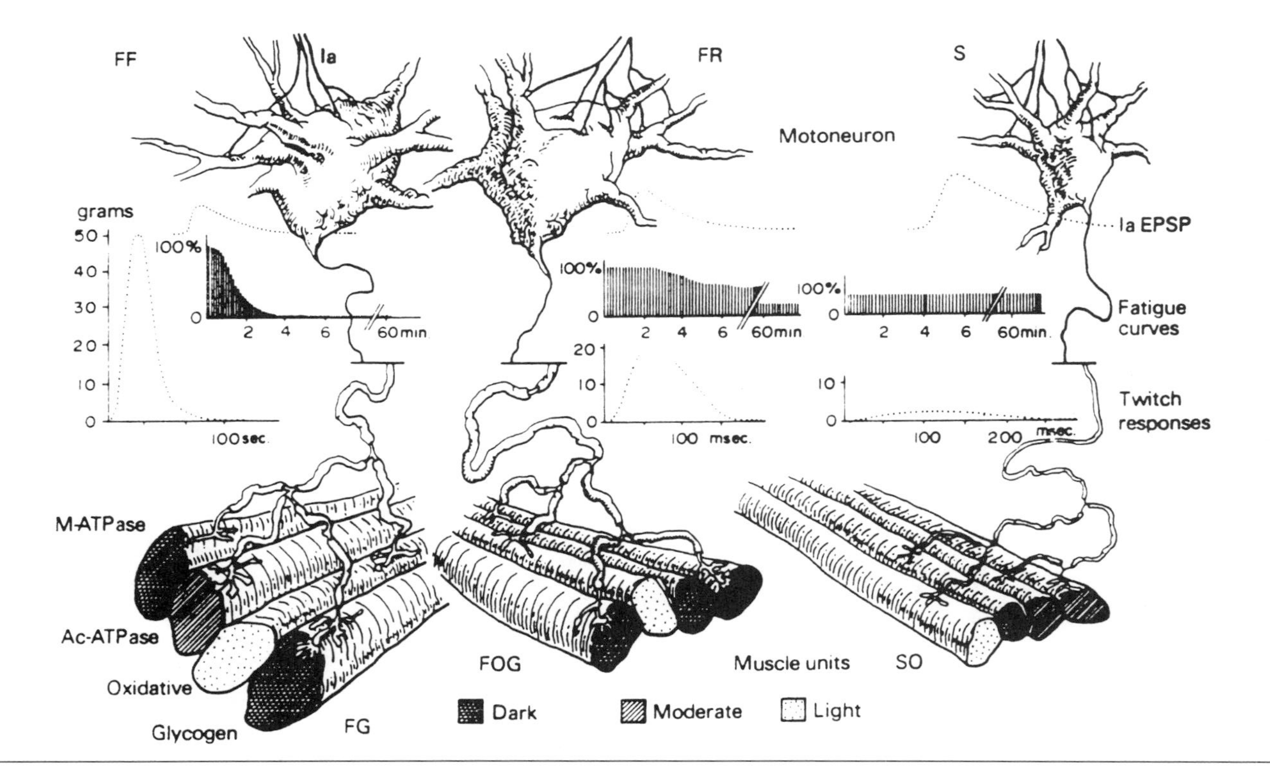

Figure 7.3 Fundamental properties of the fast-contracting fatigable (FF), fast-contracting fatigue-resistant (FR), and slow-contracting fatigue-resistant (S) motor units related to recruitment order, tension development, fatigue resistance, myosin ATPase activity, oxidative capacity, and glycogen content (FG = fast glycolytic, FOG = fast-oxidative glycolytic, SO = slow oxidative, la = afferent nerve, EPSP = excititorypost-syraptic potential). (Reprinted by permission from Edington & Edgerton, 1976.)

the quantity of contractile proteins within the muscle fiber). This variation is not thought to be a result of any differences in specific tension between fast and slow fibers, as previously noted.

There is much information on indirect estimates of motor unit output that is not discussed in this chapter. These histochemical and biochemical techniques used to measure enzymes such as NADH, SDH, and alpha-GPD add to our understanding of the aerobic and anaerobic capacity of various fibers and motor units. It is generally believed that because fibers in a motor unit have many similar properties and are activated in an all-or-none fashion, they have similar histochemical and biochemical profiles. Recent evidence, however, suggests that individual fibers within a motor unit may not necessarily be identical. We can have slight differences in enzyme profiles, for example, between fibers in the same motor unit as well as at different positions along the membrane of the same fiber within one motor unit. These types of data obviously complicate the picture of how these fibers are used in movement control. In fact, Pette and Staron (1988) reported over 50 different myosin isozyme profiles for skeletal muscle. These categories obviously are determined not only by contraction time but by a large array of myosin isoforms and variable enzyme patterns that dictate how these muscle fibers and subsequently motor units are used in the control of movement.

Compartmentalization

To this point we have discussed the structure of muscle as it relates to function and have described the traditional view of organization related to sarcomeres, fibers, orientation of fibers with respect to the pulling axis of the muscle, and structural and neural organization as they relate to the motor unit. It is now generally believed that there are identifiable compartments in the neuromuscular system, above the level of the motor unit, that can be considered functional units orchestrated by the nervous system to control movement. Windhorst, Hamm, and Stuart (1989) described levels of neuromuscular, sensory, and central partitioning as they related to the coordination of sensory, motor, and structural features in the neuromuscular system. Windhorst et al. proposed that the morphological and architectural features of skeletal muscle are sufficiently complex to require some form of partitioned control. For example, English and Weeks (1984) described four distinct regions in the cat lateral gastrocnemius muscle to illustrate neuromuscular partitioning. Several examples of sensory partitioning are supported by reports of nonuniform distribution of sensory receptors, and evidence of central partitioning rests in the data on muscle spindle projections to spinal motoneurons. Simply stated, there is now enough evidence regarding motor unit distribution and architecture of skeletal muscle to conclude that the functional organization of muscle does not necessarily rest at the level of the motor unit but with the nervous system's orchestration of compartments drawn from several muscles in the effective control movement. Use of appropriate resources based on updated sensory input will accomplish efficient movement control.

Fundamental Muscle Mechanics

During the past several decades, many classic studies have investigated the effects of muscle length and shortening velocity on muscle force production. Some studies (e.g., Goslow, Reinking, & Stuart, 1973) even report methods for calculating muscle-tendon lengths during normal movement. Although valid attempts have been made to refute these early findings, related to length, velocity, and force, the classic theories on length-tension and force-velocity relationships in skeletal muscle remain robust. One such observation concerns the position or length that muscle assumes that when if maximally activated will produce maximum force. This position is traditionally referred to as $\mathbf{L}_0$ and corresponds to optimal overlap between actin and myosin filaments (Gordon, Huxley, & Julian, 1966). Tension produced in a uniform series of sarcomeres then appears to be proportional to the number of available binding sites within each sarcomere. Tension is the same for all sarcomeres between 2.00 and 2.25 μm and decreases as the sarcomere is either lengthened or shortened from this position (see Figure 7.4a). At $\mathbf{L}_0$, given maximum excitation of all motor units, the muscle produces what is referred to as its *maximum isometric tension*, namely, $\mathbf{P}_0$. Typically, $\mathbf{L}_0$ is defined as the position of the muscle when $\mathbf{P}_0$ is achieved.

The term *mechanical stiffness* defines the instantaneous dependence of tension on length and is the derivative of force with respect to length, $d\mathbf{P}/dl$. The stiffness of a number of sarcomeres in series, a fiber, can be measured by changing length and recording tension. Unstimulated muscle fibers have very low stiffness, or large compliance ($dl/d\mathbf{P}$), with stiffness rising with the same time course as tension. Essentially, muscle fiber stiffness is directly related to filament overlap and cross-bridge attachment via muscle activation. Even during isometric contractions cross-bridges are cyclically attached, detached, and re-attached again; therefore, movements of thick and thin filaments are restrained by neighboring sarcomeres, supporting the fact that myofilaments have a high degree of stiffness and adjacent sarcomeres vary in strength. The summation of this process yields a very dynamic muscle fiber constantly responding to inter- and intra-sarcomere dynamics at varying levels of activation.

To evaluate sarcomere dynamics, Lieber and Baskin (1983) studied the contraction dynamics of the end and center regions of single fibers during a fixed-end tetanus. While resting sarcomere length varied between the end and center regions, the end regions contracted almost twice as fast as the center region sarcomeres contracted during a very rapid rise in force. During a slow rise in force, however, the velocity of contraction of the end region was almost four times the velocity of stretch of the center region. In further studies by Mai and Lieber (1990), sarcomere lengths were measured by laser diffraction in passive muscle during hip and knee rotation in the frog. Given this information, muscle sarcomere length, force, and hip and knee torque were predicted. Model output indicated that during a single coordinated movement, such as a hop, each sarcomere experienced a period of shortening followed by a period of lengthening. This

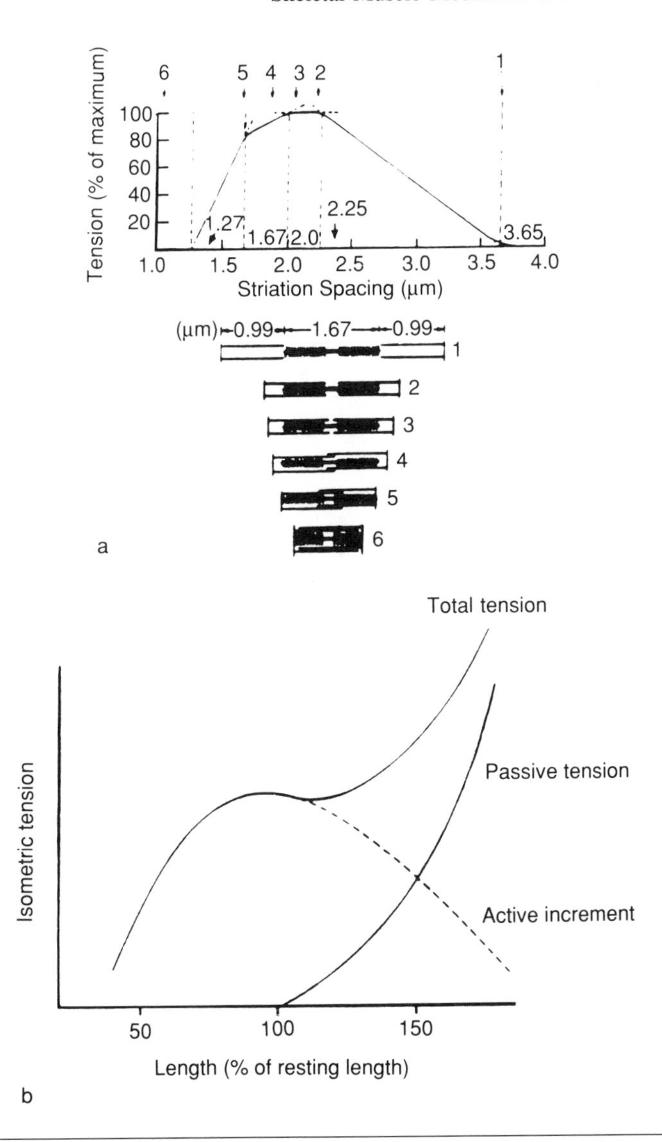

Figure 7.4 The relationship between filament overlap, sarcomere length, and force production (a). Schematic diagram of the active length-tension curve, passive length-tension curve, and total tension curve for the muscle-tendon unit (b). (Reprinted by permission from McMahon, 1984.)

stretch-shorten phenomenon at the level of the muscle fiber is similar to that observed in whole muscle during normal movement and is further discussed in a subsequent section of this chapter.

This discussion of length and tension concludes by recognizing that activation of the contractile element is not the only process employed by muscle to increase tension. An isolated muscle fiber at a sarcomere length greater than $\mathbf{L}_0$ will exert a measurable force called *passive tension.* Also, whole muscle includes a substantial amount of connective tissue (e.g., a sarcolemma, endomysium, perimysium, and epimysium), and these connective tissue structures generally respond like stiff elastic bands. When stretched they also exert a passive force that combines with the generally lower forces from the passive muscle fibers. These passive forces combined with the active contribution from cross-bridge activation result in what is referred to as the muscle's *total length-tension curve* (see Figure 7.4b). The complex interaction of both active and passive compartments in the muscle remains the focus of current research.

Force-Velocity During Shortening

Although cross-bridge overlap and muscle length changes are important to the understanding of muscle function, the mechanical output of muscle does not respond to isometric loads very often during everyday activities. Total body movements involve the acceleration and deceleration of limbs during which rotary movements about the various joints must be controlled. The actual magnitude of force a muscle produces depends on how, and whether, its ends are restrained. A muscle's function as a stabilizer, agonist, neutralizer, or antagonist depends on if, and how fast, it is allowed to change length. In reality muscles do not act in isolation but rather in a system in which load sharing among all tissues (e.g., muscle, tendon, ligament, and bone) must be done efficiently to maximize output and minimize injury. Consequently, a second important parameter related to muscle function is its ability to produce force at different velocities and accelerations under either constant or varying activation.

During the 1920s Hill's interest in muscle function and human activity stimulated many experiments involving, for example, sprint running and cycling. In their quest to answer the questions generated by these experiments, Hill and his colleagues performed several classical experiments on isolated amphibian muscle. A major outcome of the early studies was his classic paper (Hill, 1938), describing the relationship between force and the velocity of shortening as part of a rectangular hyperbola (see Figure 7.5, a & b):

$$(\mathbf{P} + a)(\mathbf{V} + b) = (\mathbf{P}_0 + a)b$$

where $\mathbf{P}$ is the force during shortening at velocity $\mathbf{V}$, $\mathbf{P}_0$ is maximum isometric tension, and a and b are constants.

(Power curves for the two muscles in Figure 7.5a are presented in Figure 7.5b.)

Hill originally thought that constants a and b could be obtained from heat as well as from mechanical measures, giving his equation fundamental significance. This has subsequently been demonstrated to be incorrect. In studies by Edman, Mulieri, and Scubon-Mulieri (1976), for example, using both after-loaded and quick-release techniques on single fibers and fiber bundles, some muscles systematically deviated from a hyperbola. Although it appears that under a range of conditions the force-velocity curve may not be hyperbolic, the original descriptive characterization by Hill and his fundamental ideas do not need any serious revision.

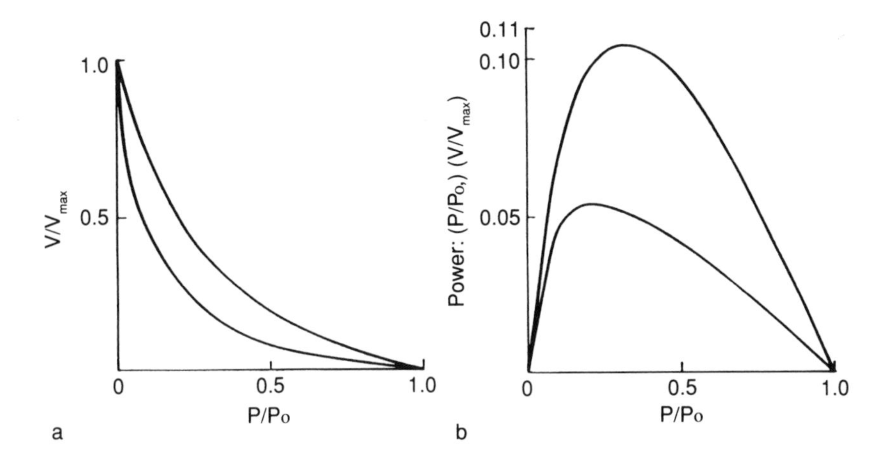

Figure 7.5 Force-velocity curves calculated from the normalized form of the Hill equation for two different muscles (a); power-force curve for those muscles (b). (Reprinted by permission from Woledge, Curtin, & Homsher, 1985.)

Major features and interpretations of the classical force-velocity relationship include the following:

- $\mathbf{P_0}$ depends on the number of cross-bridges attached, assuming the force per attached cross-bridge is constant under isometric conditions.
- $\mathbf{V_{max}}$ reflects the maximum rate of cross-bridge turnover but is independent of the number of cross-bridges that are operating.
- Experiments performed to study this property are designed to reflect the contractile component (CC) of muscle and to eliminate the effects of the series elastic component (SEC).
- Experiments are usually conducted under maximal stimulation, beginning the contraction from $\mathbf{L_0}$ in a controlled laboratory environment.

Force-Velocity During Lengthening

It has been known since the early 1920s that stimulated muscle can produce more force during stretch than during either isometric or shortening actions. Despite this early interest, little is understood about the factors that govern force production during muscle lengthening. This fact is quite surprising because we employ cycles of lengthening and shortening constantly during everyday activities. The general response of muscle under varying velocities of lengthening is presented in Figure 7.6. It should be noted that the peak tension increases above $\mathbf{P_0}$ in a range between 1.2 and 1.3 $\mathbf{P_0}$, has been studied in loads up to 1.8 $\mathbf{P_0}$, and is

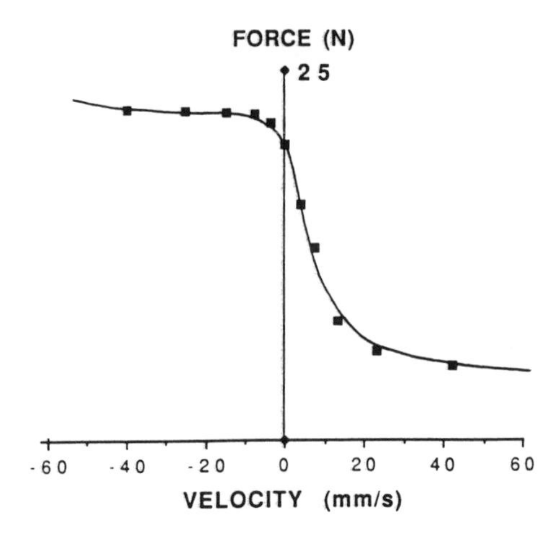

Figure 7.6 Schematic of the force-velocity profile for both lengthening and shortening contractions. This profile represents the force output of the maximally active contractile element in skeletal muscle.

considered to reach a relatively constant value regardless of additional increases in lengthening velocity. Limitations to our understanding of the force-velocity relationship during lengthening are apparent in a subsequent section of this chapter on the application of known properties of skeletal muscle to movement mechanics.

In summary, factors affecting a muscle's ability to produce force include

- length,
- velocity,
- fiber type related to myosin ATPase activity,
- physiological cross-sectional area, and
- activation through a combination of recruitment and rate coding.

This last point is discussed next.

Muscle Activation and EMG

At the level of the motor unit, it appears the nervous system has two major options available for varying force production: (a) recruiting active motor units and (b) increasing the firing rate of units already active. These two strategies are typically referred to as *motor unit recruitment* and *rate coding*, respectively. It is generally believed that motor units are recruited in an orderly sequence based on the size of the alpha motoneuron. This phenomenon, referred to as the *size principle*, suggests that the motor unit with the smallest motoneuron is recruited first and the one with the largest motoneuron is recruited last. For example, if a pool of motoneurons experience a common electrical drive, the smaller motoneurons will, in general, be recruited first. These typically are motor units in the Type S (or SO) category that produce relatively low maximum tension and have long contraction times. These muscle fibers are used on a daily basis, typically for postural control and stability.

In studies of motor units in a controlled environment generating a slow ramp contraction, a smooth increase is usually observed in force as a result of the orderly recruitment of small to large motoneurons and slow to fast motor units. In a ramp contraction with a negative slope, there is an orderly derecruitment of these same motor units, and this sequence of derecruitment is the opposite of that observed during increased force requirements. Recent reports, however, indicate that the threshold for activation of motoneurons is not dependent solely on the size of the motoneuron but can be influenced by factors related to input into a motoneuron pool. This input varies depending on changes in sensory input. Motoneuron size appears to be a significant, but not the only, factor regulating recruitment order of individual motor units. Also, it is necessary to emphasize that motor unit types do not exist as discrete populations relative to their excitation threshold. There appears to be considerable overlap between groups, especially between Type S (or SO) and Type FR (or FOG) motor units. Consequently, there is not necessarily such discrete packaging in which only Type S, then FR, and then FF units are systematically recruited to generate a smooth increase in tension. It is quite possible to activate a muscle such that Type S motor units are recruited with Type FR units. This does not contradict the size principle and also allows for almost synchronous activation of a large population of fibers necessary, for example, in high-power-output activities.

Whereas recruitment is one strategy the nervous system employs to increase force production in the muscle, modulation of motor unit firing rate is a second. Several studies in the literature have compared firing rate and force increase. Several decades ago it was reported that the upper limit of motor unit firing rate in humans was approximately 50 pulses per second. At the lower end of the frequency spectrum, the relationship between force and firing rate is nonlinear. For example, the increase in force due to the increase in firing rate from 5 to 10 Hz is not the same as that due to increases in rate between 20 Hz and 25 Hz. The force-firing rate relationship is sigmoidal, with the greatest increase in force occurring at lower rates (e.g., 3 to 10 Hz). The exact relationship depends to some degree on muscle length because the curve shifts to the left for longer muscle lengths and to the right for shorter muscle lengths. Although the relationship remains sigmoidal, the frequency for producing the greatest change in force is approximately 3 Hz to 7 Hz for a long muscle and 10 Hz to 20 Hz for shorter muscle lengths.

Similar to motor unit recruitment, the extent to which rate coding is used appears to be muscle-dependent. For example, if a muscle has the capability to recruit additional motor units, then that strategy seems to dominate. If motor unit recruitment in a given muscle is essentially complete at 30% maximum force, then subsequent increases in force must come from increases in firing rate. This carries over to muscle size because larger muscles have more motor units available for recruitment, and smaller muscles have fewer motor units and are, to some degree, dependent on increases in firing rate. For example, consider the small muscles in the hand that have recruitment rates of about 9 pulses per second. The peak rate at 40% maximum voluntary contraction for these muscles is about 25 pulses per second, whereas the peak rate at 80% maximum voluntary contraction is 42 pulses per second. In contrast, larger muscles (e.g., deltoid) have a recruitment rate of about 13 pulses per second, a peak rate at 40% maximum voluntary contraction of about 26 pulses per second, and a peak rate at 80% maximum voluntary contraction of only 29 pulses per second. Consequently, motor units in smaller muscles are recruited at lower firing rates than in larger muscles and have a greater range of firing rates available to increase force from 40% to 80% maximum voluntary contraction. Increases in muscle force, however, are not due exclusively to increases in either recruitment or firing rate because the two are undoubtedly acting concurrently. Basmajian and DeLuca (1985) presented a more detailed account of how both recruitment and rate coding strategies are employed by the nervous system to modulate tension.

A final mechanism available to the nervous system to increase force in a muscle concerns the temporal relationship between the action potentials of different motor units. Normally, the action potentials of different motor units are assumed to be discharging at different points in time (i.e., asynchronous firing). Recent studies, however, showed that motor units of a muscle subjected to a strength training program tended to discharge action potentials synchronously. Since these reports, it has been assumed, although never proven, that motor unit synchronization results in greater muscle force. It also appears that synchronization is more evident when a muscle is fatigued. This strategy then appears to be influenced by the metabolic properties of motor units and by sensory feedback to the motor output system. Strategies employed by the nervous system to increase force in muscle can focus on both recruitment and firing rate, but the domain in which these are used remains open to discussion.

The window to the nervous system in studies related to movement control is the electromyogram (EMG). Whereas voluntary muscular activity recorded in the EMG signal increases in magnitude with increases in tension, many variables can influence the resultant EMG signal at any point in time. The characteristics of the observed EMG are functions of both the instrumentation used to collect the signal and the electrical current produced by the muscle fibers. Instrumentation typically involves a series of filters, all of which affect the amplitude and frequency characteristics of the EMG. Also, coupling the EMG to the dynamics of muscle function is requisite to a complete understanding of how muscles are used during movement. For example, while sampling EMGs, one should collect information related to muscle action (i.e., lengthening and shortening velocities, joint moments, and any possible reflex activity that may influence the EMG). The objective in this section is to briefly discuss the instrumentation used to collect the EMG and to relate current methods of describing features of interest in the EMG pattern.

Instrumentation

Instrumentation typically employed to measure EMG involves a bipolar pair of either surface or indwelling electrodes, appropriate connectors, an AC wideband differential amplifier, appropriate filters, and recorders necessary for storing data. A schematic of the components in this processing and recording system is presented in Figure 7.7. As illustrated, body tissues and fluids act as filters and result in a significant decline in signal intensity as distance between the recording electrode and the motor unit increases. Grieve (1975) describes this as the *inverse square law*. Once the biopotential is generated, the electrodes transform the ion flow in the tissue to electron flow in the recording wire. Surface electrodes, which typically consist of silver or silver chloride cupped discs varying from 5 mm to 10 mm in diameter, are designed to hold the amount of electrode gel necessary for the transduction of biological signals. These electrodes act as high-pass filters whose surface area will affect the frequency content of the signal (i.e., the larger the surface area, the lower the frequency content). Surface electrodes detect average muscle activity, usually from superficial muscles. However, if the inter-electrode distance is too great (e.g., > 2 cm), the likelihood of detecting measurable amplitudes from adjacent and deep muscles increases. This *cross-talk* is a major concern of researchers because the information collected by a pair of electrodes may actually come from more than the one muscle under study. Also, the magnitude of the EMG or action potential from a single motor unit recorded at the electrode wire is a function of (a) the size of the motor unit producing the action potential (i.e., the number of fibers in that motor unit) and (b) the distance the recording electrode is from the particular motor unit under study. One cannot conclude, regardless of the electrode type, that a large action potential necessarily

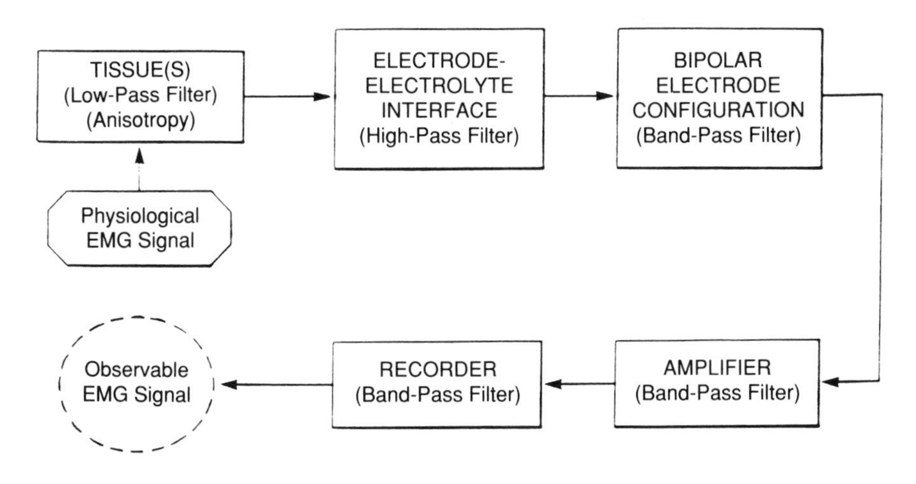

Figure 7.7 Diagram of the primary steps in data acquisition, processing, and recording of the electromyogram (EMG). (Reprinted by permission from Basmajian & DeLuca, 1985.)

comes from a large motor unit. Given the previous discussion, however, surface electrodes remain desirable for studying movement because they do not produce a significant amount of movement artifact and provide fairly reproducible results.

Whereas small discs may be used on muscles in children and smaller muscles in adults, intramuscular, or indwelling, electrodes are required to accurately assess single motor unit activity and activity from muscles deep beneath the surface of the skin. The intramuscular electrodes contain a bipolar pair of 25- to 50-µm, insulated wires placed beneath the skin via a hypodermic needle. For example, surface electrodes are appropriate for collecting activity in the vastus lateralis muscle, whereas fine wire are more appropriate for collecting activity produced by the thenar muscles in the hand. In a clinical setting, needle electrodes have been employed but have limited use in situations requiring considerable body movement. Basmajian and DeLuca (1985, pp. 22-50) present a detailed account of the different types of electrodes and connectors that use either surface, needle, or fine-wire preparations.

Although a detailed description of amplifiers and associated instrumentation is not within the domain of this chapter, several features must be considered in any discussion of electromyography. Major considerations in amplifier specifications concern

- gain and dynamic range,
- input impedance,
- frequency response and filtering capabilities, and
- common mode rejection ratio.

Detailed accounts of these characteristics can be found in instrumentation books, but sufficient for now are these basic facts:

- The amplifier should have an appropriate range of gain. A gain of 1,000x is very common for both surface and fine-wire preparations.
- Input impedance should be very high in comparison to the impedance of the electrodes. The electrodes should allow signals to pass relatively unaffected, and all major filtering should be done at the level of the amplifier.
- The frequency response of the amplifier should be appropriate for the frequency content of the EMG signal. For example, peak power of an EMG signal based on power spectral density analyses is below 100 Hz for surface electrodes and 200 Hz for indwelling electrodes. Ninety-five percent of the energy in a signal obtained using surface electrodes has a frequency content below 400 Hz, whereas indwelling electrodes have most of their frequency content below 500 Hz. Amplifier settings must be made so that signals with frequency content in this range pass through the system unattenuated, and frequencies of nonbiological origin, outside this range, are attenuated to the maximum degree. The frequency response of the amplifier should be capable of attenuating low-frequency movement artifact (< 3 Hz), high-frequency noise from nonbiological origin (> 1000 Hz), and 60-Hz noise from power lines.

• The common mode rejection ratio should be very high to insure that signals common to both electrodes be markedly attenuated. Differential signals between the two electrodes, usually of biological origin, should be allowed to pass relatively unattenuated (3 db).

EMG Processing

Several procedures are employed to process the EMG; they vary from simple evaluation of the bipolar waveform, or interference pattern, to full-wave rectification and integration. Typically, the types of information gained from the EMG signal involve a time-domain analysis yielding information related to such parameters as burst duration (BD) and interburst intervals, with amplitudes quantified using a smoothed rectified signal, a mean rectified signal, a total integral of the EMG (IEMG), the mean EMG over a certain period of time (MEMG = IEMG/BD), and the root-mean-square (RMS) value. The specific objectives of any study dictate how the EMG is processed and how it is related to other kinematic and kinetic variables (i.e., force, acceleration, or velocity). Winter (1990) gave a detailed account of kinesiological applications of electromyography and provided insight into the variety of ways this signal can be processed in the study of movement.

Much of the literature on EMG evaluation focuses on its relationship to muscle force. In a controlled environment, an increased force requirement dictates an increase in the activation because the force-generating unit of the muscle (i.e., the contractile component) is the element that generates the EMG. In dynamic situations, however, where movements require lengthening and shortening of muscle and acceleration of body segments, relationships between the EMG and the associated force or joint torque are more complex. For example, one of the first considerations when associating the EMG to muscle force is that the signals are not coincident in time. As illustrated in Figure 7.8, there can be a considerable delay between the onset of the EMG burst and the beginning of muscle force production. This time difference, commonly referred to as *electromechanical delay* (EMD), varies depending on a number of variables. Recent evidence reported by Norman and Komi (1979) indicates the delay ranges from 26 to 42 ms, depending on the speed of contraction, the type of contraction, and the type of muscle involved. It appears, for example, that the delay is shorter during a lengthening muscle action than when the muscle experiences a shortening action. Delays of only 5 ms have been reported during high-velocity lengthening actions of the fast gastrocnemius muscle in the cat hindlimb. Some researchers argue the delay is more consistent, but the literature indicates a great deal of variability in EMDs.

Models that employ EMG as a driving function to predict muscle force rely on several assumptions made during calibration of the EMG in a controlled environment. Norman (1977), for example, established a relationship between the EMG of elbow flexors and their percent contribution to the joint moment.

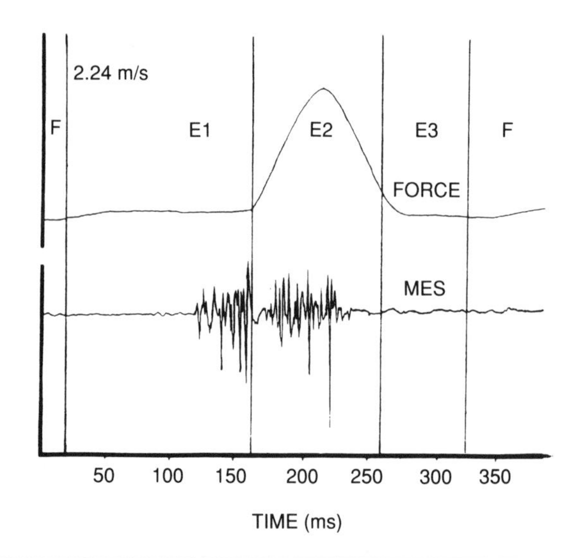

Figure 7.8 Force and EMG records from the cat medial gastrocnemius during a single step cycle at a treadmill speed of 2.24 m/s. The delay between EMG and force is approximately 50 ms (MES = myoelectric signal). (Reprinted by permission from Sherif, Gregor, Liu, Roy, & Hager, 1983.)

Essentially, the net flexor moment was distributed according to the PCSA of the major agonist muscles. That value was then divided by the measured moment arm of each muscle, leaving a value of force for that specific agonist. If it is assumed that all muscles were maximally active during the calibration, this force value was related to the EMG recorded via surface electrodes, and a relationship was expressed for each muscle in N/V. Under dynamic conditions, this relationship was modified according to the estimated force-velocity and length-tension properties of the muscle as well as to compliance associated with tendon excursion. Hof and van den Berg (1977) followed a similar argument in distributing the joint moment at the ankle within the tricep surae complex and subsequently (1983) established a more sophisticated relationship in which force was predicted from EMG during slow walking. Although models of this type have some inherent assumptions, in some situations they are quite promising.

Muscle Stiffness

When studying function of the muscle-tendon unit during movement, one must understand the stiffness of the muscle and tendon operating in series (Praske & Morgan, 1984; Rack & Westbury, 1974). I previously stated that muscle stiffness is directly related to cross-bridge activation and muscle length. Static stiffness increases as tension and length increase until a maximum value is attained and failure occurs. Dynamic stiffness is either incremental, as measured under transient conditions, or it is instantaneous and related to the slope of the length-tension curve. Stiffness is also under reflex control, which when combined with central nervous system control yields a relatively complicated picture. The actual regulation of stiffness involves both force feedback components and length feedback components that constantly update the muscle to accommodate load and length perturbations.

Tendon, in series with the muscle fibers, has its own mechanical properties and is anatomically divided into a portion external to the muscle (i.e., free tendon), and a portion internal to the muscle, (e.g., the aponeurosis). Data suggest that similar strains are experienced throughout both internal and external portions. It is also convenient to assume that the material properties (i.e., the stress-strain relationship in both the external and internal tendon) are similar. Stress is the relationship between force and tendon cross-sectional area (**F**/*A*), and strain is the relationship between a change in length over an initial length (*L*/*L*). For all parts of the tendon to experience the same strain, each must experience the same stress. Figure 7.9 shows the mechanical properties of isolated tendon, indicating a toe region in which the tendon tangent modulus of elasticity increases with strain at low strains, a linear region in which the proportion is constant, and ultimate failure at higher strains, which occurs at approximately 10% strain. The toe region begins where a tendon is stretched to approximately 2% strain.

The time history of repetitive loading, however, also needs to be considered. For instance, the tendon must be stretched to consistently longer lengths during the first few cycles of repetitive loads to develop even a small amount of stress. It is also believed that exercise can increase the tendon tangent modules in the toe region. The stress-strain response is thought to be independent of strain rate during locomotion, and it is agreed that when the tendon shortens at physiological rates, it can lose between 6% and 11% of its energy.

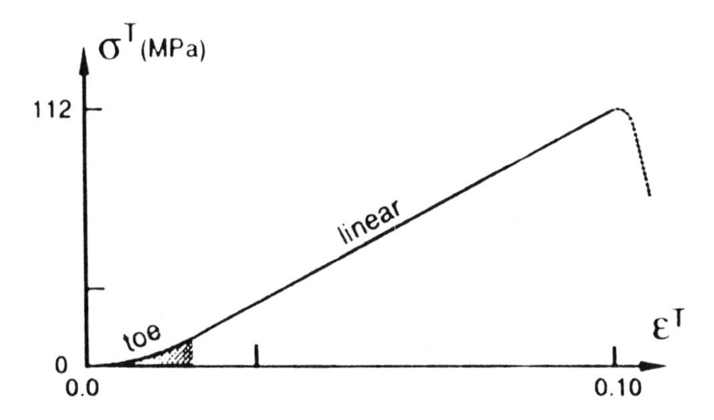

Figure 7.9 Material properties of tendon described by a nominal stress-strain curve (o^T = nominal stress, ε^T = nominal strain, ^{m}Pa = megapascals). (Reprinted by permission from Zajac, 1989.)

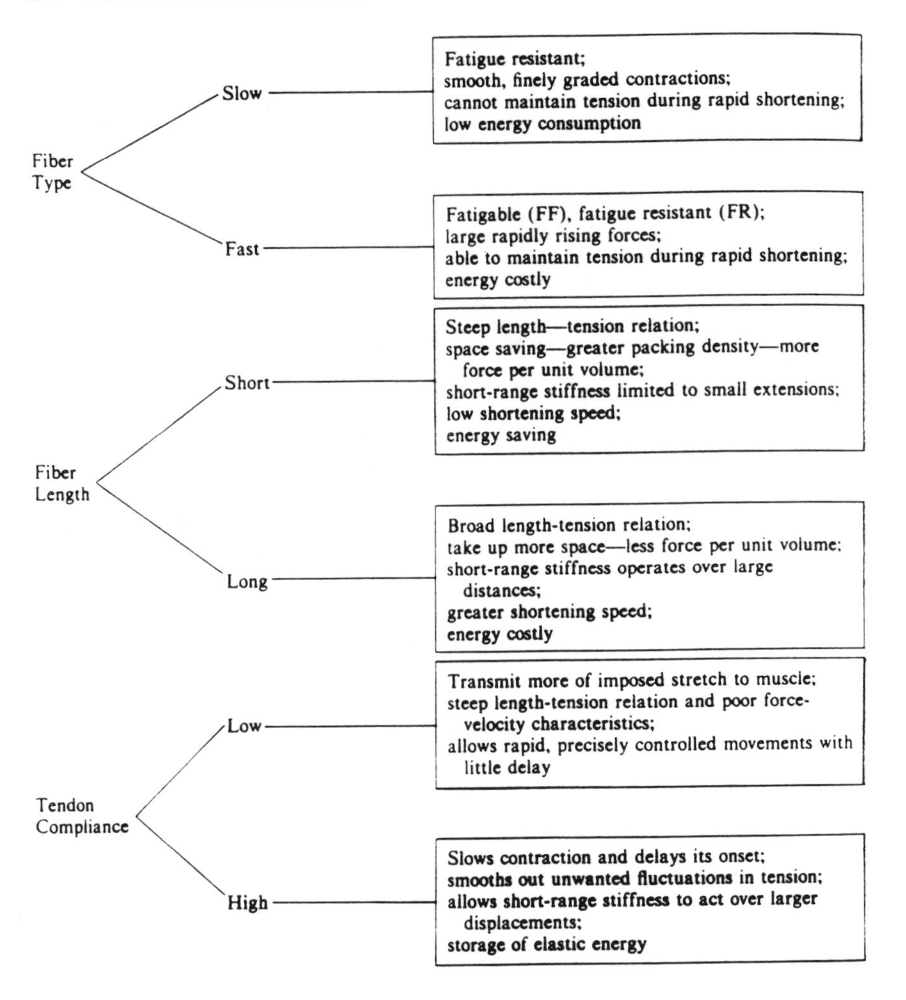

Figure 7.10 Relationship among major factors influencing the performance of the muscle-tendon unit in movement control. (Reprinted by permission from Walmsley & Proske, 1981.)

When evaluating the design of muscle, one must consider several points about muscle fiber and tendon lengths. Figure 7.10 summarizes various factors in muscle design and tendon compatibility and lists their advantages and disadvantages. This figure, from an article by Walmsley and Proske (1981), is useful in understanding the issues presented in this chapter related to muscle fiber type, architecture, and tendon compliance. These considerations are referred to in the final section of this chapter regarding muscle function in vivo.

Muscle Mechanics and Normal Movement

The mechanical properties of muscle and tendon typically are studied in isolation under conditions in which temperature, blood supply, length, and load are strictly controlled. Because these variables cannot be strictly regulated in vivo and the performance of the muscle-tendon unit is greatly influenced by the way it functions with its synergists, mechanical output of the muscle-tendon unit measured during normal movement is difficult to interpret. Also, certain muscles are better suited to a given task than others, and the manner in which one muscle responds to environmental demands affects how other muscles in the same group respond. For example, for any given load at the ankle (e.g., those generated in response to ground reaction forces), there is a differential response by the fast biarticular gastrocnemius and the slow uniarticular soleus. Studying how these two muscles work together (i.e., load sharing) is more interesting than knowledge of their performance alone. In fact, knowledge of how effectively all resource tissues, both active and passive, are used is important to our understanding of joint control. Muscles do not act in isolation.

Knowledge of the forces produced by the muscle-tendon unit (i.e., the interface between the central nervous system and the articulated skeleton), is requisite to a thorough analysis of movement. Thus far, we have discussed the morphological, histochemical, and mechanical properties of muscle and tendon in isolation and how these two compartments might behave. Our task now is to describe how these fundamental mechanical properties relate to data obtained during normal movement.

Currently, three methods are employed to estimate, or measure, muscle forces in vivo. They are (a) forward solution techniques using the calibrated EMG to predict muscle force (Sherif, Gregor, Liu, Roy, & Hager, 1983), (b) indirect estimates using principles in inverse dynamics and optimization, and (c) direct measurement of muscle forces using implanted force transducers (Whiting, Gregor, Roy, & Edgerton, 1984; Gregor, Roy, Whiting, Hodgson, & Edgerton, 1988). A major challenge to predictive modeling (i.e., methods a and b) is the validation of model output using experimental data. An objective of the third method presented above (i.e., direct force measurements) could be the validation of other noninvasive techniques. Results of this type of validation experiment have been presented recently and indicate surprising success. Norman et al. (1988) compared forces predicted from an EMG-driven model to forces measured directly from the cat soleus muscle during locomotion across a range of treadmill speeds. In most cases model output was very close to the directly measured force.

Data obtained from implanted tendon transducers have broad application to questions in muscle mechanics, motor control, and musculoskeletal biomechanics (see Figure 7.11). In the remainder of this chapter, three examples are presented in which forces are measured directly from cat hindlimb or human lower extremity

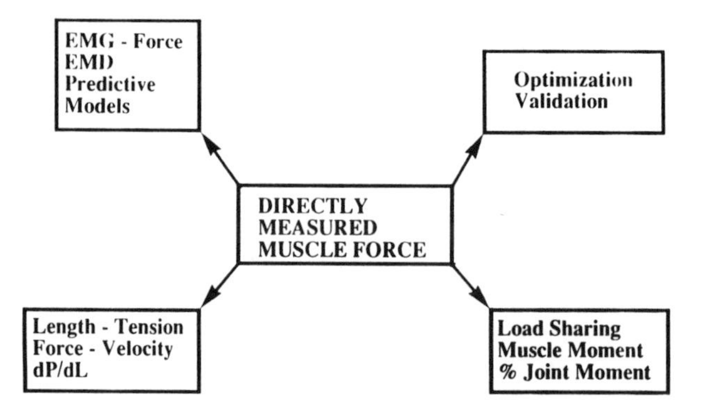

Figure 7.11 Block diagram illustrating the applications of tendon force data to questions in motor control, muscle mechanics, optimization modeling, and joint kinetics (EMD = electromechanical delay; P = force; L = length).

muscles. Each example is designed to illustrate a specific point regarding load sharing and the compatibility of data measured in vivo with known properties of skeletal muscle measured in isolation. Relevance of these data to the three previously mentioned areas of research should become apparent.

Load Sharing Between Medial and Lateral Heads of the Gastrocnemius Muscle

There are three major ankle extensors in the cat hindlimb, each with different physiological and mechanical characteristics. Load sharing among these three muscles (i.e., gastrocnemius, soleus, and plantaris) has been studied, and results indicate that as kinetic demands increase, that is, requirements increase for plantarflexor torques, individual muscle output increases in a manner related to the muscle's architecture and histochemistry (Fowler, Gregor, Hodgson, Roy, & Broker, 1989). To further explore the distribution of effort in the ankle extensor group, Tjoe, Gregor, Perell, and Roy (1991) considered separately the function of the medial gastrocnemius (MG) and lateral gastrocnemius (LG) during normal movements in the cat. For the most part, the functional role of these two muscle heads has been derived using electromyography, and although each head has a common distal insertion on the calcaneus, published results imply they are recruited differently. This fact has been discussed previously in our section on compartmentalization, in particular, regarding the lateral head of the gastrocnemius. To support the functional role of the MG and LG based on EMG, data on the histochemical, morphological, and isometric force production capabilities are presented in Table 7.2 (Tjoe et al.).

It may be concluded from these data comparing LG and MG capabilities that the lateral head appears capable of producing more force—because of its physiological cross-sectional area and, possibly, because of its high percentage

Table 7.2 Architectural, Histochemical, and Isometric Force-Production Capabilities of the Medial and Lateral Heads of the Gastrocnemius Muscle in the Cat

Variable under study	Lateral gastrocnemius	Medial gastrocnemius
Fiber length (mm)	24.5 ± 0.6	20.9 ± 0.3
Fiber angle (degrees)	17	21
Physiological cross-sectional area (cm^2)	4.58	4.01
Estimated P_0 (N)	105.3	92.2
S fibers (%)	25	18
FR fibers (%)	14	16
FF fibers (%)	61	66

of large fast-twitch fibers. It also appears capable of higher power output because the fiber length to muscle length ratio is greater than in the MG, a feature directly related to $\mathbf{V}_{max}$. One might infer then, given the differential recruitment of the two muscle heads and the data presented in Table 7.2, that as extensor requirements increase (e.g., from walking to vertical jumping), the neuromuscular system meets the increased power demands by employing additional resources—first with the MG and subsequently with the LG, which is better suited to meet the higher power demands.

To substantiate these inferential data, the distal tendons of both the medial and lateral heads of the gastrocnemius were implanted with tendon transducers capable of measuring forces in the freely moving cat (see Figure 7.12). Postoperatively, each cat stood in a quiet, quadruped posture; walked on a level and on a 12% inclined surface; and jumped vertically to a height of 0.65 m. Ground reaction forces were measured beneath each hindlimb using piezoelectric load washers. Average peak muscle force for both MG and LG, as well as percent of maximal isometric tension, are presented for each condition in Table 7.3 (Tjoe et al., 1991). Peak ground reaction force for each condition is also presented.

In general, muscle force data appear consistent with predictions made from the histochemical and morphological features of each muscle. Although absolute values were low and similar between the two heads during quadruped standing, differences became more apparent during level and incline walking. To meet the increased demands during walking and vertical jumping, the following scenario appears plausible. The MG meets the initial demands for increased force by generating consistently higher forces than the LG generates during walking. The force production demands can be met by using this resource first while the LG, to some degree, appears to be held in reserve. During the vertical jump, higher power output is required, and whereas both LG and MG respond with similar

Table 7.3 Average Peak Muscle Forces, Both Absolute and as a Percent of Maximum Tetanic Tension

Condition	Average peak muscle force (N) MG[a]	Average peak muscle force (N) LG[b]	% P_0 MG	% P_0 LG	Average peak GRF[c] (N)
Quadruped standing	3.2	2.6	2.9	2.1	12.1
Level walking	9.2	6.0	8.2	4.7	17.1
Incline walking	14.6	8.6	13.2	6.8	20.0
Vertical jump	89.5	89.5	96.0	82.6	22.5

Note. [a]MG = medial gastrocnemius.
[b]LG = lateral gastrocnemius.
[c]GRF = ground reaction force.

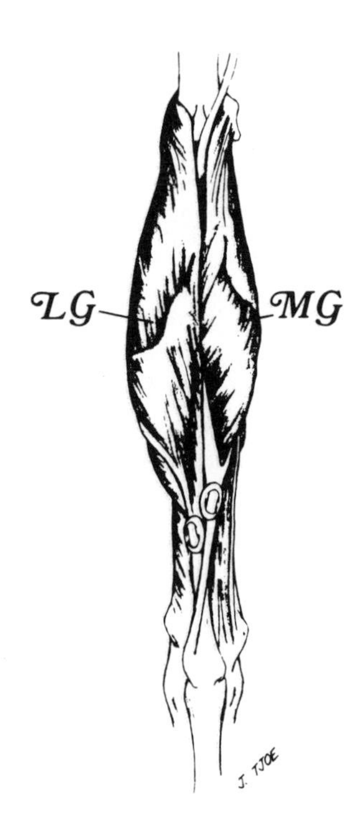

Figure 7.12 Illustration of the medial (MG) and lateral (LG) heads of the gastrocnemius muscle, depicting the placement of separate force transducers onto the individual tendons of each head. (Illustration provided by Ms. Judy Tjoe.)

absolute forces (approximately 89 N), the MG still contributes more in a relative sense because it performs at a greater percent of its maximum capability. The LG displays a larger increase in absolute force than the MG does from walking to jumping but maintains an 18% reserve based on its maximum isometric tension at the height of 0.65 m. This reserve would presumably be employed to meet the demands of jumps to even higher heights. Both LG and MG meet the increased demands, but the sharing of effort seems consistent with their different morphological and neuromuscular profiles.

The issue of compartmentalization has been raised in this chapter, and it is quite plausible that there is a further division of effort within subdivisions of both MG and LG. The limitation of this transducer technology is that the sampling is from the entire muscle head and not from any identified compartments within the MG or LG. In summary, two heads of a single muscle appear to respond differently to increased load. The response of each head, as inferred through its morphological and histochemical profiles, is supported by direct force measurements across a spectrum of environmental demands.

In Vivo Versus In Situ Force-Velocity Characteristics

Skeletal muscle has a unique force-velocity relationship that describes the force-production capabilities of the contractile element in relation to its shortening velocity. Application of this relationship to the study of muscle function during normal movement involves a complete understanding of muscle-tendon length changes during the same movement. Several algorithms permit the calculation of muscle-tendon length changes using either high-speed cinematography or video and an appropriate anatomic model (e.g., Brand, Crowninshield, Wittstock, Pederson, Clark, & Kriekin, 1982; Frigo & Pedotti, 1978; Grieve, Pheasant, & Cavanagh, 1978; Goslow, 1973). A major limitation of these algorithms, however, is their inability to dissect the total musculotendon length into its composite elements, which are a function of (a) the change in length of the sarcomeres and subsequently the fibers, (b) the change in angulation of the fibers with respect to the line of pull of the muscle force, and (c) the change in length of free tendon.

The change in length of the muscle fibers under maximal stimulation is described by the active length-tension curve; however, length changes under submaximal activation are less understood. The distribution of length among the sarcomeres within a fiber has already been discussed, and it would be naive to think that a more complex relationship does not exist between tendon and muscle strains as environmental demands and neural activation change. Reports from studies using rigid-body modeling techniques on human subjects (Bobbert & Ingen Schenau, 1990; Hoy, Zajac, & Gordon, 1990) and in animal models using instrumentation implanted on superficial muscle fibers (Griffiths, 1987) indicated that variable length changes between free tendon and muscle commonly exist. In fact, Hoffer, Caputi, Pose, and Griffiths (1989) reported that muscle fibers may actually shorten while the muscle-tendon unit is observed to lengthen. These findings have significant impact on experimental results describing the stretch-shorten cycle of the muscle-tendon unit (Komi, 1984). Implied in these experiments is that the lengthening and shortening of the muscle fibers are in phase with the lengthening and shortening of the whole muscle-tendon unit. Clearly, this may not be the case, and it appears that the entire concept of the stretch-shorten cycle, as commonly presented in the literature, needs revisiting.

Angulation of fibers in human skeletal muscle has been presented with data indicating that, at rest, fiber angulation can vary, for example, from 0° in the long head of the biceps femoris to approximately 25° in the short head of the same muscle. Research on the cat hindlimb indicates that the soleus muscle, for example, has a fiber angulation of only a few degrees, whereas the medial head of the gastrocnemius shows angulations of approximately 20° and a change in angulation from approximately 20° at rest to 45° at maximum shortening. The vast majority of muscles reported in the literature, however, have very few degrees of angulation (approximately 5°) at rest, which minimizes the contribution of fiber angulation changes to muscle shortening.

The third piece of information needed to understand length changes in the muscle-tendon unit is tendon compliance (i.e., how much tendon stretches under normal loading conditions and how it interacts with muscle compliance). The significance of the interaction between muscle and tendon has long been recognized, as evidenced by the classic studies of Hill (1938) in which the force-velocity properties of the contractile element were described using methodology that eliminated the series elastic element. This fundamental property of isolated muscle, measured in situ, should then be related to the fact that tendon interacts with the contraction process of muscle tissue (not the activation process) and the combined force they produce controls the movement of body segments. In turn, the mass and moment characteristics of the body segments affect the length and velocity of the muscle-tendon unit, which, of course, affect its force-production capability. This dynamic coupling between the muscle-tendon unit and the environment is a significant concept to understand in musculoskeletal mechanics and in application of the classic force-velocity relationship to movement control.

Zajac (1989) presented the concept of the musculotendon *actuator* to describe how muscle and tendon function as one entity, not as two independent tissues. This concept is similar to the schema presented by Walmsley and Proske (1981) (see Figure 7.10). Zajac described the function of different systems containing various combinations of high and low muscle and tendon compliance. For example, in a system where tendon compliance is high, the muscle fibers may change length very little in comparison with the total length change of the muscle-tendon unit. Zajac also presented the concept of tendon slack length, which is the length at which force just begins to develop. Furthermore, Zajac formulated a dimensionless property of his muscle actuator, relating tendon slack length to muscle fiber length. A high ratio infers high compliance not simply because a muscle has a long tendon, but because it has a long tendon in comparison with fiber length. The important aspect of this concept is the emphasis on the muscle-tendon relationship, rather than considering just the muscle or the tendon separately.

Compliance in the series elastic element of the muscle-tendon unit, which can be in the tendon as well as in the muscle cross-bridges and in other tendinous elements in the muscle, also affects the coupling between the muscle's EMG and its subsequent force production. The (EMD) can vary from a few milliseconds to over 100 ms, depending on the type of contraction. The concept of EMD is reintroduced at this point because the EMG represents the contractile element, and it is the contraction process of the muscle that interacts with tendon in responding to environmental demands.

Another example focuses on a comparison of the force-velocity pattern obtained for the cat soleus muscle in situ with the pattern obtained during an experiment in vivo while the cat was walking and running on a motor drive treadmill. The instrumentation used to collect these data were high-speed cinematography and specialized transducers surgically implanted to measure the force in an individual muscle (see Figure 7.12). The proximal and distal insertions of the soleus muscle were marked and identified, and once position data were collected, muscle-tendon length changes were calculated during the entire step cycle. Velocities of lengthening and shortening were then calculated as the first time derivative of the original length calculations. As shown in Figure 7.13, a, b, c, and d, the soleus muscle, during normal locomotion, can produce forces higher than those forces measured in situ. (The limitations of the in situ preparation are discussed in this chapter.) In contrast, in vivo data indicate the soleus muscle experiences a lengthening action immediately after ground contact; the velocity of stretch on the muscle-tendon unit is indicated to the left of the zero-velocity line. As observed, the velocity changes very little as force increases to a peak value around midstance. One of these animals actually produced peak forces during stance that exceeded $\mathbf{P}_0$. As positive velocities increase during the shortening phase (i.e., move to the right of the zero line) forces produced at certain velocities exceed the maximal force-production capacity of the contractile element of that muscle at essentially the same shortening velocity. Essentially, this represents an *enhancement* of the force-production capabilities of the soleus muscle-tendon unit because the forces are greater than those produced by the contractile element alone. Therefore, this muscle-tendon unit has the potential to store and reuse elastic strain energy, even at the relatively slow walking speeds used in this experiment. Similar statements have been previously made by Cavagna (1978).

It has been clearly demonstrated in the literature that the muscle-tendon can perform a greater amount of positive work during a shortening contraction if that shortening contraction is preceded by active lengthening. It has also been demonstrated that this enhanced force or work output is evident even if the shortening contraction is preceded by an active isometric contraction. From a work-energy standpoint, the increase in the work performed requires an increase in energy expenditure. Because the series elastic component is considered more important than the parallel elastic component in its potential for storage of elastic energy when an active muscle is stretched, tendon may be the first place to consider as a potential source of the additional work. A general scenario may be as follows: An external load stretches the series elastic component during a

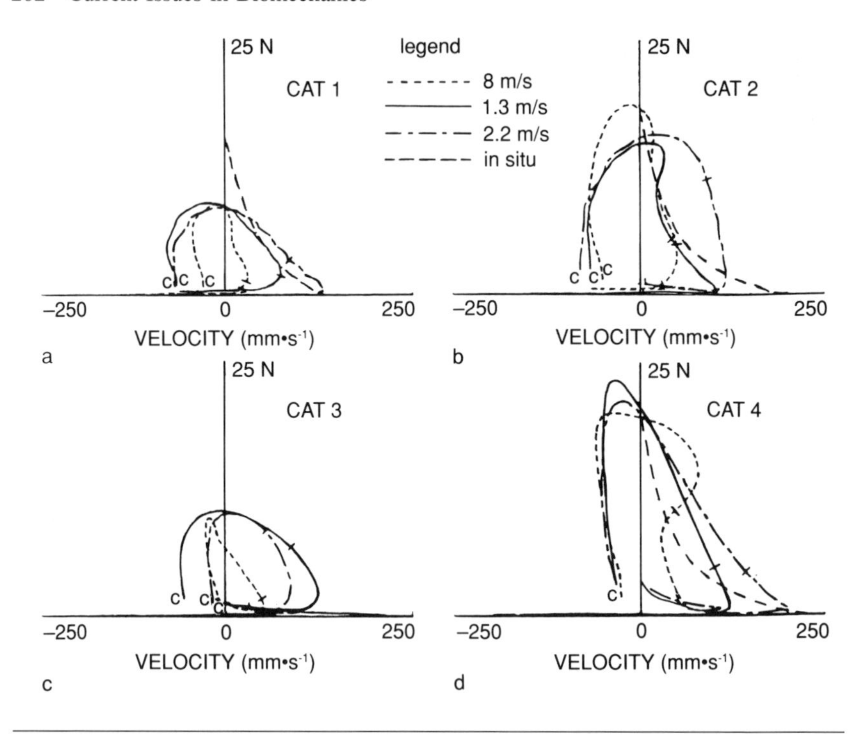

Figure 7.13 Force-velocity patterns within a step cycle for four cats—Cat 1(a), Cat 2(b), Cat 3(c), and Cat 4(d)—at treadmill speeds of 0.8, 1.3, and 2.2 m/s. Patterns represent an average of five step cycles. In situ force-velocity curves are also presented for Cats 1, 2, and 4. (Reprinted by permission from Gregor, Roy, Whiting, Hodgson, & Edgerton, 1988.)

lengthening contraction, which can be envisioned as a transfer of energy from the load to the series elastic component. Essentially, the environment is doing work on the muscle-tendon unit, and energy is stored in the series elastic component. Once the system is released, the molecular structures in the various tissues return to their original shape and the muscle shortens. The energy received from that stored in the tendon is then used in the shortening contraction. As previously considered in Figure 7.10, however, the relationship between active muscle compliance and tendon compliance must always be considered in any attempt to distribute load in the musculotendon system.

The ability to use stored elastic energy is affected by three variables: time, magnitude of stretch, and velocity of stretch. It is currently believed that there should be a minimum time delay between the lengthening and shortening contractions; otherwise most of the stored energy would be lost as heat. The actual magnitude of this delay remains open to discussion, but the shorter the delay, the greater the enhancement during shortening. With regard to the magnitude of stretch, if the lengthening contraction of the muscle is too great, fewer cross-bridges will remain attached following the stretch (i.e., more will be mechanically broken and hence less elastic energy would be stored in the cross-bridges). If the cross-bridges remain attached and the velocity of stretch is high, the percent contribution by the cross-bridges will be higher, and the storage of elastic energy will potentially be greater.

Recent evidence suggests that the tendon is actually recoiling during the shortening phase, producing what is referred to as an enhanced velocity. If this were the case, the actual velocity seen by the muscle-tendon unit would be dominated by the velocity seen in the tendon. The velocity of the contractile element then would be lower than that of the whole muscle-tendon unit and would place the contractile element on a different portion of its force-velocity curve than indicated by evaluating the in vivo locomotion data. This argument is appealing in light of the previous discussion on length distribution within the musculotendon unit and might explain the shift to the right of the in vivo force-velocity curve with respect to the force-velocity curve observed in isolated muscle (see Figure 7.13). The basic points of these data are that

- it appears that forces can be produced at certain velocities that exceed the capabilities of the contractile element, and
- much more information is needed on the response of the total muscle-tendon unit in normal movements.

Individual Muscle Moments and the Joint Moment

It has previously been stated that muscles do not act alone in the control of movement. Studies of the mechanical output of skeletal muscle typically consider a muscle in isolation to simplify the task of identifying elements related to its mechanical output. Studying the same muscle's function within a group of muscles during normal locomotion provides insight into the complexity of how muscles work together. For example, in the discussion on the mechanical output of the soleus, the output measured was the muscle's response to environmental demands in concert with its reaction to the output of its synergists and antagonists. The muscle must respond to external and internal loads produced by other surrounding muscles.

An indirect method of evaluating muscle forces and joint loads commonly reported in the literature involves modeling the body as a linked system of rigid bodies and employing principles in inverse dynamics. As a result of this analysis, researchers calculate segmental velocities and accelerations, joint reaction forces, and joint moments. This *joint moment* represents the summation of all active, passive, and external forces acting on the joint. Other terms reported in the literature are *residual muscle moment*, which is that moment produced by all active and passive biological structures crossing the joint, and *net intersegmental moment*, which is the result of a very similar calculation. Dissection of the residual muscle moment is a primary concern of scientists interested in musculoskeletal biomechanics. Although many reports in the literature used the two indirect methods described in this chapter, several recent reports have emerged in which forces were directly measured and compared to the joint moment. For example, Landjerit, Maton, and Peres (1988) reported the use of implanted transducers to measure muscle forces in two elbow flexors in the monkey and related these force values to the static flexor torque calculated at five separate elbow positions. Results, however, were limited to static equilibrium conditions and were not easily applied to dynamic movements.

The next example involves directly measured muscle forces in both an animal and human model during normal movement. The major questions addressed by these experiments focused on the relationship between the moment of force produced by individual or groups of muscles and the residual muscle moment at the ankle calculated using inverse dynamics. Requisite to the completion of these experiments was the development of methodology to

- measure muscle forces,
- calculate instant centers of rotation in the ankle and subsequent moment arms to muscles under study, and
- employ an appropriate rigid body model for use in the inverse solution.

A series of experiments was conducted by Fowler et al. (1989) to develop these methodologies and apply them to the study of cat hindlimb kinetics during the stance phase of overground locomotion at 0% and 12% grades. The results of these experiments are illustrated by the exemplar data in Figure 7.14, a and b. Individual muscle moments for the soleus (SOL, Figure 7.14a) and medial gastrocnemius (MG, Figure 7.14b) were calculated as the product of the tendon force impulse and its changing moment arm throughout the range of motion at the ankle during the stance phase of walking. The ankle generalized muscle moment (GMM) was calculated using ground reaction forces measured by means of special force platforms concealed in the floor of a plexiglass enclosed walkway and an appropriate rigid body model of the cat hindlimb.

The illustrated data present two interesting features:

- The soleus moment seems to comprise a significant portion of the GMM during stance at slow speeds of walking.
- The MG moment comprises very little of the GMM generated during the stance phase of slow walking.

Whereas the data in Figure 7.14, a and b, were not taken from the same trial, they represent the very consistent response of these fast and slow synergists to the demands of walking. Results obtained from all major ankle extensors from over 50 trials at various speeds and at a 12% grade indicate that each plantar-flexor muscle contributed to the increased joint in a manner highly related to their muscle fiber type. That is, the soleus contributed the most at the slower speeds, followed by the plantaris and medial and lateral gastrocnemius muscles, respectively. The range of speeds employed were very low, however, and it is conceded that at higher power output requirements the picture is not as clear. This conclusion is supported by the discussion on the role played by the LG and MG during

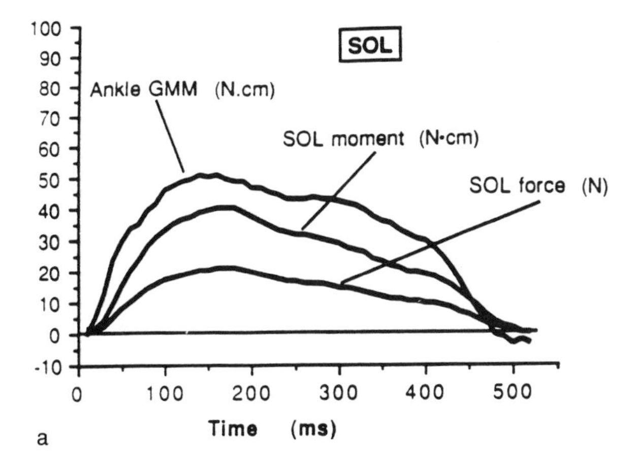

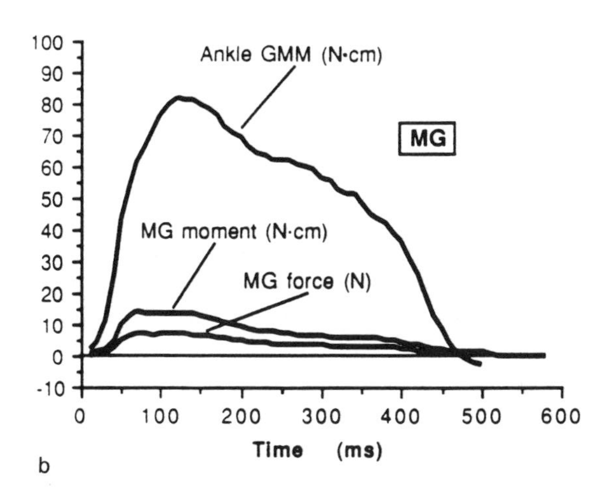

Figure 7.14 The generalized muscle moment (GMM), soleus (SOL) force and moment, and medial gastrocnemius (MG) force and moment during the stance phase of overground locomotion. SOL (a) and MG (b) data are from separate trials and separate hindlimbs.

jumping in which the responsibility for increased power output falls predominantly on the two heads of the gastrocnemius.

In a similar experiment conducted in humans, Gregor, Komi, Browning, and Jarvinen (1991) described the forces produced by the triceps surae complex during cycling. In order to dissect the residual muscle moment at the ankle, a specially designed transducer, similar to the design used by Fowler et al. (1989), was surgically implanted on the Achilles tendon. Data were then collected during cycling at three separate power outputs: 90 RPM at 1 Kp (kilopons) (90 W), 60 RPM at 3 Kp (180 W), and 90 RPM at 3 Kp (270 W). Data presented in

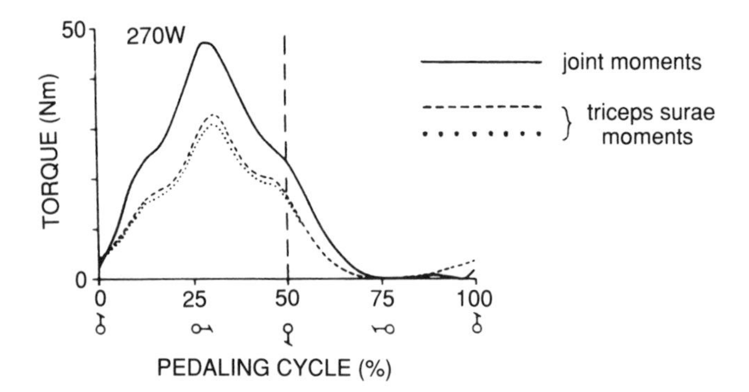

Figure 7.15 Exemplar torque (moment) pattern at the ankle for one male subject riding at a power output of 270 W on a stationary bicycle. Joint moments and triceps surae moments are illustrated throughout the pedaling cycle.

Figure 7.15 illustrate the response at the 270-W condition. Subsequent to the experimental data session, moment arms were calculated in a separate set of experiments using serial magnetic resonance images of the ankle for the entire range of motion observed during the cycling task. Major findings include the following:

- Peak Achilles tendon force occurred at approximately the same point in the pedaling cycle at all power outputs.
- Peak Achilles tendon force increased with increased load but not with increased cadence.
- The moment produced by the triceps surae complex is temporally in phase with the joint moment.
- The angular impulse calculated for the Achilles tendon moment represented approximately 65% of the angular impulse calculated for the residual muscle moment regardless of power output requirements.

The most significant limitation in this experiment was that data were collected for the entire triceps surae and not for individual muscles, as was the case in the experiments by Fowler et al. (1989). Whereas the relatively constant relationship between the triceps surae moment and the residual muscle moment is useful to understanding joint mechanics, the contribution made by the gastrocnemius and soleus muscles remains unclear. Coupling EMG and muscle-tendon length change data with the tendon force data, however, does support the following tentative arguments:

- Based on the appearance of an isometric contraction and increased EMG activity, the soleus muscle appears to contribute more than gastrocnemius does to the tendon moment calculated during the initial portion of the pedaling cycle (0° to 90°).

• Based on muscle-tendon length change calculations, both muscle-tendon units appear to experience a lengthening-shortening cycle during the power phase of cycling, thus suggesting the potential for the storage and reuse of elastic strain energy.

• Based on the same length calculations, each muscle-tendon unit exhibits a unique length change pattern as pedaling cadence and cycling kinematics change.

These statements are also based on the facts that the two muscles are different in their fiber composition and that they may have different force-velocity patterns.

In summary, it appears in both animal and human models, within the domain of each experiment presented, that muscles contribute to increased environmental demands and joint moments in a manner consistent with their fiber type. This conclusion must be taken with caution, however, because each experiment was conducted in a relatively narrow domain of kinetic demands, possibly precluding the obvious effects of other muscle properties (i.e., muscle architecture).

Conclusion

I have evaluated both the structural and functional units within skeletal muscle and described the most recent findings regarding their mechanical and physiological profiles. I have identified various parameters that affect the muscle's ability to produce force and have associated length changes, velocity changes, activation patterns, fiber type, and previous history to the mechanical output of skeletal muscle. The mechanical properties are discussed in this chapter in an attempt to understand the factors that affect muscle force production capability from the cellular to the total joint level. Also, three examples have shown the need to think in a more thorough way about the relevance of muscle-tendon properties studied in isolation to the function of the muscle-tendon unit in vivo. Results from research using both direct and indirect methods of analysis are converging, but much work remains to be done.

References

Basmajian, J.V., & DeLuca, C. (1985). *Muscles alive: Their function revealed by electromyography* (5th ed.). Baltimore: Williams & Wilkins.

Bobbert, M.F., & Ingen Schenau, G.J. van (1990). Isokinetic plantar flexion: Experimental results and model calculations. *Journal of Biomechanics*, **23**, 105-120.

Bodine, S.C., Garfinkel, A., Roy, R.R., & Edgerton, V.R. (1988). Spatial distribution of motor unit fibers in the cat soleus and tibialis anterior muscles: Local interactions. *Journal of Neuroscience*, **8**, 2142-2152.

Bodine, S.C., Roy, R.R., Meadows, D.A., Zernicke, R.F., Sacks, R.D., Fournier, M., & Edgerton, V.R. (1982). Architectural, histochemical and contractile characteristics of a unique biarticular muscle: The cat semitendinosus. *Journal of Neurophysiology*, **48**, 192-210.

Brand, R.A., Crowninshield, R.D., Wittstock, C.E., Pederson, D.R., Clark, C.R., & Kriekin, F.M. (1982). A model of lower extremity muscular anatomy. *Journal of Biomechanical Engineering*, **104**, 304-310.

Cavagna, G.M. (1978). Storage and utilization of elastic energy in skeletal muscle. In R.S. Hutton (Ed.), *Exercise and Sport Science Reviews* (pp. 89-130). Baltimore: Williams & Wilkins.

Edgerton, V.R., Roy, R.R., Gregor, R.J., & Rugg, S. (1986). Morphological basis of skeletal muscle power output. In N.L. Jones, N. McCautney, & A.J. McComes (Eds.), *Human Muscle Power* (pp. 43-58). Champaign, IL: Human Kinetics.

Edman, K.A.P., Mulieri, L.A., & Scubon-Mulieri, B. (1976). Non-hyperbolic force-velocity relationship in single muscle fibers. *Acta Physiologica Scandinavica*, **98**, 143-156.

English, A.W., & Weeks, O.I. (1984). Compartmentalization of single muscle units in the cat lateral gastrocnemius. *Experimental Brain Research*, **56**, 361-368.

Fowler, E.G., Gregor, R.J., Hodgson, J., Roy, R.R., & Broker, J.P. (1989). The contribution of individual muscles to the ankle moment produced in the cat hindlimb. *Journal of Biomechanics*, **22**, 1011.

Friederich, J.A., & Brand, R.A. (1990). Muscle fiber architecture in the human lower limb. *Journal of Biomechanics*, **23**, 91-95.

Frigo, C., & Pedotti, A. (1978). Determination of muscle length during locomotion. In E. Asmussen & K. Jorgensen (Eds.), *Biomechanics VI-A* (pp. 355-360). Baltimore: University Park Press.

Garrett, W., & Tidball, J.G. (1987). Myotendinous junction: Structure, function and failure. In S.L.Y. Woo & J.A. Buckwalter (Eds.), *Injury and repair of the musculoskeletal soft tissues* (pp. 171-207). Park Ridge, IL: American Academy of Orthopaedic Surgeons.

Gordon, A.M., Huxley, A.F., & Julian, F.J. (1966). The variation in isometric tension with sarcomere length in vertebrate muscle fibers. *Journal of Physiology*, **184**, 170-192.

Goslow, G.E., Reinking, R.M., & Stuart, D.G. (1973). The cat step cycle: Hindlimb joint angles and muscle lengths during unrestrained locomotion. *Journal of Morphology*, **141**, 1-42.

Gregor, R.J., Komi, P.V., Browning, R.C., & Jarvinen, M. (1991). A comparison of the triceps surae and residual muscle moments at the ankle during cycling. *Journal of Biomechanics*, **24**, 287-297.

Gregor, R.J., Roy, R.R., Whiting, W.C., Hodgson, J.A., & Edgerton, V.R. (1988). Mechanical output of the cat soleus during treadmill locomotion: In-vivo vs in-situ characteristics. *Journal of Biomechanics*, **21**, 721-732.

Grieve, D.W. (1975). Electromyography. In H.T.A. Whiting (Ed.), *Techniques for the analysis of human movement* (The Human Movement Series, pp. 109-149). London: Lepus Books.

Grieve, D.W., Pheasant, S., & Cavanagh, P.R. (1978). Prediction of gastrocnemius length from knee and ankle joint posture. In E. Asmussen & K. Jorgensen (Eds.), *Biomechanics VI-A* (pp. 405-412). Baltimore: University Park Press.

Griffiths, R.I. (1987). Ultrasound trainsit-time gives direct measurement of muscle fibre length in vivo. *Journal of Neuroscience Methods*, **21**, 159-165.

Hill, A.V. (1938). The heat of shortening and the dynamic constants of muscle. *Proceedings of the Royal Society of London: Series B: Biological Sciences*, **126**, 136-195.

Hof, A.L., Geelen, B.A., & van den Berg, J. (1983). Calf muscle moment, work and efficiency in level walking: Role of series elasticity. *Journal of Biomechanics*, **16**, 523-537.

Hof, A.L., & van den Berg, J.W. (1977). Linearity between the weighted sum of the EMGs of the human triceps surae and the total torque. *Journal of Biomechanics*, **10**, 529-539.

Hoffer, J.A., Caputi, A.A., Pose, I.E., & Griffiths, R.I. (1989). Role of muscle activity and load on the relationship between muscle spindle length and whole muscle length in the freely walking cat. In J.H.J. Allum & M. Hulliger (Eds.), *Progress in Brain Research* (Vol. 80, pp. 75-85). New York: Elsevier Science Publishers.

Hoy, M.G., Zajac, F.E., & Gordon, M.E. (1990). A musculoskeletal model of the human lower extremity: The effect of muscle, tendon, and moment arm on the moment-angle relationship of musculotendon actuators at the hip, knee and ankle. *Journal of Biomechanics*, **23**, 157-170.

Ingen Schenau, G.J. van (1989). From rotation to translation: Constraints on multi-joint movements and the unique action of bi-articular muscles. *Human Movement Science*, **8**, 301-337.

Komi, P.V. (1984). Physiological and biomechanical correlates of muscle function: Effects of muscle structure and stretch shortening cycle on force and speed. In R.H. Terjung (Ed.), *Exercise and Sport Science Reviews* (pp. 81-122). Baltimore: Williams & Wilkins.

Landjerit, B., Maton, M., & Peres, G. (1988). In-vivo muscular force analysis during the isometric flexion on a monkey's elbow. *Journal of Biomechanics*, **21**, 577-584.

Lieber, R.L., & Baskin, R.L. (1983). Intersarcomere dynamics of single muscle fibers during fixed-end tetani. *Journal of General Physiology*, **82**, 347-364.

Mai, M.T., & Lieber, R.L. (1990). A model of semitendinosus muscle sarcomere length, knee and hip joint interaction in the frog hindlimb. *Journal of Biomechanics*, **23**, 271-279.

Norman, R. (1977). The use of electromyography in the calculation of joint torque. Unpublished doctoral dissertation, Pennsylvania State University, University Park, PA.

Norman, R., Gregor, R.J., & Dowling, J. (1988). The prediction of cat tendon force from EMG in dynamic muscular contractions. *Proceedings of the Fifth Biennial Conference of Human Locomotion Symposium of the Canadian Society of Biomechanics* (pp. 120-121). London, ON: Spodym Publishers.

Norman, R.W., & Komi, P.V. (1979). Electromechanical delay in skeletal muscle under normal movement conditions. *Acta Physiologica Scandinavica*, **106**, 241-248.

Ounjian, M., Roy, R.R., Eldred, E., & Edgerton, V.R. (1987). Muscle fiber lengths within single motor units in the cat tibialis anterior muscles. *Society for Neuroscience Abstracts*, **13**, 1213.

Pette, D., & Staron, R.S. (1988). Molecular basis of the phenotypic characteristics of mammalian muscle fibers. In D. Evered & J. Whalen (Eds.), *Plasticity of the Neuromuscular System: Ciba Foundation Symposium 138*, pp. 22-30. Chichester: J. Wiley.

Proske, U., & Morgan, D.L. (1984). Stiffness of cat soleus muscle and tendon during activation of part of muscle. *Journal of Neurophysiology*, **52**, 459-468.

Rack, P.M.H., & Westbury, D.R. (1974). The short range stiffness of active mammalian muscle and its effect on mechanical properties. *Journal of Physiology*, **240**, 331-350.

Sacks, R.D., & Roy, R.R. (1982). Architecture of the hind limb muscles of cats: Functional significance. *Journal of Morphology*, **173**, 185-195.

Sherif, M.H., Gregor, R.J., Liu, L.M., Roy, R.R., & Hager, C.L. (1983). Correlation of myoelectric activity and muscle force during selected cat treadmill locomotion. *Journal of Biomechanics*, **16**, 691-701.

Spector, S.A., Gardiner, P.F., Zernicke, R.F., Roy, R.R., & Edgerton, V.R. (1980). Muscle architecture and force-velocity characteristics of cat soleus and medial gastrocnemius: Implications for motor control. *Journal of Neurophysiology*, **44**, 951-960.

Tjoe, J., Gregor, R.J., Perell, K.L., & Roy, R.R. (1991). A comparison of the forces produced by the lateral and medial heads of the gastrocnemius in the cat across a continuum of postural and movement demands. *Journal of Biomechanics*, **24**, 267.

Walmsley, B., & Proske, U. (1981). Comparison of stiffness of soleus and medial gastrocnemius muscles in cats. *Journal of Neurophysiology*, **46**, 250-259.

Whiting, W.C., Gregor, R.J., Roy, R.R., & Edgerton, V.R. (1984). A technique for estimating mechanical work of individual muscles in the cat during treadmill locomotion. *Journal of Biomechanics*, **17**, 685-694.

Wickiewicz, T.L., Roy, R.R., Powell, P.L., & Edgerton, V.R. (1983). Muscle architecture of the human lower limb. *Clinical Orthopaedics and Related Research*, **179**, 317-325.

Wickiewicz, T.L., Roy, R.R., Powell, P.L., Perrine, J.J., & Edgerton, V.R. (1984). Muscle architecture and force velocity relationships in humans. *Journal of Applied Physiology*, **57**, 435-443.

Windhorst, U., Hamm, T.M., & Stuart, D.G. (1989). On the function of muscle and reflex partitioning. *Behavioral and Brain Sciences*, **12**, 629-681.

Winter, D.A. (1990). *Biomechanics and motor control of human movement* (2nd ed.). New York: J. Wiley.

Woittiez, R.D., Huijing, P.A., Boom, H.B.K., & Rozendal, R.H. (1984). A three dimensional muscle model: A quantified relationship between form and function of skeletal muscles. *Journal of Morphology*, **182**, 95-11.

Zajac, F.E. (1989). Muscle and tendon: Properties, models, scaling, and application to biomechanics and motor control. In J.R. Bourne (Ed.), *CRC Critical Reviews in Biomedical Engineering* (Vol. 17, pp. 359-411). Boca Raton, FL: CRC Press Inc.

Part III

Neuromotor Elements

The two chapters in this final section of the book emphasize the neuromotor elements of biomechanics. As a follow-up to the chapter on skeletal muscle mechanics, chapter 8 presents some of the neuromotor factors that contribute to restrictions on the maximum voluntary force potential of muscle contraction. Because the success of many motor activities depends on the skillful expression of maximum muscular force, the principles in this chapter should be of interest. Chapter 9 reviews research into the interaction between knee joint ligaments and skeletal muscle contraction in humans. Because elementary musculoskeletal anatomy courses may not present the neuroanatomy of ligaments, students in such courses may be unaware of the significant impact on muscle excitation patterns of forces experienced by joint and capsular ligaments.

Chapter 8

Neuromuscular Basis of the Maximum Voluntary Force Capacity of Muscle

Roger M. Enoka and Andrew J. Fuglevand

Biomechanists often aim to improve technique, or motor skill execution, by improving the athlete or improving the sport-related implements. Considerable research has been devoted to improving the athlete through muscle strengthening. Historically, it also has been a key clinical variable used in deciding when a person may resume activity following injury.

In chapter 8, Roger Enoka and Andrew Fuglevand discuss several neuromotor mechanisms that underlie the expression of maximum muscular force. Measurement of maximum effort is a commonly used method, both experimentally and clinically, but Enoka and Fuglevand argue that these measurements do not, in fact, represent maximum force. The chapter contends that neuromuscular mechanisms normally preclude maximal neural activation of muscle. Techniques are available, however, to supplement the output of the motor neuron pool.

The maximal voluntary force that an individual can exert with a muscle, or group of muscles, depends on the ability to provide a sufficient neural drive to the muscles involved in the task, the mechanical characteristics of the muscles, and the geometric arrangement of the muscles in the body. The neural factors refer to motor unit activation, including recruitment and discharge rate and pattern, and the coordination of activity among the muscles involved in the task. The muscular factors are muscle length, the rate of change in muscle length, and the size of the muscles, as indicated by cross-sectional area. The main geometric factor is the arrangement of the muscle with respect to the joint axis of rotation, whereby the muscle exerts a force on the skeleton at a distance from the joint and the net effect is a muscle torque about the joint.

Because the maximal voluntary torque can be influenced by so many physiological and biomechanical factors, it has become common practice to standardize the protocols that are used for its assessment. The procedures used to measure handgrip strength (Costa, Borghi, Mussi, & Ambrosioni, 1987; Martin, Neale, & Elia, 1985; Niebuhr & Marion, 1987) and to determine a maximum voluntary contraction (MVC) (Aitkens, Lord, Bernauer, Fowler, Lieberman, & Berck, 1989; Kroemer & Marras, 1980) are examples of attempts to standardize the assessment of the maximal voluntary torque. These standardized procedures minimize the strategies that an individual can use to execute the task, and therefore they limit the number of variables that influence the measurement. An MVC, for example, is typically performed as a 3- to 5-s isometric contraction with the subject secured to an apparatus. This procedure standardizes three factors, one neural and two mechanical, that could confound the interpretation of the measurement. An isometric contraction, such as an MVC, minimizes the influence of the component of the neural command that is associated with coordination of the muscles involved in the task (Rutherford & Jones, 1986) and the mechanical effects of joint angle and its rate of change on the MVC (Jones, Rutherford, & Parker, 1989; Kitai & Sale, 1989; Kulig, Andrews, & Hay, 1984; Thépaut-Mathieu, van Hoecke, & Maton, 1988). For these reasons, we will focus on the MVC as a standardized example of the maximal voluntary torque that an individual can exert.

Mechanical and electromyographic (EMG) measurements obtained during MVCs have been used as a means to evaluate the global effect of experimental, clinical, and training interventions (e.g., fatigue, immobilization, beta-blocker medication, strength training) on the neuromuscular apparatus. In addition, torque, force, and EMG measurements are frequently normalized by being expressed as a percentage of the respective values obtained during an MVC. The implicit rationale in this procedure is that the MVC is a reliable and valid criterion. In this sense, the MVC is regarded as a standard unit of neuromuscular system behavior to which experimental effects can be gauged. The purpose of this chapter is to review the experimental basis of the often-expressed assumption that the MVC is a reliable and valid measure of the maximum force capacity of muscle. Support for this assumption seems to be widespread, as indicated by Thomas, Woods, and Bigland-Ritchie (1989, p. 1835): "The central nervous system (CNS) remains capable of fully activating all motor units to respond with maximum force under conditions of extreme contractile failure. . . . This capability has been well established for all muscles so far examined."

For a given volume of muscle and a particular limb position, the force exerted during an MVC is largely determined by the neural drive supplied to the muscle by the motor neurons. The nature of this input to muscle can be characterized by studying the behavior of individual motor units or a population of motor units. For technical reasons, little is known about behavior of single motor units during an MVC. In contrast, more is known about the association between force and interference EMG, where the latter represents the summated electrical activity of many motor units. In this chapter, we review what is known about motor unit behavior during an MVC, discuss the limitations of using the EMG as an index of neural input during an MVC, and consider evidence that suggests an MVC does not yield the absolute maximum force that a muscle can exert.

Motor Unit Behavior

Since the time of Sherrington (1925), we have known that muscle force is graded in terms of motor unit activity. This gradation of force is accomplished by variable combinations of recruitment and activation rate, where the latter refers to both the average rate at which action potentials are discharged and the pattern of discharge. The experimental evidence suggests that muscles are generally activated with one of two strategies; either all the motor units are recruited by 50% MVC or recruitment continues up to about 90% MVC. The behavioral significance of this distinction is that for muscles in which recruitment is completed by 50% MVC, the remaining 51% to 100% MVC must be elicited by manipulating discharge rate alone, whereas the force in other muscles can be varied by concurrent recruitment and modulation of discharge rate. Hand muscles, such as adductor pollicis and first dorsal interosseus, seem to have all their motor units recruited by about 50% MVC (DeLuca, LeFever, McCue, & Xenakis, 1982; Kukulka & Clamann, 1981). This conclusion, however, should be regarded as somewhat tentative because it is difficult to determine experimentally the upper limit of recruitment as this involves observing the recruitment of all motor units in a muscle. Because experimental techniques allow only the assessment of an exiguous fraction of the motor unit population, the upper limit of recruitment is difficult to determine conclusively. Simulated behavior of motor units in first dorsal interosseus (Fuglevand, Winter, Patla, & Stashuk, 1989) has indicated that recruitment would need to operate beyond 70% MVC for the relationship between surface EMG and force to be linear, and hence consistent with experimental observation.

In muscles such as biceps brachii (Kukulka & Clamann, 1981), deltoid (DeLuca et al., 1982), toe extensors (Grimby & Hannerz, 1977), and tibialis anterior (Hannerz, 1974), recruitment continues up to about 90% MVC. Grimby and Hannerz (1977) have also indicated that a fraction of the motor unit population may not be capable of sustained activity during maximum contractions. These motor units may discharge briefly during the development of force, particularly if the rate of force increase is high, but cease to discharge after only a few action potentials. The inability of these units to remain active during a sustained effort suggests that not all units are recruited throughout the duration of an MVC. The early transient peak torque or force observed during a rapid MVC, which then

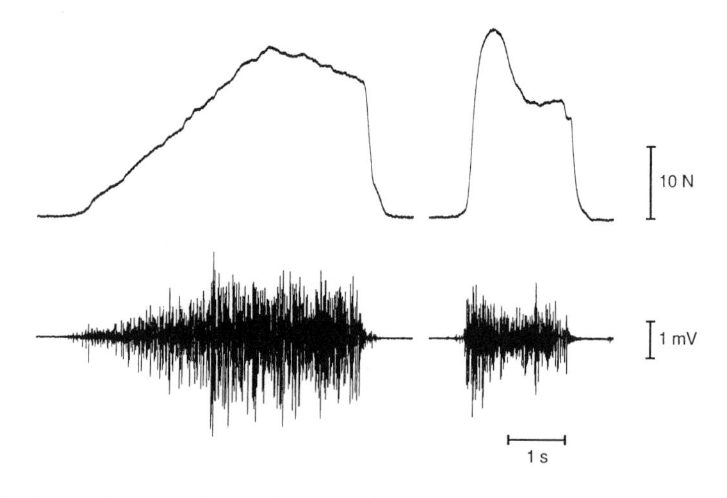

Figure 8.1 Index finger abduction force (upper trace) and first dorsal interosseus surface electromyogram (EMG) (lower trace) during a standard, graded maximum voluntary contraction (left panel) and during a rapid maximum voluntary contraction (right panel). The subject was instructed to smoothly increase force to maximum over a 3-s ramp for the contraction in the left panel and was instructed to exert a maximum force as quickly as possible for the contraction depicted in the right panel. The enhanced maximal force during the rapid contraction may reflect a transient contribution of motor units that were incapable of sustained activity during the slow, graded contraction depicted in the left panel.

declines to a lower steady state, may reflect the brief contribution and subsequent derecruitment of these units (see Figure 8.1). We have found that subjects (n=8) can exert a significantly (p<0.05) greater peak index finger abduction force (mean increase = 5%) when the MVC is performed rapidly compared to 3- to 5-s increase in force (A.J. Fuglevand & R.M. Enoka, unpublished observations, April 1990).

The force that a motor unit exerts varies in a sigmoidal manner with the discharge rate so that for equal increments in discharge rate the increase in force is not constant over the entire operating range (Bigland & Lippold, 1954; Rack & Westbury, 1969). The specific shape of the force-discharge rate curve depends on the contractile characteristics of the motor unit. The relatively sluggish properties of slow-twitch units allow the individual twitch profiles to summate more readily at lower discharge rates compared with fast-twitch units. For this reason, the force-discharge rate curve for slow-twitch units rises more steeply at comparatively lower rates and attains tetanic fusion (i.e., the plateau in the sigmoid) at lower rates than does the curve for fast-twitch units. Burke, Rudomin, and Zajac (1976) have shown that the slope of the sigmoid is maximal when the time between stimuli is equivalent to the rise time of the motor unit twitch. These data have been used to model the nonlinear force-discharge rate behavior of individual motor units (Fuglevand, 1989). Simulated force-discharge rate relationships of a slow-twitch motor unit (twitch contraction time [CT] = 90 ms) and a fast-twitch

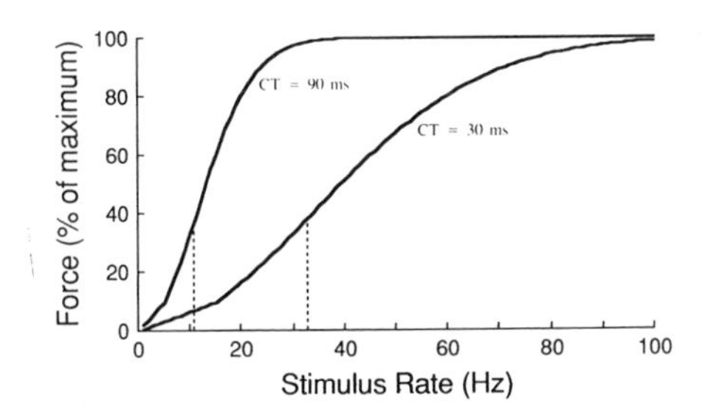

Figure 8.2 The simulated relationship between motor unit force and discharge rate for two motor units. The left trace is the simulated force-discharge rate behavior for a slow-twitch motor unit (contraction time [CT] = 90 ms). The right trace represents the force-discharge rate relationship for a fast-twitch motor unit (CT = 30 ms). The vertical lines indicate the stimulus rate at which the interpulse interval durations were equivalent to the CT of each motor unit. (Reprinted by permission from Fuglevand, 1989.)

motor unit (CT = 30 ms) are shown in Figure 8.2. The slow-twitch unit attained the maximum tetanic tension at a stimulus rate of approximately 30 Hz, whereas the fast-twitch unit required a rate of 80 to 100 Hz to achieve maximum force.

Although it is technically difficult to record identifiable single motor unit potentials at high muscle forces, some investigators have been able to measure discharge rate during sustained efforts at 80% to 100% MVC. Marsden, Meadows, and Merton (1983) reported peak discharge rates of 100 Hz by single motor units in adductor pollicis during a 60-s MVC. DeLuca et al. (1982) found rates of 41 ± 10 Hz for first dorsal interosseus and 29 ± 3 Hz for deltoid. Kukulka and Clamann (1981) found 25 to 30 Hz for adductor pollicis and 20 to 25 Hz for biceps brachii, a range similar to that for extensor digitorum communis (20 to 25 Hz) (Monster & Chan, 1977). Tanji and Kato (1973) reported maximum rates of 25 to 35 Hz for motor units of abductor digiti minimi. Discharge rates greater than 35 Hz were never observed in first dorsal interosseus or extensor indicis, even when the rate of increase in force was high (Freund, Büdingen, & Dietz, 1975). Based on population averages that were obtained with a moving tungsten microelectrode, Bellemare, Woods, Johansson, and Bigland-Ritchie (1983) found no difference between adductor pollicis (30 ± 9 Hz) and biceps brachii (31 ± 10 Hz) but a substantially lower average discharge for soleus (11 ± 3 Hz), a slow-twitch muscle, during an MVC. Interestingly, Bellemare et al. also reported quite large differences in discharge rate (35 ± 8 Hz vs. 23 ± 7 Hz) for adductor pollicis of 2 subjects.

The observed discharge rates during an MVC, with the exception of Marsden et al. (1983), appear to be inconsistent with those necessary to elicit artificially the maximum tetanic force in all motor units of a muscle. In cat soleus, a muscle composed almost exclusively of slow-twitch motor units that have an average twitch contraction time of about 100 ms (Burke, 1981), a stimulus rate of 25 to

30 Hz is necessary to achieve maximal tetanic force (Rack & Westbury, 1969). The average contraction times of motor units in human muscle are generally less than those of cat soleus: 52 ms for biceps brachii; 45 ms for triceps brachii and tibialis anterior; 74 ms for soleus (Buchthal & Schmalbruch, 1970); 55 ms for first dorsal interosseus (Milner-Brown, Stein, & Yemm, 1973); 65 ms for adductor digiti minimi (Burke, Skuse, & Lethlean, 1974); 74 ms for abductor pollicis brevis (Thomas, Ross, & Calancie, 1987). By comparison, the observed maximum discharge rates during an MVC (20 to 40 Hz) seem to be sufficient to produce the maximum force in only those motor units that have relatively long contraction times (see Figure 8.2). Furthermore, some data indicate that earlier recruited, slow-twitch motor units attain higher discharge rates during near-maximum contractions than do the fast-twitch units that are recruited later (DeLuca et al., 1982; Monster & Chan, 1977; Tanji & Kato, 1973), which suggests that the higher threshold motor units may not be discharging at rates sufficient for maximal force.

In contrast to these studies, Grimby and Hannerz (1977) have distinguished two types of motor units: those that can discharge continuously during a sustained maximum effort and those that discharge briefly at a high rate. The continuous motor units (about 50% of the units in the short toe extensors) tend to have low thresholds and discharge at 20 to 30 Hz, whereas the intermittent motor units (about 10% of the units) can reach 100 Hz for the short toe extensors (Grimby & Hannerz) and 65 Hz in tibialis anterior (Hannerz, 1974). Similarly, DeLuca et al. (1982) examined the behavior of single motor units during a force-tracking task and reported high discharge rates (60 Hz) for high threshold units in first transiently dorsal interosseus when the force was about 80% MVC. The inability of high threshold units to remain active during a sustained maximal effort suggests that not all units are maximally activated during an MVC. Furthermore, during rapid rhythmic movements, the intermittent units are active and discharge at high discharge rates (up to 100 Hz) (Grimby & Hannerz). The contribution of intermittent units to rapid changes in force suggests that they may contribute to the early transient peak force in a rapid MVC, as mentioned previously (see Figure 8.1).

This apparent distinction between motor units that discharge continuously and those that discharge intermittently is underscored by the observations of Nardonne, Romanò, and Schieppati (1989) on the task-dependent behavior of motor units. Subjects were asked to make slow plantar-flexion movements in which the voluntarily activated triceps surae were either shortened or lengthened. Nardonne et al. found that motor units in soleus and gastrocnemius were recruited either during the shortening phase, the lengthening phase, or the transition between the two phases. Interestingly, the motor units that were recruited during the slow lengthening phase were not recruited during the slow shortening phase but could be activated during rapid shortening movements and during high-force isometric contractions. Although these data do not directly address the issue of motor unit activity during an MVC, they do suggest an experiment to determine whether units that are active during high-force lengthening contractions are recruited during an MVC.

One feature of motor unit activity that has been difficult to characterize is the pattern of action potential discharge. It is known that the pattern of discharge can have a profound effect on the force exerted by muscle (Burke et al., 1976; Zajac & Young, 1980) and that voluntarily activated motor units can exhibit irregular discharge patterns (Bawa & Calancie, 1983; Enoka et al., 1989). Three pattern effects may be important:

- Doublet—the presence of short intervals (< 20 ms) between the action potentials discharged by one motor unit
- Synchronization—the near-coincident timing of action potentials from different motor units
- Force-frequency relation—the force elicited in a motor unit with a particular stimulus frequency

The force-frequency relation is not unique but depends on the activation history (Binder-Macleod & Clamann, 1989). Apart from the report of brief intervals in both adductor pollicis and biceps brachii when the force was greater than 44% MVC (Kukulka & Clamann, 1981) and the suggestion that strength training can increase motor unit synchronization (Milner-Brown, Stein, & Lee, 1975), little is known about the significance of discharge pattern during an MVC.

Taken together, these observations on the voluntary control of motor unit behavior suggest that not all motor units may be fully activated during an MVC. This conclusion is underscored by a comparison of the observed motor unit discharge rates to the stimulus rates that are necessary when a muscle is activated artificially with electrical stimulation. The stimulation rates used to elicit maximum force ranged from 50 to 120 Hz in both normal human muscle (Davies, Dooley, McDonagh, & White, 1985; Duchateau & Hainaut, 1987, 1988; Marsden et al., 1983; Miller, Mirka, & Maxfield, 1981; Rutherford & Jones, 1988) and paralyzed human muscle (Carroll, Triolo, Chizeck, Kobetic, & Marsolais, 1989). These stimulation rates are higher than the average discharge rates during an MVC, but they appear consistent with the rates necessary to attain maximum tetanic force in motor units. Since high threshold units tend to have brief contraction times and undoubtedly exert large forces, it would seem critical that these units be activated with reasonably fused tetani in order to elicit the maximal force—hence the need for rates of 50 to 120 Hz.

Despite the limited data available on motor unit behavior at high forces, the reports of (a) unsustainable activity in some high threshold motor units and (b) the discrepancy between the observed discharge rates during an MVC and the stimulus rates necessary to drive many motor units to maximum tetanic force suggest that a subject is unlikely to be able to exert the maximum force during a slow, sustained contraction. These observations suggest caution in interpreting motor unit activity during an MVC as maximal.

EMG-Force Relationship

Largely due to the ease with which an EMG can be recorded, it is used frequently as a window into the nervous system. The intensity of the EMG signal reflects, indirectly, the summated activity of both the recruitment and discharge rates of those motor units within the detection range of the recording electrodes. The exact manner, however, by which motor unit activity influences either EMG or muscle force is not well understood because of the difficulty in quantifying the neural input (viz., number of units recruited and their discharge rates) to muscle.

The complex voltage pattern recorded with surface electrodes during a muscle contraction results from the temporal summation of randomly occurring action potentials generated by a number of motor units. The amplitude and shape of an individual motor unit action potential depends on a number of factors that include

- the number and size of fibers innervated by the motor unit,
- the spatial orientation of the fibers relative to the electrode,
- the electrode configuration and dimensions,
- the conduction velocity of the fiber action potential,
- the spatial relationship of the electrode to the innervation zone, and
- the length of the muscle fibers (Fuglevand, 1989).

The relationship between EMG and isometric muscle force has been described as linear for some muscles (e.g., soleus, adductor pollicis, first dorsal interosseus) but parabolic for other muscles (e.g., biceps brachii, brachioradialis, triceps brachii) (Bigland-Ritchie, 1981). The relatively simple form of the EMG-force relationship (i.e., linear or parabolic) should not be misinterpreted as indicative of a simple relationship between neural drive and the EMG or between neural drive and muscle force. The effect of neural input on EMG and on muscle force is shaped by many nonlinear processes (Calvert & Chapman, 1977; Fuglevand, 1989; Milner-Brown & Stein, 1975). A principal nonlinearity affecting muscle force is the sigmoidal relationship between motor unit force and discharge rate (see Figure 8.2). The amount of force increase due to a given discharge rate increase varies over the entire operating range. Similarly, one factor that contributes to nonlinear variation in EMG with neural input is signal cancellation due to the overlap of motor unit action potentials. The random nature of the discharge of many motor units within the detection range of the electrodes will cause some superposition of the action potentials emanating from different motor units. Thus, at any instant in time, the negative phase of an action potential of one motor unit may coincide with the positive phase of an action potential of another motor unit. As motor unit recruitment and discharge rates increase, the EMG does not increase linearly because signal cancellation due to overlap also increases (Moore, 1967). Apparently, the complex, nonlinear effects of neural input on EMG and on muscle force happen to occur in the same direction during isometric contractions so that when EMG is compared to force, the relationship appears to have a simple form (Fuglevand; Milner-Brown & Stein).

Measurements of EMG during an MVC are known to be somewhat unreliable (Bigland & Lippold, 1954; Calvert & Chapman, 1977; Chapman & Belanger, 1977; Lawrence & DeLuca, 1983). In an assessment of this unreliability, Yang and Winter (1983) had subjects perform five MVCs on separate days with the triceps brachii muscle. Yang and Winter found that the averaged rectified EMG had a coefficient of variation (standard deviation/mean) of 9.1% on a given day and 16.4% between days. Similarly, Howard and Enoka (1991) had subjects perform three knee-extensor MVCs and noted that the average EMG for vastus lateralis could vary substantially, whereas the force exerted by the leg remained constant (see Figure 8.3a).

Although the trial-to-trial variability in the averaged, rectified EMG associated with the MVCs may reflect intertrial variations in neural drive, two lines of

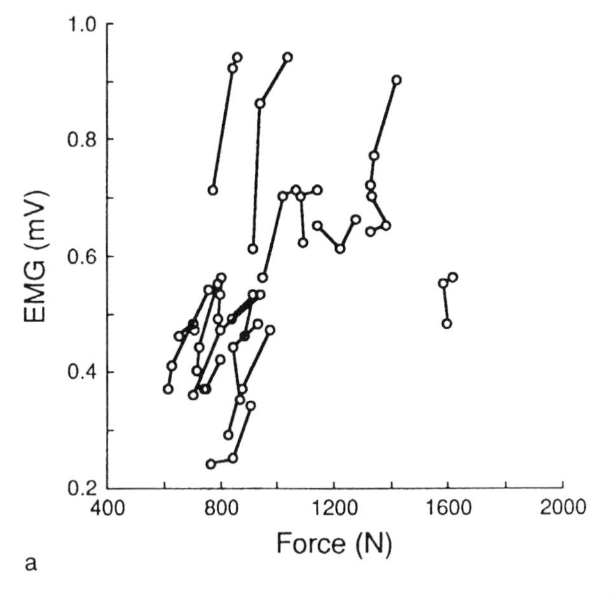

(continued)

Figure 8.3 Relationship between surface EMG and isometric force during repeated maximum voluntary contraction (MVC) trials. Isometric knee extensor force and average rectified surface EMG of vastus lateralis for three MVC trials by 18 subjects. The three trials of each subject are connected by lines. Force varied little from trial to trial, whereas there was considerable variation in the EMG (a). (Reprinted by permission from Howard & Enoka, 1991.) The rectified and averaged EMG (AEMG) and isometric force from five different simulations of a maximum force contraction (b). (Reprinted by permission from Fuglevand, 1989.) The simulation period for each trial was 2 s. The neural input (number of motor units recruited and mean discharge rate of each unit) was identical from trial to trial. The only difference between trials was the specific time of occurrence of the action potentials; this was based on a Gaussian renewal process. Trial-to-trial variability in surface EMG during MVCs is influenced by the random nature of the motor unit discharge; it is not simply a reflection of intertrial variations in neural drive to the muscle.

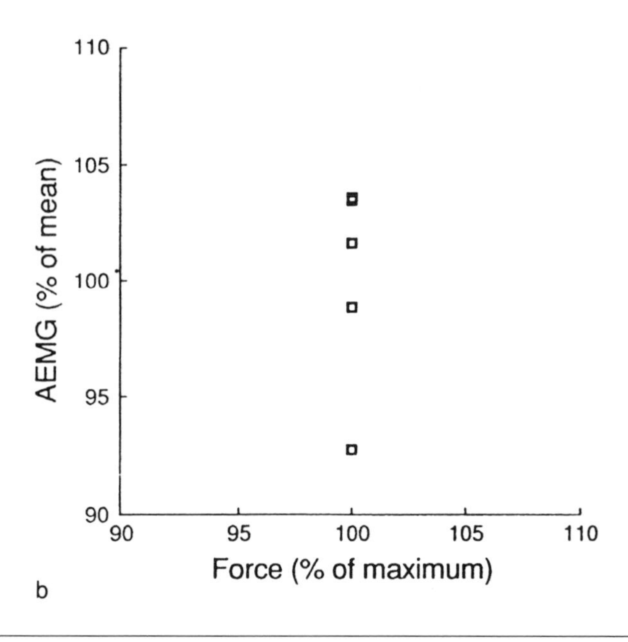

Figure 8.3 *(continued)*

evidence suggest that another factor accounts for much of the variability. First, little variation is seen in the torque or force measurements from trial to trial (see Figure 8.3a), which may indicate that variation in neural input is minimal. Second, repeated simulations of surface EMG in which the neural input (i.e., number of motor units recruited and their mean discharge rates) was identical from trial to trial yielded a substantial variability in the rectified and averaged EMG (Fuglevand, 1989) (see Figure 8.3b). For the data shown in Figure 8.3b, the simulation period was 2 s for each trial, and the simulated muscle force was essentially identical across trials. The discharge of motor units was modeled as a random process with interpulse intervals having a Gaussian probability distribution and a coefficient of variation of 20%. Thus, the only difference in the neural input pattern across trials was the specific time of occurrence of the action potentials. The randomness of the interpulse intervals, therefore, appears to contribute to the trial-to-trial variability in the rectified average EMG. This variability was greater when simulating large muscle forces. Taken together, these observations caution against using the EMG as a direct index of the magnitude of the neural drive to muscle at high forces, such as an MVC.

An MVC Is Not the Maximal Force

Three lines of evidence suggest that the MVC is not the maximal force that a muscle, or muscle group, can exert. This evidence consists of training-induced facilitation of muscle force without changes in muscle size, the demonstration that the neural input to muscle can be insufficient for maximum force, and the observation that an MVC can be increased by supplementing the neural input to the motor neuron pool.

Neural Adaptations With Training

Significant gains in muscle force have been reported following short-term strength training regimens that are generally regarded as too brief to induce gross morphological changes in muscle (Häkkinen, Alèn, & Komi, 1985; Moritani & deVries, 1979; Rutherford & Jones, 1986). For example, after 4 to 5 weeks of quadriceps femoris strength training, subjects increased their MVC torque but did not exhibit any significant changes in muscle enzyme activities, fiber size, or mitochondrial properties (Eriksson, Haggmark, Kiessling, & Karlsson, 1981). Young, McDonagh, and Davies (1985) reported a 27% increase in MVC following an isometric training program, but no change in the maximal force that could be artificially elicited. From this evidence, it would seem that the increase in force that is associated with short-term strength training results from an increased ability to activate the available muscle tissue rather than from an alteration in the quantity of muscle. When training does cause increases in cross-sectional area, the increase in MVC is often disproportionately larger than the increase in muscle size (Narici, Roi, Landoni, Minetti, & Cerretelli, 1989).

Along similar lines, several investigators (see Enoka, 1988) have reported interactions between limbs that influence the MVC a subject can achieve. One example is the increase in strength of an untrained limb contralateral to a limb that has undergone strength training. The magnitude of this phenomenon, termed *cross-education*, can be quite substantial. For example, Moritani and deVries (1979) reported an increase in MVC force of 36% in isometrically trained elbow flexor muscles of 15 subjects and a 25% increase in the MVC force of the contralateral, untrained limb. These changes in MVC of the untrained limb occur without alterations in muscle fiber areas or enzyme activities (Houston, Froese, Valeriote, Green, & Ranney, 1983). The amplitude of the EMG (normalized to the maximum evoked motor response) associated with the MVC, however, does increase in the untrained limb (Butler & Darling, 1990). Consequently, it is probable that the cross-education phenomenon also reflects a neural adaptation. Furthermore, the recent observations of an increase in the MVC and associated EMG following imaginary MVC training (Yue & Cole, 1992) suggest a role for neural factors in strength gains. Another example of an interlimb effect is the dependence of the MVC on whether one or two limbs have been activated. Most subjects exhibit a lower single-limb MVC when both limbs are active concurrently; this has been termed the *bilateral deficit* (Howard & Enoka, 1991; Schantz, Moritani, Karlson, Johansson, & Lundh, 1989; Secher, Rube, & Elers, 1988). These observations on the enhanced force in muscle that has undergone short-term strength training, on cross-education, on imaginary training, and on the bilateral deficit provide compelling, although indirect, evidence that voluntary neural drive is inadequate to fully activate muscle during an MVC in normal, untrained subjects.

Neural Insufficiency

Whether the maximal force that an individual can exert is indeed maximum essentially depends on the ability of the individual to provide the necessary neural drive. In order to test the ability of individuals to achieve this goal, investigators have used two techniques:

- The superimposed twitch method involves imposing a single electrical stimulus on a muscle during an MVC to see if the stimulus elicits a detectable (interpolated) twitch response.
- The force elicited by supramaximal, high frequency electrical stimulation is compared to that exerted during an MVC.

Both methods test the null hypothesis that the neural drive is insufficient and provide an index of the degree to which the maximal force can be increased by artificial activation of the muscle; the inability to provide an adequate neural drive can be called the *neural insufficiency hypothesis*.

Because the twitch superimposition technique is less disagreeable for the subject, it tends to be used more frequently to test the neural insufficiency hypothesis. In a relaxed muscle, a single electrical stimulus administered percutaneously across either the nerve or the muscle will elicit a twitch response of a given magnitude. As the voluntary force increases in magnitude, the size of the superimposed twitch will decrease in proportion to the muscle force until it cannot be detected at maximal force (Belanger & McComas, 1981; Merton, 1954). The final state where the twitch cannot be detected has been referred to as *occlusion of the twitch* (Denny-Brown, 1928). If the central nervous system does not fully activate muscle, then the amplitude of the twitches superimposed on the MVC will reflect this deficiency (Woods, Furbush, & Bigland-Ritchie, 1987).

Based on the twitch superimposition test, a number of investigators have concluded that motivated subjects are able to fully activate a variety of muscles (Belanger & McComas, 1981; Bellemare et al., 1983; Gandevia & McKenzie, 1985, 1988; Jones & Rutherford, 1987; Merton, 1954; Rutherford, Jones, & Newham, 1986; Woods et al., 1987). In a slightly more conservative interpretation of the twitch superimposition observations, Hales and Gandevia (1988) suggested that 99% of the true maximal force can be achieved voluntarily, perhaps with 95% of the muscle receiving maximal neural drive and the remaining 5% being activated at 80% of maximum.

At least four issues affect the validity and reliability of the twitch superimposition test. First, the efficacy of superimposing a twitch on a maximal force is greatest when the stimulus can activate the entire muscle or muscles contributing to the force. Often, however, it is not possible to activate all the involved musculature due to the size (e.g., quadriceps femoris) or the distributed nature of the muscles (e.g., elbow flexors—biceps brachii, brachialis, brachioradialis). If the stimulus does not excite the entire muscle, then the test will reveal only whether the stimulated fibers are maximally activated (Merton, 1954). The degree of confidence in the conclusion of maximal neural activation depends on the proportion of the muscle that can be activated by the electrical stimulation. Second, the gain of the force record is usually so low that it is difficult to detect twitches that are less than 2% to 3% of the control values (Belanger & McComas, 1981; Hales & Gandevia, 1988). One way to improve the resolution is to use sample-and-hold amplification of the force signal, which will allow the detection of twitches as small as 0.02% of the control value (Hales & Gandevia, 1988). Another approach is to replace the single stimulus with a train of stimuli, such as four stimuli at 50 Hz (Hales & Gandevia) or five stimuli at 100 Hz (Garland, Garner, & McComas, 1988). Third, the stimulus should not activate muscles that generate torque that is antagonistic to the test muscle or muscles. This is particularly important in nerve stimulation because some peripheral nerves carry axons to an agonist-antagonist muscle pair. For example, the ulnar nerve innervates both first dorsal interosseus and first palmar interosseus, an abductor and adductor of the index finger, respectively. Thus, the index finger abduction torque recorded during ulnar nerve stimulation is the resultant of the opposing actions of these two muscles and is not equivalent to the maximum capacity of either. Fourth, fatigue-related impairment of neuromuscular propagation may render some muscle fibers incapable of responding to either neural activation or the interpolated stimulus.

The alternative approach to test the neural insufficiency hypothesis has been to tetanize the muscle or muscles with a supramaximal electrical stimulus and to compare the elicited force ($\mathbf{P}_0$) to that obtained in an MVC. Ratios ($\mathbf{P}_0$/MVC) greater than 100% indicate that the neural drive was insufficient for maximal activation of the muscle, whereas ratios less than 100% suggest that the stimulus (current and frequency) was inadequate to achieve complete activation of the muscle. This comparison has been characterized by divergent results, with some investigators finding $\mathbf{P}_0$/MVC ratios less than 100% and others reporting ratios equal to or greater than 100%.

Davies and colleagues used conventional stimulation procedures (brief rectangular pulses) with rates of 40 Hz to 60 Hz and reported $\mathbf{P}_0$/MVC ratios of approximately 72% for biceps brachii, 80% for triceps surae, and 75% for first dorsal interosseus (Davies et al., 1985; McDonagh, Hayward, & Davies, 1983; Young et al., 1985). Similarly, Duchateau and Hainaut (1987, 1988) found ratios of 80% to 93% for adductor pollicis when the ulnar nerve was stimulated with subcutaneous electrodes at 100 Hz. Rutherford and Jones (1988) also stimulated the ulnar nerve at 100 Hz and noted that $\mathbf{P}_0$ and MVC for first dorsal interosseus were essentially equivalent. This is surprising given that percutaneous stimulation of the ulnar nerve would also result in activation of the antagonist. Because these ratios are generally less than 100%, these results indicate that the stimulus was inadequate to achieve complete activation of the muscle.

In contrast, Delitto, Brown, Strube, Rose, and Lehman (1989) performed extensive longitudinal tests on a highly motivated subject and reported that electrical stimulation elicited, on average, a quadriceps femoris force that was 112% of MVC (range = 95% to 126% MVC). Several features of this interesting observation were unusual, including the bilateral application of the stimulus to the legs, the use of a clinical electromyostimulation unit designed to minimize the discomfort (see also Selkowitz, 1985), the delivery of a substantial current (200 mA), and the involvement of a subject who could tolerate the sensations associated with electrical stimulation. The observation by Delitto et al. of a $\mathbf{P}_0$/MVC ratio greater than 100% indicates that it is possible to artificially elicit a force in a large muscle mass that is greater than the maximum force the subject can exert voluntarily.

Neural Supplementation

The rationale underlying the tests of the neural insufficiency hypothesis is to determine whether the descending command impinging on the motor neuron pool is adequate for maximal activation of the muscle. An alternative approach is to consider whether the descending command can be supplemented by input from other sources, such as sensory feedback. Under these conditions, investigators have subjects provide the supposed maximum descending command (i.e., perform an MVC) and then manipulate afferent feedback to determine if additional input to the motor neuron pool can increase the maximum force exerted by the subject. The few studies that have employed this paradigm indicate that low-frequency (20-Hz) vibration (Samuelson, Jorfeldt, & Ahlborg, 1989) and low-frequency (2-Hz) transcutaneous nerve stimulation (Kaada, 1984) do not influence an MVC, but that cutaneous stimulation (Howard & Enoka, 1987, 1991; Matyas, Galea, & Spicer, 1986) and a brief stretch (Lagasse, 1974; Morris, 1974) can increase an MVC.

Howard and Enoka (1987, 1991) examined the effect of right-leg stimulation on the MVC that the knee extensors of the left leg could exert. Stimulation of the right leg involved the percutaneous application of a current (1 to 100 mA) that consisted of a 2500-Hz sine wave modulated at 50 bursts per second with a 50% duty cycle. In addition to activating the quadriceps femoris of the right leg, the stimulus was associated with a strong cutaneous sensation. The stimulus elicited a right-leg force that averaged 49 ± 15% MVC. For all subjects (n=12), the right-leg stimulation caused a facilitation of the left-leg MVC (mean ± SD=10.8 ± 7.5%, range 0.7% to 23.5% MVC). Similarly, Matyas et al. (1986) found that the therapeutic procedure of rapid brushing, which provides a cutaneous sensation, caused an increase in the EMG associated with an MVC in both quadriceps femoris and biceps femoris. These observations are consistent with reports of increases in the H reflex associated with contralateral cutaneous stimulation (Pierrot-Deseilligny, Bussel, Sideri, Cathala, & Castaigne, 1973; Robinson, McIlwain, & Hayes, 1979), which suggest that the facilitation of the MVC was due to supplementation of the descending command with afferent feedback.

Similar effects have been elicited by stretching the contralateral muscles. Morris (1974) had subjects perform a bilateral reciprocal MVC with the knee musculature. For this task, the subjects exerted a right-leg knee extensor MVC and a left-leg knee flexor MVC. While performing this task, the knee angle for the right leg was rapidly decreased by 15°, from 145° to 130°, thereby stretching the knee extensor muscles. The left-leg MVC was not different with the right knee joint at 145° and 130°. With the stretch, however, the MVC exerted by the knee extensors of the right leg increased by 25.2%, and the MVC exerted by the knee flexors of the left leg increased by 8%. Lagasse (1974) used a similar paradigm to examine the effect on homologous muscles and found that when the ipsilateral knee angle was suddenly changed by 15°, the knee-extensor MVC increased by an average of 8% and the contralateral knee-extensor MVC decreased by 18%. These observations underscore the effect of stretch on the contralateral MVC; presumably this effect is mediated by neural mechanisms.

Conclusion

The literature on neural adaptations with training, the neural insufficiency hypothesis (i.e., the twitch superimposition test and the comparison of tetanic force to MVC), and the facilitation of an MVC by manipulating sensory feedback provide an adequate rationale to conclude that an MVC does not represent the maximum capacity of the muscle-joint system. This conclusion, however, does not preclude the utility of the MVC as a standardized measure of performance. Rather, the conclusion invites efforts to determine the mechanisms underlying the apparent inability of the central nervous system to fully activate muscle during an MVC and underscores the need for a greater appreciation of the interaction between descending drive and sensory feedback in the output of the motor system.

References

Aitkens, S., Lord, J., Bernauer, E., Fowler, W.M., Lieberman, J.S., & Berck, P. (1989). Relationship of manual muscle testing to objective strength measurements. *Muscle & Nerve*, **12**, 173-177.

Bawa, P., & Calancie, B. (1983). Repetitive doublets in human flexor carpi radialis muscle. *Journal of Physiology (London)*, **339**, 123-132.

Belanger, A.Y., & McComas, A.J. (1981). Extent of motor unit activation during effort. *Journal of Applied Physiology*, **51**, 1131-1135.

Bellemare, F., Woods, J.J., Johansson, R., & Bigland-Ritchie, B. (1983). Motor-unit discharge rates in maximal voluntary contractions of three human muscles. *Journal of Neurophysiology*, **50**, 1380-1392.

Bigland, B., & Lippold, O.C.J. (1954). The relation between force, velocity and integrated electrical activity in human muscles. *Journal of Physiology (London)*, **123**, 214-224.

Bigland-Ritchie, B. (1981). EMG/force relations and fatigue of human voluntary contractions. In D.I. Miller (Ed.), *Exercise and sport sciences reviews* (Vol. 9, pp. 75-117). Philadelphia: Franklin Institute.

Binder-Macleod, S.A., & Clamann, H.P. (1989). Force output of cat motor units stimulated with trains of linearly varying frequency. *Journal of Neurophysiology*, **61**, 208-217.

Buchthal, F., & Schmalbruch, H. (1970). Contraction times and fibre types in intact human muscle. *Acta Physiologica Scandinavica*, **79**, 435-452.

Burke, D., Skuse, W.F., & Lethlean, A.K. (1974). Isometric contraction of the abductor digiti minimi muscle in man. *Journal of Neurology, Neurosurgery, and Psychiatry*, **37**, 825-834.

Burke, R.E. (1981). Motor units: Anatomy, physiology, and functional organization. In V.B. Brooks (Ed.), *Handbook of physiology* (Sec. 1, Vol. 2, P. 1, pp. 345-422). Washington, DC: American Physiological Society.

Burke, R.E., Rudomin, P., & Zajac, F.E., III. (1976). The effect of activation history on tension production by individual muscle units. *Brain Research*, **109**, 515-529.

Butler, A.J., & Darling, W.G. (1990). Reflex changes accompanying isometric strength training of the contralateral limb. *Society for Neuroscience Abstracts*, **16**, 884.

Calvert, T.W., & Chapman, A.E. (1977). Relationship between the surface EMG and force transients in muscle: Simulation and experimental studies. *Proceedings of the IEEE*, **65**, 682-689.

Carroll, S.G., Triolo, R.J., Chizeck, H.J., Kobetic, R., & Marsolais, E.B. (1989). Tetanic responses of electrically stimulated paralyzed muscle at varying interpulse intervals. *IEEE Transactions on Biomedical Engineering*, **36**, 644-653.

Chapman, A.E., & Belanger, A.Y. (1977). Electromyographic methods of evaluating strength training. *Electromyography and Clinical Neurophysiology*, **17**, 265-280.

Costa, F.V., Borghi, C., Mussi, A., & Ambrosioni, E. (1987). Reproducibility of pressor response to handgrip: Influence of time intervals, strength and duration of exercise. *Clinical and Experimental Pharmacology and Physiology*, **14**, 587-595.

Davies, C.T.M., Dooley, P., McDonagh, M.J.N., & White, M.J. (1985). Adaptation of mechanical properties of muscle to high force training in man. *Journal of Physiology (London)*, **365**, 277-284.

Delitto, A., Brown, M., Strube, M.J., Rose, S.J., & Lehman, R.C. (1989). Electrical stimulation of quadriceps femoris in an elite weight lifter: A single subject experiment. *International Journal of Sports Medicine*, **10**, 187-191.

DeLuca, C.J., LeFever, R.S., McCue, M.P., & Xenakis, A.P. (1982). Behavior of human motor units in different muscles during linearly varying contractions. *Journal of Physiology (London)*, **329**, 113-128.

Denny-Brown, D. (1928). On inhibition as a reflex accompaniment of the tendon jerk and of other forms of active muscular response. *Proceedings of the Royal Society of London: Series B: Biological Sciences*, **103**, 321-336.

Duchateau, J., & Hainaut, K. (1987). Electrical and mechanical changes in immobilized human muscle. *Journal of Applied Physiology*, **62**, 2168-2173.

Duchateau, J., & Hainaut, K. (1988). Training effects of sub-maximal electrostimulation in a human muscle. *Medicine and Science in Sports and Exercise*, **20**, 99-104.

Enoka, R.M. (1988). Muscle strength and its development: New perspectives. *Sports Medicine*, **6**, 146-168.

Enoka, R.M., Robinson, G.A., & Kossev, A.R. (1989). Task and fatigue effects on low-threshold motor units in human hand muscle. *Journal of Neurophysiology*, **62**, 1344-1359.

Eriksson, E., Haggmark, T., Kiessling, K.H., & Karlsson, J. (1981). Effect of electrical stimulation on human skeletal muscle. *International Journal of Sports Medicine*, **2**, 18-22.

Freund, H.J., Büdingen, H.J., & Dietz, V. (1975). Activity of single motor units from human forearm muscles during voluntary isometric contractions. *Journal of Neurophysiology*, **38**, 933-946.

Fuglevand, A.J. (1989). *A motor unit pool model: Relationship of neural control properties to isometric muscle tension and the electromyogram.* Unpublished doctoral dissertation, University of Waterloo, Ontario.

Fuglevand, A.J., Winter, D.A., Patla, A.E., & Stashuk, D. (1989). Motor unit recruitment and rate coding influences on muscle tension and EMG. *Society for Neuroscience Abstracts*, **15**, 523.

Gandevia, S.C., & McKenzie, D.K. (1985). Activation of the human diaphragm during maximal static efforts. *Journal of Physiology (London)*, **367**, 45-56.

Gandevia, S.C., & McKenzie, D.K. (1988). Activation of human muscles at short muscle lengths during maximal static efforts. *Journal of Physiology (London)*, **407**, 599-613.

Garland, S.J., Garner, S.H., & McComas, A.J. (1988). Reduced voluntary electromyographic activity after fatiguing stimulation of human muscle. *Journal of Physiology*, **401**, 547-556.

Grimby, L., & Hannerz, J. (1977). Firing rate and recruitment order of toe extensor motor units in different modes of voluntary contraction. *Journal of Physiology (London)*, **264**, 865-879.

Häkkinen, K., Alèn, M., & Komi, P.V. (1985). Changes in isometric force- and relaxation-time electromyographic and muscle fibre characteristics of human skeletal muscle during strength training and detraining. *Acta Physiologica Scandinavica*, **125**, 573-585.

Hales, J.P., & Gandevia, S.C. (1988). Assessment of maximal voluntary contraction with twitch interpolation: An instrument to measure twitch responses. *Journal of Neuroscience Methods*, **25**, 97-102.

Hannerz, J. (1974). Discharge properties of motor units in relation to recruitment order in voluntary contractions. *Acta Physiologica Scandinavica*, **91**, 374-384.

Houston, M.E., Froese, E.A., Valeriote, S.P., Green, H.J., & Ranney, D.A. (1983). Muscle performance, morphology and metabolic capacity during strength training and detraining: A one leg model. *European Journal of Applied Physiology*, **51**, 25-35.

Howard, J.D., & Enoka, R.M. (1987). Enhancement of maximum force by contralateral-limb stimulation. *Journal of Biomechanics*, **20**, 980.

Howard, J.D., & Enoka, R.M. (1991). Maximum bilateral contractions are modified by neurally mediated interlimb effects. *Journal of Applied Physiology*, **70**, 306-316.

Jones, D.A., & Rutherford, O.M. (1987). Human muscle strength training: The effects of three different regimes and the nature of the resultant changes. *Journal of Physiology (London)*, **391**, 1-11.

Jones, D.A., Rutherford, O.M., & Parker, D.F. (1989). Physiological changes in skeletal muscle as a result of strength training. *Quarterly Journal of Experimental Physiology*, **74**, 233-256.

Kaada, B. (1984). Improvement of physical performance by transcutaneous nerve stimulation in athletes. *International Journal of Acupuncture and Electro-Therapeutic Research*, **9**, 165-180.

Kitai, T.A., & Sale, D.G. (1989). Specificity of joint angle in isometric training. *European Journal of Applied Physiology*, **58**, 744-748.

Kroemer, K.H.E., & Marras, W.S. (1980). Towards an objective assessment of the 'maximal voluntary contraction' component in routine muscle strength measurements. *European Journal of Applied Physiology*, **45**, 1-9.

Kukulka, C.G., & Clamann, H.P. (1981). Comparison of the recruitment and discharge properties of motor units in human brachial biceps and adductor pollicis during isometric contractions. *Brain Research*, **219**, 45-55.

Kulig, K., Andrews, J.G., & Hay, J.G. (1984). Human strength curves. In R.L. Terjung (Ed.), *Exercise and sport sciences reviews* (Vol. 12, pp. 417-466). Lexington, MA: Collamore.

Lagasse, P.P. (1974). Muscle strength: Ipsilateral and contralateral effects of superimposed stretch. *Archives of Physical Medicine and Rehabilitation*, **55**, 305-310.

Lawrence, J.H., & DeLuca, C.J. (1983). Myoelectric signal versus force relationship in different human muscles. *Journal of Applied Physiology*, **54**, 1653-1659.

Marsden, C.D., Meadows, J.C., & Merton, P.A. (1983). 'Muscular wisdom' that minimizes fatigue during prolonged effort in man: Peak rates of motoneuron discharge and slowing of discharge during fatigue. In J.E. Desmedt (Ed.), *Motor control mechanisms in health and disease* (pp. 169-211). New York: Raven Press.

Martin, S., Neale, G., & Elia, M. (1985). Factors affecting maximal momentary grip strength. *Human Nutrition: Clinical Nutrition*, **39C**, 137-147.

Matyas, T.A., Galea, M.P., & Spicer, S.D. (1986). Facilitation of the maximum voluntary contraction in hemiplegia by concomitant cutaneous stimulation. *American Journal of Physical Medicine*, **65**, 125-134.

McDonagh, M.J.N., Hayward, C.M., & Davies, C.T.M. (1983). Isometric training in human elbow flexor muscles: The effects on voluntary and electrically evoked forces. *Journal of Bone and Joint Surgery*, **65B**, 355-358.

Merton, P.A. (1954). Voluntary strength and fatigue. *Journal of Physiology (London)*, **123**, 553-564.

Miller, R.G., Mirka, A., & Maxfield, M. (1981). Rate of tension development in isometric contractions of human hand muscle. *Experimental Neurology*, **73**, 267-285.

Milner-Brown, H.S., & Stein, R.B. (1975). The relation between the surface electromyogram and muscular force. *Journal of Physiology (London)*, **246**, 549-569.

Milner-Brown, H.S., Stein, R.B., & Lee, R.G. (1975). Synchronization of human motor units: Possible roles of exercise and supraspinal reflexes. *Electroencephalography and Clinical Neurophysiology*, **38**, 245-254.

Milner-Brown, H.S., Stein, R.B., & Yemm, R. (1973). The orderly recruitment of human motor units during voluntary isometric contractions. *Journal of Physiology (London)*, **230**, 359-370.

Monster, A.W., & Chan, H. (1977). Isometric force production by motor units of extensor digitorum communis muscle in man. *Journal of Neurophysiology*, **40**, 1432-1443.

Moore, A.D. (1967). Synthesized EMG waves and their implications. *American Journal of Physical Medicine*, **46**, 1302-1316.

Moritani, T., & deVries, H.A. (1979). Neural factors versus hypertrophy in the time course of muscle strength gain. *American Journal of Physical Medicine*, **58**, 115-130.

Morris, A.F. (1974). Myotatic reflex effects on bilateral reciprocal leg strength. *American Corrective Therapy Journal*, **28**, 24-29.

Nardonne, A., Romanò, C., & Schieppati, M. (1989). Selective recruitment of high-threshold human motor units during voluntary isotonic lengthening of active muscles. *Journal of Physiology (London)*, **409**, 451-471.

Narici, M.V., Roi, G.S., Landoni, L., Minetti, A.E., & Cerretelli, P. (1989). Changes in force, cross-sectional area and neural activation during strength training and detraining of the human quadriceps. *European Journal of Applied Physiology*, **59**, 310-319.

Niebuhr, B.R., & Marion, R. (1987). Detecting sincerity of effort when measuring grip strength. *American Journal of Physical Medicine*, **66**, 16-24.

Pierrot-Deseilligny, E., Bussel, B., Sideri, G., Cathala, H.P., & Castaigne, P. (1973). Effect of voluntary contraction on H reflex changes induced by cutaneous stimulation in normal man. *Electroencephalography and Clinical Neurophysiology*, **34**, 185-192.

Rack, P.M.H., & Westbury, D.R. (1969). The effects of length and stimulus rate on tension in the isometric cat soleus muscle. *Journal of Physiology (London)*, **204**, 443-460.

Robinson, K.L., McIlwain, J.S., & Hayes, K.C. (1979). Effects of H-reflex conditioning upon contralateral alpha motoneuron pool. *Electroencephalography and Clinical Neurophysiology*, **46**, 65-71.

Rutherford, O.M., & Jones, D.A. (1986). The role of learning and coordination in strength training. *European Journal of Applied Physiology*, **55**, 100-105.

Rutherford, O.M., & Jones, D.A. (1988). Contractile properties and fatiguability of the human adductor pollicis and first dorsal interosseus: A comparison of the effects of two chronic stimulation patterns. *Journal of the Neurological Sciences*, **85**, 319-331.

Rutherford, O.M., Jones, D.A., & Newham, D.J. (1986). Clinical and experimental application of the percutaneous twitch superimposition technique for the study of human muscle activation. *Journal of Neurology, Neurosurgery, and Psychiatry*, **49**, 1288-1291.

Samuelson, B., Jorfeldt, L., & Ahlborg, B. (1989). Influence of vibration on endurance of maximal isometric contraction. *Clinical Physiology*, **9**, 21-25.

Schantz, P.G., Moritani, T., Karlson, E., Johansson, E., & Lundh, A. (1989). Maximal voluntary force of bilateral and unilateral leg extension. *Acta Physiologica Scandinavica*, **136**, 185-192.

Secher, N.H., Rube, N., & Elers, J. (1988). Strength of two- and one-leg extension in man. *Acta Physiologica Scandinavica*, **134**, 333-339.

Selkowitz, D.M. (1985). Improvement in isometric strength of the quadriceps femoris muscle after training with electrical stimulation. *Physical Therapy*, **65**, 186-196.

Sherrington, C.S. (1925). Remarks on some aspects of reflex inhibition. *Proceedings of the Royal Society of London: Series B: Biological Sciences*, **97**, 519-545.

Tanji, J., & Kato, M. (1973). Firing rate of individual motor units in voluntary contraction of abductor digiti minimi muscle in man. *Experimental Neurology*, **40**, 771-783.

Thépaut-Mathieu, C., van Hoecke, J., & Maton, B. (1988). Myoelectrical and mechanical changes linked to length specificity during isometric training. *Journal of Applied Physiology*, **64**, 1500-1505.

Thomas, C.K., Ross, B.H., & Calancie, B. (1987). Human motor-unit recruitment during isometric contractions and repeated dynamic movements. *Journal of Neurophysiology*, **57**, 311-324.

Thomas, C.K., Woods, J.J., & Bigland-Ritchie, B. (1989). Impulse propagation and muscle activation in long maximal voluntary contractions. *Journal of Applied Physiology*, **67**, 1835-1842.

Woods, J.J., Furbush, F., & Bigland-Ritchie, B. (1987). Evidence for a fatigue-induced reflex inhibition of motoneuron firing rates. *Journal of Neurophysiology*, **58**, 125-137.

Yang, J.F., & Winter, D.A. (1983). Electromyography reliability in maximal contractions and submaximal isometric contractions. *Archives of Physical Medicine and Rehabilitation*, **64**, 417-420.

Young, K., McDonagh, M.J.N., & Davies, C.T.M. (1985). The effects of two forms of isometric training on the mechanical properties of the triceps surae in man. *Pflügers Archiv*, **405**, 384-388.

Yue, G., & Cole, K.J. (1992). Strength increases from the motor program: A comparison of training with maximal voluntary and imagined muscle contractions. *Journal of Neurophysiology*, **67**, 1114-1123.

Zajac, F.E., & Young, J.L. (1980). Properties of stimulus trains producing maximum tension-time area per pulse from single motor units in medial gastrocnemius muscle of the cat. *Journal of Neurophysiology*, **43**, 1206-1220.

Acknowledgments: We thank Peter Worden for technical assistance and Amy Tyler, Jeffrey Jasperse, and Drs. Robin Callister and Douglas Seals for editorial comments on a draft of the manuscript. Work described in this chapter was supported by National Institutes of Health grants NS 07309 (to the University of Arizona Motor Control Group) and NS 20544 (to Douglas G. Stuart and Roger M. Enoka). Andrew J. Fuglevand was supported by a National Institutes of Health Postdoctoral Fellowship (NS 08634).

Chapter 9

Ligamentous Mechanoreceptors and Knee Joint Function: The Neurosensory Hypothesis

Mark D. Grabiner

In the final chapter of the book, Mark Grabiner addresses injury to the anterior cruciate ligament, the most commonly injured ligament of the human knee. This area is of increasing interest to both clinicians and basic scientists. The recognition of the presence of mechanoreceptors within the substance of ligaments, especially the anterior cruciate ligament, has revived interest in mechanoreceptors relative to their knee joint function. In this area the human-related research lags behind research that uses animal models. The animal work, which is not emphasized in the chapter, presents strong evidence that these mechanoreceptors provide static and dynamic afferent information regarding ligament strain and that they can exert considerable influence on reflexive and voluntary muscular contractions. The work on humans has been far less convincing. However, this chapter highlights research opportunities in an area that has strong implications for participation in sport, rehabilitation from sport-related knee ligament injury, and surgical interventions to restore the mechanical and neuromotor function to the ligamentously deficient knee.

Traditionally, ligaments have been thought of as passive collagenous cords. The mechanical properties of ligaments such as load deformation and viscoelasticity are well suited for the primary task that they perform, which is passively restricting undesirable motions of the joints they span. Along with the geometry of the articulating surfaces of the bones that form the joint, ligaments are a primary source of joint stability.

In addition to this passive role of restricting undesirable joint kinematics, ligaments may also actively influence skeletal muscle excitation patterns (Brand, 1986). This active role has been referred to as the neurosensory hypothesis of ligament function (Frank et al., 1988). Injury to knee joint ligaments can result in joint instabilities that restrict function; such injuries have also been implicated in the long-term degeneration of the articular cartilage. An injury to a ligament caused by an elongation beyond its elastic limit will cause a permanent deformation and can therefore result in changes in the joint kinematics. Changes in joint kinematics can subsequently alter the strain milieu, or environment, of the ligament during joint motion. The neurosensory hypothesis can be used to relate adaptations of knee joint muscular synergies to an altered ligament strain milieu following ligamentous injury and associated joint instability. This chapter will integrate the recent findings of ligament mechanics and neurophysiological studies that have spurred a renewed orthopedic interest in the neurosensory hypothesis of ligament function as it relates to the knee joint.

Ligamentous Mechanoreceptors

The key physical components of the neurosensory hypothesis are mechanoreceptors within the substance of the ligament. These mechanoreceptors differ relative to the stimuli to which they are sensitive and in their ability to partially or completely adapt to the stimulus as a function of time. Afferent firing during a length change of a ligament is its dynamic response, and firing during no length change is its static response. Slowly adapting mechanoreceptors transmit impulses for as long as the stimulus is present. This continuous stream of information to the central nervous system can provide instantaneous appraisal of joint position, muscle contraction intensity, tensile force, and pain. Rapidly adapting mechanoreceptors are sensitive to the rate at which the change occurs. These mechanoreceptors therefore transmit impulses during a change in stimulus status such as a change in length or tension.

Four types of mechanoreceptors have been identified within the ligaments and joint capsule of the human knee joint:

- Ruffini endings
- Pacinian corpuscles
- Mechanoreceptors similar to Golgi tendon organs
- Free nerve endings

Ruffini endings—found in the human medial meniscus, the joint capsule, and the anterior cruciate ligament (ACL)—are slowly adapting and have low mechanical thresholds. Additionally, these mechanoreceptors have been characterized as static and dynamic (depending on their location) and are sensitive to static joint position, intra-articular pressure, movement amplitude, and velocity.

Pacinian corpuscles, rapidly adapting mechanoreceptors having low mechanical thresholds, have been found in the human knee joint capsule, medial meniscus, and the ACL. These mechanoreceptors are regarded as dynamic and are inactive in an immobile joint or a joint moving at constant velocity. They become active during acceleration and deceleration of the joint.

The largest articular mechanoreceptors are similar to Golgi tendon organs and have been found in human collateral ligaments, cruciate ligaments, and the medial meniscus. These mechanoreceptors are slowly adapting and have high mechanical thresholds. Golgi organs are inactive in the immobile joint. Because of their high mechanical threshold, it has been suggested that Golgi organs sense ligament tension, especially at the extreme ranges of motion.

The articular nociceptive system is comprised of free nerve endings. These mechanoreceptors are inactive but become active in the presence of extreme mechanical deformation or various chemical agents.

These four types of morphologically different mechanoreceptors, demonstrating different stimulus response profiles, provide information to the central nervous system. Information from these types of mechanoreceptors reach all levels of the central nervous system (Sjolander, 1989; Tracey, 1980). This information may be used to detect and subsequently to respond to signals reflecting joint kinematics and kinetics.

Influence of Capsular Mechanoreceptors on Knee Joint Musculature

Studies on the effects of mechanical stimulation of mechanoreceptors within the medial collateral ligaments and joint capsule have been a cornerstone for similar study of the ACL. Stimulation of human knee joint mechanoreceptors can have significant effects on muscle excitation, force output of knee joint musculature, and proprioceptive performance (Wood, Ferrell, & Baxendale, 1988).

A knee joint effusion model has often been used to examine mechanoreceptor influence on knee muscle excitation patterns. Knee joint trauma often involves an effusion, or escape, of body fluids into the joint and manifests itself as swelling. The effusion increases intra-articular pressure and subsequently stimulates mechanoreceptors embedded within the stretched joint capsule. An inverse relation between knee joint effusion and muscle excitability was reported by deAndrade, Grant, and Dixon (1965). Progressive distension of the knee joint with blood plasma was found to decrease both knee joint extension strength and associated vastus lateralis electromyogram (EMG) amplitude. This clinical observation was later confirmed by Stratford (1981).

Kennedy, Alexander, and Hayes (1982) and Spencer, Hayes, and Alexander (1984) further investigated the relationship between joint effusion and muscle excitability. Hoffmann (H) reflexes from the vastus medialis, vastus lateralis, and rectus femoris were elicited during progressive, experimentally induced knee joint distensions. H-reflex inhibitions of 30% to 60% were observed with infusions of 60 ml of saline. Notably, the H-reflex inhibition did not demonstrate time-dependency characteristics; that is, the level of inhibition did not decrease during maintenance of a 60-ml infusion for 30 min. Notably, infusion of an anesthetic (lidocaine) prior to saline abolished the previously observed reflexive inhibition. Ostensibly, the effect of the lidocaine was to anesthetize the capsular mechanoreceptors, thereby rendering them unable to sense the distension of the knee.

Baxendale and Ferrell (1987) demonstrated that infusion of the knee joint with as little as 20-ml saline (one third the maximum amount used by Kennedy et al., 1982) significantly affected joint position sense, primarily at the extremes of knee joint range of motion. In this experiment, consistent with the experiment by Kennedy et al., injection of anesthetic into the knee joint subsequently improved the joint position sense. In contrast, knee joint anesthesia was not found to have an effect on joint position sense in noneffused knees.

Many types of musculoskeletal disorders are associated with chronic joint effusions, making this joint condition a subject of study and comparison. Geborek, Moritz, and Wollheim (1989) reported progressive inhibition of knee joint extensor torque and muscle excitation with increased joint effusion in arthritic patients. D.W. Jones, D.A., Jones, and Newham (1987) investigated chronically effused knee joints—associated with conditions such as inflammatory synovitis, osteoarthritis, and rheumatoid arthritis—and the effects of joint aspiration (removal by suction). Excitability of the quadriceps femoris muscle was investigated using the interpolated twitch technique.

The interpolated twitch technique consists of comparing the torque generated with maximum voluntary effort to the torque generated during maximal voluntary effort supplemented with electrical stimulation of either the involved musculature or nerve supplying the muscle. In normal muscle, increased muscular output with electrical stimulation is associated with decreased excitability of the musculature.

Contrary to experiments conducted on acutely effused knees, D.W. Jones et al. (1987) reported that joint aspiration of chronically effused knees had no effect on the quadriceps femoris inhibition. This observation was attributed to changes in joint capsule compliance caused by the long-term distension and, presumably, an adaptation of the mechanoreceptors to the distension. In contrast to the findings of D.W. Jones et al., Fahrer, Rentsch, Gerber, Beyeler, Hess, and Grunig (1988) found an approximately 13.6% improvement in isometric knee extension strength following joint aspiration. Associated increases in the EMG signal were also observed.

The previously mentioned studies are particularly relevant to rehabilitation of knee joint function following knee injuries. The improvements in muscular strength, power, and (to a limited extent) endurance are key elements of therapy and rely on the ability to cause a muscle group to be functionally overloaded. The reduction in skeletal muscle excitability associated with joint effusion influences the degree to which a muscle group can be overloaded.

The Neurosensory Hypothesis and the Anterior Cruciate Ligament

The ACL is the primary ligamentous restraint to anterior movement of the tibia below the femur, providing up to 80% of the required force (Butler, Noyes, & Grood, 1980). The ACL also restrains some types of rotary motion (Noyes, Grood, Suntay, & Butler, 1983) and can serve as a restraint to medial-lateral joint opening (Grood, Noyes, Butler, & Suntay, 1981).

An increasing orthopedic interest has focused on the neurosensory hypothesis using animal and human ACL models. This reflects, in part, (a) a general increase in the number of orthopedically serious ACL injuries and (b) the relationship between injury, treatment, and rehabilitation on long-term knee joint function. The knee joint, the largest and one of the most complex joints in the body, is quite susceptible to recreational and sport-related injury. A 7-year retrospective study by DeHaven and Lintner (1986) of injury data on 4,551 patients revealed that the knee joint was the most commonly injured area of the body for males (42.1%) and females (59.2%). Over 40% of these injuries were associated with some type of ligamentous injury.

Of the major knee joint ligaments, the ACL is the most commonly injured (Tibone, Antich, Fanton, Moynes, & Perry, 1986). Injuries to this ligament range from partial to complete rupture and result in varying degrees of knee joint instability. If untreated, these injuries can become more serious. Noyes, Mooar, Moorman, and McGinniss (1989) reported that the extent to which partial ACL tears become more extensively injured is significantly and linearly related to the extent of the original tear.

There is a dichotomy of opinion, however, on the prognosis for long-term function in the ACL-deficient knee. Knee joint instability has been related to significant progressive damage to articular components such as the articular cartilage, the menisci, and the secondary capsular and ligamentous restraints. These joint changes may ultimately lead to early onset osteoarthritis (Noyes & McGinniss, 1985). On the other hand, Brand (1986) has written that existing clinical evidence does not conclusively demonstrate a relationship between mechanical instability, functional deficits, and degenerative changes.

ACL Strain Milieu: Effect of Muscle Excitation

Clinically, the mechanical strain behavior of the ACL during normal knee joint motion is of interest to the orthopedic surgeon and the physical therapist. The repair, reconstruction, and rehabilitation of ACL injuries require specification of the expected strains so that extremes may be avoided. If the maximum strain of a repaired or reconstructed ACL is exceeded, then subsequent damage is inevitable. On the other hand, if the strain environment of the ligament is not maintained, then the ligament can experience changes of its mechanical properties.

Investigators of the neurosensory hypothesis must take into consideration the expected strain milieu of the ACL, including consideration of ACL strain as a function of muscle excitation patterns. Muscle excitation patterns are coupled with joint stability requirements (Andriacchi, Andersson, Ortengren, & Mikosz, 1984). Deviations from expected muscle excitation patterns may be interpreted relative to possible ligament strains and feedback from ligamentous and capsular mechanoreceptors.

Arms, M.H. Pope, Johnson, Fischer, Arvidsson, and Eriksson (1984), using in situ instrumented cadaveric ACLs, documented the relationship between the ACL strain and the knee flexion angle during passive motion with and without simulated quadriceps femoris contraction. Based on the control strain at full knee extension, passive knee flexion decreased ACL strain to a minima between 30° and 35° of knee flexion. Further knee flexion increased ACL strain to an observed maximum at 120° of knee flexion. Quadriceps femoris contraction was simulated using a load (400 N) applied through the quadriceps tendon. Isometric and eccentric contraction resulted in a statistically significant increase in ACL strain during the initial 45° of knee flexion. After 75° of knee joint flexion, simulated isometric and eccentric quadriceps femoris contraction resulted in less ACL strain than did passive flexion (see Figure 9.1).

The quadriceps contraction-related increases in ACL strain observed with these knee flexion arcs are associated with changes in the sagittal plane patellar ligament angle. Figure 9.2 illustrates how the patellar ligament angle influences the shear force applied to the tibia, causing subsequent ACL strain. The transition point at

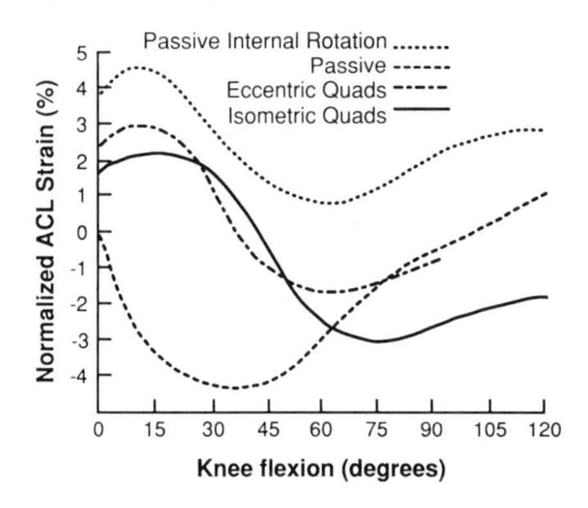

Figure 9.1 Strain measured in the anterior cruciate ligament (ACL) as a function of knee joint flexion angle and conditions of simulated eccentric and isometric quadriceps femoris contraction. Note how passive internal rotation of the tibia dramatically increases ACL strain relative to a neutral tibial position. Not shown is the effect of external tibial rotation of the ACL strain, which was less than that of internal rotation, greater than that of passive knee position at flexion angles less than approximately 60°, and greater than passive knee position at angles greater than approximately 60°. (Adapted by permission from Arms et al., 1984.)

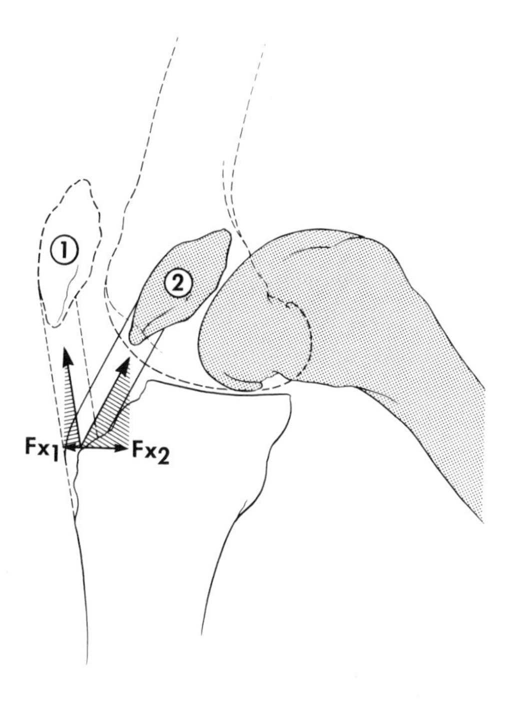

Figure 9.2 Illustration of the effect of knee flexion angle on the angle of the patellar ligament in the sagittal plane. As the knee joint flexes, the anterior shear component of quadriceps femoris contraction force is reduced to zero and subsequently directed posteriorly. Fx_1 and Fx_2 represent the horizontal components of the force vector in the patellar ligament.

which the angle changes the directed force from anterior to posterior has been reported as ranging from 65° to 85° of knee flexion (Draganich, Andriacchi, & Andersson, 1987; Yamaguchi & Zajac, 1989; Yasuda & Sasaki, 1987).

Arms et al. (1984) demonstrated that quadriceps contraction has a significant impact on ACL strain. Renstrom, Arms, Stanwyke, Johnson, and M.H. Pope (1986) subsequently investigated the effect of simulated isometric hamstrings contraction on the quadriceps-femoris-induced ACL strains. Hamstrings co-contraction decreased ACL strain below that associated with passive motion between 0° and 120° of knee flexion. Furthermore, the decrease was statistically significant between 75° and 105° of knee flexion. Simulated quadriceps femoris-hamstrings co-contraction increased ACL strain significantly between 0° and 30° of knee flexion relative to that associated with passive knee motion. However, this ACL strain was less than that of the isolated quadriceps contraction observed in this experiment (see Figure 9.3).

These results provide evidence that hamstrings contraction serves as a strain relief for the ACL. However, the decrease in strain is knee-angle-dependent. From 0° to 30° of knee flexion, the hamstrings appear unable to protect the ACL.

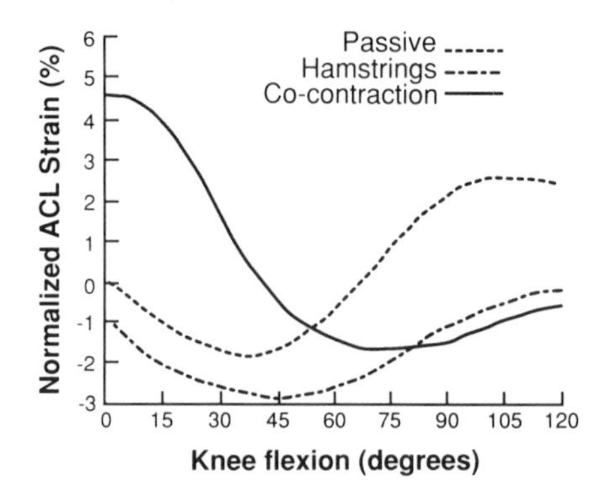

Figure 9.3 The effect of simulated hamstrings contraction and hamstrings-quadriceps femoris co-contraction on ACL strain as a function of knee flexion angle. (Adapted by permission from Renstrom, Arms, Stanwyck, Johnson, & M.H. Pope, 1988.)

Draganich and Vahey (1990) reexamined the question of mechanical hamstrings-ACL synergy using a different method of obtaining the control zero-strain condition and the muscular loading conditions. They reported findings consistent with those of Arms et al. (1984) and Renstrom et al. (1986) relative to quadriceps-induced increases in ACL strain. However, in contrast to the earlier studies, Draganich and Vahey found that the hamstrings could significantly reduce ACL strain at 10° and 20° of knee flexion. This difference was attributed to the order in which the simulated muscle forces were applied. The sequence of quadriceps-hamstrings activation has been shown to affect the final strain of the ACL (Kain, McCarthy, Arms, & Manske, 1987). When hamstrings contraction precedes quadriceps contraction, ACL strain will be larger than if quadriceps femoris activation occurs initially.

The hamstrings muscles may play a significant role relative to rotary stability of the knee. Insofar as internal rotation of the tibia can dramatically increase ACL strain, control of rotary stability may be a crucial function of the hamstrings (see Figure 9.3). Draganich (1985) reported progressive increases in the contribution of the biceps femoris (long and short head) to external tibial rotation with increases in knee flexion angle. Indeed, from approximately 45° to 90° of knee flexion, the biceps femoris reportedly functioned primarily as an external tibial rotator rather than as a knee flexor. However, the contribution of the medial hamstrings, semimembranosus and semitendinosus, to internal tibial rotation was small and constant up to 40° of knee flexion. The internal tibial rotation component of the medial hamstrings increased between 40° and 90° of knee flexion, but the maximum contribution remained smaller (approximately 34%) than the external rotation contribution of the biceps femoris. Excluding factors such as muscle cross-sectional area and neural drive, this suggests a potential for rotary imbalance caused by the hamstrings on the tibia. The difference in the magnitude of the tibial rotation functions of the medial and lateral hamstrings is probably related to the degree of excursion these muscles experience with knee joint flexion (Haynes, Pepe, Feinstein, & Hungerford, 1990).

ACL strain is significantly influenced by passive knee joint flexion angle but is also significantly affected by contraction of the quadriceps femoris and hamstrings musculature. At knee flexion angles less than 65° to 80°, contraction of the quadriceps femoris significantly increases ACL strain by imparting an anteriorly directed shear force to the tibia. In contrast, hamstrings contraction may reduce ACL strain by applying a posteriorly directed shear force to the tibia. Hamstrings strength, therefore, has been a traditional rehabilitation concern for individuals with ACL injuries, ostensibly because the hamstrings provide enhanced knee joint stability and a mechanism by which an injured or repaired ACL may be protected against strain associated with daily activities such as walking, running, and stair-climbing—not to mention athletic activities. Although rehabilitation techniques emphasizing increased knee joint stability as a function of hamstrings strength relative to quadriceps strength (hamstrings:quadriceps ratio) have been reported as clinically successful (Giove, Miller, Kent, Sanford, & Garrick, 1983; Kannus, 1988; Seto, Orofino, Morrissey, Medeiros, & Mason, 1988), these results have not been universal (Tibone et al., 1986; Walla, Albright, McAuley, Martin, Eldridge, & El-Khoury, 1985). This clinical inconsistency, similar to the differences demonstrated in long-term knee joint function subsequent to serious ACL injury, has generated interest and support for therapeutic interventions that emphasize the control of muscular coordination.

ACL Mechanoreceptors and Knee Stability in ACL Deficiency

The presence of mechanoreceptors in the ACL and evidence that mechanical stimulation of these mechanoreceptors may elicit reflexive hamstrings contraction have led to speculation regarding their neurosensory role. This speculation includes contributions to control of knee joint musculature and functional adaptation following injury. Solomonow et al. (1987) used a feline model to demonstrate an ACL-hamstrings reflex arc. External fixation of the tibia and femur and a wire looped around and secured to the ACL allowed loading and subsequent deformation of the ACL. The intramuscular EMG showed considerable hamstrings excitation during ACL loading. Solomonow et al. concluded that there is an afferent-driven synergy between the ACL and hamstrings and that the hamstrings may provide posterior knee joint stability as required and signaled by the ACL mechanoreceptors.

D.F. Pope, Cole, and Brand (1989) designed a study intended to confirm the results of Solomonow et al. (1987). However, D.F. Pope et al. elected to strain the ACL in a manner that more resembled physiological loading, by anterior displacement of the tibia below the femur. The magnitude and rate of anterior tibial displacement were controlled using a Materials Test System, and the intramuscular hamstrings EMG was recorded simultaneously. Additionally, the wire loop technique described by Solomonow et al. was replicated in four animals. Contrary to the findings of Solomonow et al., neither the anterior displacement nor the wire loop method generated a reflexive hamstrings excitation.

D.F. Pope, Cole, and Brand (1990a, 1990b) subsequently reported on the findings using a different preparation. In this experiment, loads were applied to the isolated ACL and recordings were made directly from the posterior articular nerve, which is thought to innervate the feline ACL. Contrary to their previous results (D.F. Pope et al., 1989), afferents with endings in the ACL were found to be sensitive to low-level mechanical probing (2 g), step increases in tensile loading, and vigorous twisting of the ACL. These findings suggest that simple ACL strain may not be the only mechanical stimulus to which the ligament is sensitive.

The in situ animal model has an in vivo human counterpart. Investigation has been conducted on whole body performance functions such as gait and on isolated knee joint function. Andriacchi, Kramer, and Landon (1985) compared subjects with an ACL injury, referred to as ACL-deficient, with subjects without knee injury, using a 90° side-step cutting maneuver. This particular performance test can elicit the signs of knee joint instability in a manner similar to that of clinical tests used for this purpose. The ACL-deficient patients were observed to perform the maneuver with larger knee flexion angles than the normal subjects. This posture—suggested as a compensatory, or functional, adaptation—decreases the anteriorly directed shear force associated with quadriceps femoris contraction and increases the potential hamstrings contribution to posterior stabilization of the tibia. Similar results were reported by Rezin, Birac, Bach, and Andriacchi (1990) and by Shiavi, Limbird, Frazer, Stivers, Strauss, and Abramovitz (1987).

Limbird, Shiavi, Frazer, and Borra (1988) compared the EMG patterns of ACL-deficient patients during walking gait with those of normals. Significantly higher levels of biceps femoris (but not medial hamstrings) excitation during the swing-to-stance transition were reported. This finding appears to support the suggestions of Andriacchi et al. (1985) and Rezin et al. (1990).

Other investigators have studied isolated knee joint function. Solomonow et al. (1987) reported an experiment in which subjects without knee ligament injury were compared with subjects having arthroscopically verified ACL deficiencies. The subjects performed maximum isokinetic knee extension from 90° of knee flexion to full extension at 15 deg/s. Surface EMGs from the rectus femoris and hamstrings, synchronized with the isokinetic torque signals, were used to identify excitation changes associated with knee extension torque deficits identified in most of the 12 ACL-deficient patients. The sequence of excitation and torque changes were interpreted as reflecting a response to assumed ACL strain.

The excitation pattern demonstrated by normal subjects was a typical reciprocal inhibition pattern. During knee flexion and extension, respectively, quadriceps femoris and hamstrings activities were less than 10% of their maximum observed values. However, ACL-deficient subjects demonstrated a significantly different pattern. During knee joint extension, at a knee angle of approximately 37° to 46° of flexion, approximately an 80% decrease in the magnitude of the isokinetic torque was observed. The torque deficit was associated with a simultaneous decrease in quadriceps excitation (approximately 50%) and a dramatic increase in hamstrings excitation (approximately 700%). The knee angle at which these events occurred was similar to the knee angle at which anterior subluxation has been reported to be the greatest in cadaveric specimens (Hirokawa, Solomonow, Baratta, Shoji, & D'Ambrosia, 1990), but it fell outside the range of those reported to be associated with the transition of patellar ligament angle from positive to negative (Draganich et al., 1987; Yamaguchi & Zajac, 1989; Yasuda & Sasaki, 1987), maximum in situ ACL strain (Arms et al., 1984; Renstrom et al., 1986), and maximum in vivo ACL strain (Johnson, 1989).

Draganich, Jaeger, and Kralj (1989), using normal subjects who performed unconstrained knee extension at 10 deg/s with external weight affixed to the ankle, reported somewhat different results than Solomonow et al. (1987). Knee joint angle-specific increases in hamstrings excitation were observed and attributed to an ACL-hamstrings synergy serving to reduce strain in the ligament. However, whereas Solomonow et al. reported increased hamstrings excitation between 37° and 46° of knee flexion, Draganich et al. reported that the increase in hamstrings co-contraction occurred at full knee extension.

Grabiner, Campbell, Hawthorne, and Hawkins (1989) reported on quadriceps femoris-hamstrings synergies during isometric knee extension. The knee was positioned at approximately 15° of knee flexion, and isometric knee extension contraction intensity ranged from 10% to 100%. This knee angle was selected because of the magnitude of the anterior shear applied to the tibia with quadriceps femoris contraction; this angle was also a compromise between the angles producing co-contraction for Solomonow et al. (1987) and Draganich et al. (1989). The design provided a full range of isometric contraction intensity; therefore, the shear forces on the tibia during this task may arguably be larger than those of the other experiments. It was anticipated that if an ACL-hamstrings synergy was present, it would be triggered as the contraction intensity and anterior tibial shear increased above a threshold value. The results, inconsistent with those of Solomonow et al. and Draganich et al., provided no evidence of synergy. Hamstrings excitation during isometric knee extension ranged between 9.6% and 13.0% of its maximum observed value collected during a maximum knee flexion. It was concluded that ligamentously normal knees do not require hamstrings stabilization of the tibia during this type of task.

Grabiner, Weiker, Anderson, and Bergfeld (1990) and Grabiner and Weiker (in press) expanded on the previous study by Grabiner et al. (1989) by comparing subjects with ligamentously normal knees with two groups of patients. One group of patients had documented ACL deficiency, and the second group had undergone reconstructive ACL surgery. Two knee angles, 15° and 85° of knee flexion, were used in this protocol. These knee joint angles were selected to compare hamstrings excitation between two conditions having different quadriceps-related tibial shear forces and ACL strain milieus. Quadriceps femoris contraction at the 15° knee flexion angle increases ACL strain; at 85° of knee flexion, ACL strain is reduced by quadriceps femoris contraction. It was expected that if an ACL-hamstrings synergy could be elicited it would be observed in the ACL-deficient patients at the 15° knee flexion angle and would result in a significantly different pattern of hamstrings excitation than that of the 85° flexion angle. It was also expected that patients with reconstructed ACLs, having clinically normal knee joint stability, would have hamstrings excitation patterns similar to normal subjects.

The results revealed no significant differences between the ACL-reconstructed and ACL-deficient groups relative to hamstrings excitation pattern. When the patient data from these two groups were combined, it was found that the patient group was characterized by significantly higher hamstrings excitation than the normal subjects through the entire range of contraction intensities. The hamstrings excitation pattern was not significantly different for the two knee joint angles for either the normal subjects or patients. Figure 9.4 compares the combined patient data against the normal data for the combined 15° and 85° knee flexion angles. The slopes of the regression lines were not significantly different, but a significant difference was revealed between the intercepts.

The similarity of hamstrings excitation patterns for the ACL-reconstructed and ACL-deficient groups was unexpected. The purpose of the surgical intervention is to restore normal anterior-posterior and rotary knee joint stability, and in the subjects tested this was clinically documented. The higher levels of hamstrings excitation may represent a functional compensation in response to the original instability that had not been restored to preinjury levels. This is suggested as a possible neuromotor maladaptation. With the loss of ligamentous restraint, the joint stability generated by hamstrings contraction is desirable. However, unnecessary increases in hamstrings contraction can decrease knee extension torque and

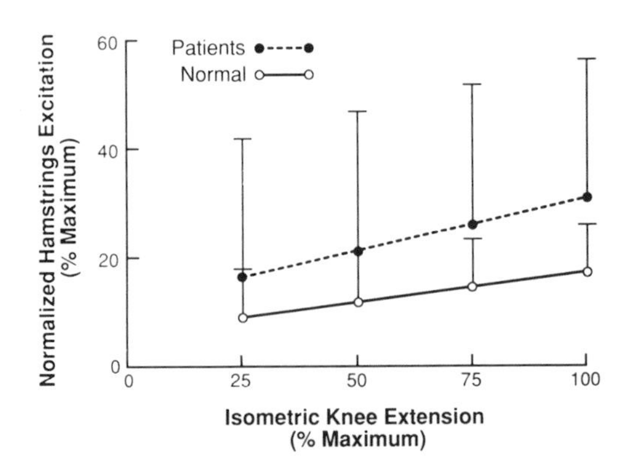

Figure 9.4 Comparison of normalized hamstrings excitation during isometric knee extension. The regression lines for each group combine two knee angles, 15° and 85° of knee flexion, which were not significantly different from one another. The regression line for the patients represents the ACL-reconstructed and ACL-deficient groups, which were not significantly different from one another. Hamstrings excitation for the patients was significantly higher than that of the normal subjects throughout the entire range of isometric contraction intensity. (Adapted by permission from Grabiner and Weiker, in press).

also increase joint reaction forces and pressures. The increase in joint reaction forces and pressures associated with unnecessary hamstrings contraction may be found to be a factor in long-term knee joint dysfunction.

A second unexpected observation was the insensitivity of the hamstrings excitation pattern to knee joint angle in the patients. Both of these findings could reflect the influence of a knee joint mechanism other than anterior tibial shear forces on hamstrings excitation. The studies by Grabiner, Weiker, Anderson, and Bergfeld (1990) and Grabiner and Weiker (in press) were unable to verify a reflexive ACL-hamstrings synergy but identified a potential chronic neuromotor adaptation affecting hamstrings behavior.

A subsequent study was conducted using normal subjects to determine if the rate at which isometric torque was developed would demonstrate an influence on reflexive hamstrings excitation (Grabiner, Miller, Campbell, & Metzger, 1989). Two knee flexion angles were used, 15° and 85°, but in this experiment, maximum knee extension torque was generated under three temporal conditions: minimum elapsed time, 1,000 ms, and 2,000 ms. Surface EMGs of the medial and lateral hamstrings were analyzed for indications of reflexive excitation in response to the imposed ACL strain rate. It was expected that if reflexive excitation was observed, it would be at the 15° knee flexion angle because of the imposed anterior tibial shear force. Analysis did not reveal reflexive hamstrings excitation. However, a significant difference was observed between the excitation of the medial and lateral hamstrings that was evident at the onset of knee extension torque, prior to imposed ACL strain. This suggests that ACL strain was not a factor in the excitation pattern.

During the maximum-torque, minimum-time condition, the lateral hamstrings excitation increased significantly (35.6%) with the knee at 85° of flexion compared with the 15° flexion angle. The medial hamstrings excitation did not change meaningfully (–5.8%) with the same knee joint angle change (see Figure 9.5). This finding supports the possibility that a knee joint mechanism other than anterior tibial shear force influences hamstrings excitation. The proposed mechanism was internal rotation of the tibia associated with quadriceps femoris contraction.

Two aspects of the knee joint support the internal rotation hypothesis. First, as the knee flexes, its torsional stiffness decreases significantly (Crowninshield, M.H. Pope, & Johnson, 1976; Markolf, Mensch, & Amstutz, 1976), thereby increasing its susceptibility to internal and external rotation moments. Second, the patellar ligament in the frontal plane is directed superiorly-medially to inferiorly-laterally, therefore, a component of quadriceps femoris force contributes to a moment that tends to internally rotate the tibia. The lateral hamstrings may functionally negate this internal rotation moment (Draganich, 1985). The observed significant increase in lateral hamstrings excitation suggests that this occurs in ligamentously normal knee joints.

The evidence to date supports the contention that knee joint instability related to ACL deficiency is associated with kinematic, kinetic, and neuromotor changes, or adaptations, in a variety of movement tasks. It is not certain if mechanoreceptors

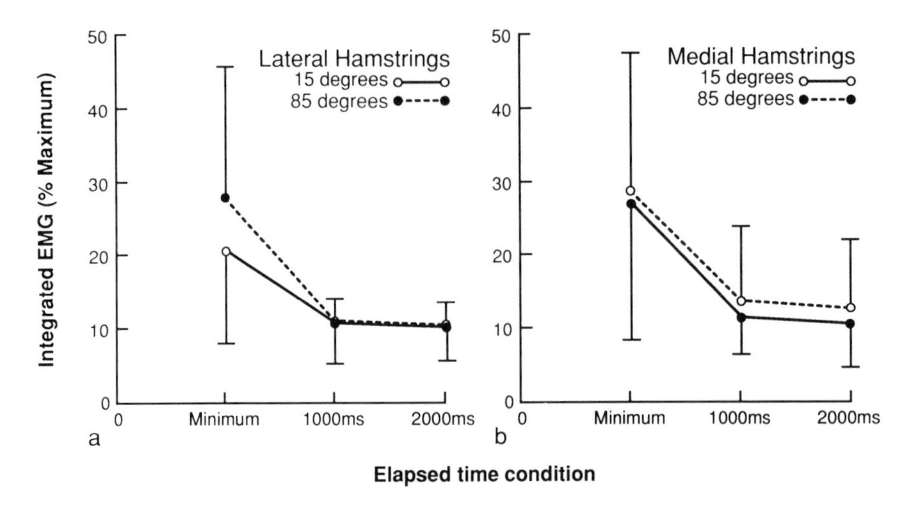

Figure 9.5 Comparison of how the rate of maximum isometric knee extension torque development and knee joint angle affected the excitation of the lateral (a) and medial (b) hamstrings. The lateral hamstrings revealed a significant increase in excitation after the knee was flexed to 85°. For each muscle, the excitation from the minimum elapsed time was greater than that from both slower conditions, which were not significantly different from one another.

are directly or indirectly involved, and therefore, an issue requiring attention is the source of the signal that triggers adaptive responses (Andriacchi, 1989). Notably, a unified consensus regarding the neuromotor patterns of the ligamentously normal knee joint also remains unspecified.

Conclusion

This chapter has presented evidence that mechanoreceptors around and within the knee joint, in general, and within the ACL, in particular, can significantly impact the neuromotor control of the knee joint musculature. Although far from conclusive, the potential neurosensory role cannot be discounted. Variables that have played a role in confounding the findings of studies to date include the variability of biological systems, structurally and functionally. Also, methodological differences in surgical preparation, afferent stimulation, surface and intramuscular EMG, and contraction modes can result in subtle or large disparities in findings. Future research in this area will probably include more complex models of the musculoskeletal system and standardized methods of data collection and analysis.

This research has both basic and applied significance, and the future of this work will necessarily incorporate both animal and human models. Basic science benefits from information about the role of ACL mechanoreceptors through an improved understanding of the structure and function of, and interaction between, various levels and elements of the nervous and musculoskeletal systems, relative to the control of human motion. Knowledge of modified sensory characteristics of injured joint systems should be applied in the development and execution of rehabilitation programs for those recovering from injury or surgery. The effect of altered sensory feedback from an injured joint on preinjury neuromotor patterns may be detrimental to healing, rehabilitation, and long-term function. The surgical technique for repairing the ruptured ACL may be improved with knowledge obtained from more in-depth research of the importance of mechanoreceptors to joint stability and function. It may be found that surgical methods that retain some of the neurosensory potential of even a completely ruptured ligament merit consideration.

References

Andriacchi, T.P. (1989). Biomechanics and orthopaedic problems. In J.S. Skinner, C.B. Corbin, D.L. Landers, P.E. Martin, & C.L. Wells (Eds.), *Future directions in exercise and sport science research* (pp. 45-56). Champaign, IL: Human Kinetics.

Andriacchi, T.P., Andersson, G.B.J., Ortengren, R., & Mikosz, R.P. (1984). A study of factors influencing muscle activity about the knee joint. *Journal of Orthopaedic Research*, **1**, 266-275.

Andriacchi, T.P., Kramer, G.M., & Landon, G.C. (1985). The biomechanics of running and knee injuries. In G. Finerman (Ed.), *American Academy of Orthopaedic Surgeons Symposium on Sports Medicine: The Knee* (pp. 23-32). St. Louis: Mosby.

Arms, S.W., Pope, M.H., Johnson, R.J., Fischer, R.A., Arvidsson, I., & Eriksson, E. (1984). The biomechanics of anterior cruciate ligament rehabilitation and reconstruction. *American Journal of Sports Medicine*, **12**, 8-18.

Baxendale, R.H., & Ferrel, W.R. (1987). Disturbance of proprioception at the human knee resulting from acute joint distension. *Journal of Physiology*, **392**, 601.

Brand, R.A. (1986). Knee ligaments: A new view. *Journal of Biomedical Engineering*, **108**, 106-109.

Butler, D.L., Noyes, F.R., & Grood, E.S. (1980). Ligamentous restraints to anterior-posterior drawer in the human knee: A biomechanical study. *Journal of Bone and Joint Surgery*, **62A**, 259-270.

Crowninshield, R., Pope, M.H., & Johnson, R.J. (1976). An analytical model of the knee. *Journal of Biomechanics*, **9**, 397-405.

deAndrade, J.R., Grant, C., & Dixon, A.St.J. (1965). Joint distension and reflex muscle inhibition in the knee. *Journal of Bone and Joint Surgery*, **47A**, 313-322.

DeHaven, K.E., & Lintner, D.M. (1986). Athletic injuries: Comparison by age, sport, and gender. *American Journal of Sports Medicine*, **14**, 218-224.

Draganich, L.F. (1985). *The influence of the cruciate ligaments, knee musculature, and anatomy on knee joint loading*. Unpublished doctoral dissertation, University of Illinois, Chicago.

Draganich, L.F., Andriacchi, T.P., & Andersson, G.B.J. (1987). Interaction between intrinsic knee mechanics and the knee extensor mechanism. *Journal of Orthopaedic Research*, **5**, 539-547.

Draganich, L.F., Jaeger, R.J., & Kralj, A.R. (1989). Coactivation of the hamstrings and quadriceps during extension of the knee. *Journal of Bone and Joint Surgery*, **71A**, 1075-1081.

Draganich, L.F., & Vahey, J.W. (1990). An in vitro study of the anterior cruciate ligament strain induced by quadriceps and hamstrings forces. *Journal of Orthopaedic Research*, **8**, 57-63.

Fahrer, H., Rentsch, H.U., Gerber, N.J., Beyeler, C., Hess, C.W., & Grunig, B. (1988). Knee effusion and reflex inhibition of the quadriceps: A bar to effective training. *Journal of Bone and Joint Surgery*, **70B**, 635-637.

Frank, C., Woo, S.L.Y., Andriacchi, T.P., Brand, R.A., Oakes, Dahners, L., DeHaven, K.E., Lewis, J., & Sabiston, P. (1988). Normal ligament: Structure, function, and composition. In S.L.Y. Woo & J.A. Buckwalter (Eds.), *Injury and repair of the musculoskeletal soft tissues* (pp. 45-101). Park Ridge, IL: American Academy of Orthopaedic Surgeons.

Geborek, P., Moritz, U., & Wollheim, F.A. (1989). Joint capsular stiffness in knee arthritis. Relationship to intraarticular volume, hydrostatic pressures, and extensor muscle function. *Journal of Rheumatology*, **16**, 1351-1358.

Giove, T.P., Miller, S.J., Kent, B.E., Sanford, T.L., & Garrick, J.G. (1983). Nonoperative treatment of the torn anterior cruciate ligament. *Journal of Bone and Joint Surgery*, **65A**, 184-192.

Grabiner, M.D., Campbell, K.R., Hawthorne, D.L., & Hawkins, D.A. (1989). Electromyographic study of the anterior cruciate ligament: Hamstrings synergy during isometric knee extension. *Journal of Orthopaedic Research*, **7**, 152-155.

Grabiner, M.D., Miller, G.F., Campbell, K.R., & Metzger, D. (1989). Effects of rate of force development on ligament-muscle synergies in normal knees. *Proceedings of the XII International Congress of Biomechanics,* 26-30 June 1989, UCLA, Los Angeles, California, 164.

Grabiner, M.D., Weiker, G.G., Anderson, T.E., & Bergfeld, J.A. (1990). Neuromotor synergies of the ligamentously injured knee. *Journal of Biomechanics*, **23**, 728.

Grabiner, M.D., & Weiker, G.G. (in press). Anterior cruciate ligament injury and hamstrings coactivation. *Clinical Biomechanics*.

Grood, E.S., Noyes, F.R., Butler, D.L., & Suntay, W.J. (1981). Ligamentous capsular constraints preventing straight and medial laxity in intact human cadaver knees. *Journal of Bone and Joint Surgery*, **63A**, 1257-1269.

Haynes, D.W., Pepe, C.L., Feinstein, J.A., & Hungerford, D.S. (1990, February). The changing excursions of the hamstring muscles during flexion, rotation, varus and valgus movements of the tibia. *Proceedings of the 36th Annual*

Meeting of the Orthopaedic Research Society, New Orleans, Louisiana, **15**, 496.

Hirokawa, S., Solmonow, M., Baratta, R., Shoji, H., & D'Ambrosia, R. (1990, February). The hamstrings do prevent anterior subluxation of the tibia. *Proceedings of the 36th Annual Meeting of the Orthopaedic Research Society, New Orleans, Louisiana*, **15**, 497.

Johnson, R.J. (1989). The in vivo biomechanics of the anterior cruciate ligament. *Proceedings of the 1st International Olympic Committee World Congress on Sport Sciences*, 162-166.

Jones, D.W., Jones, D.A., & Newham, D.J. (1987). Chronic knee effusions and aspiration: The effect on quadriceps inhibition. *British Journal of Rheumatology*, **26**, 370-374.

Kain, C.C., McCarthy, J.J., Arms, S., & Manske, P.R. (1987). An in vivo study of the effect of transcutaneous electrical muscle stimulation on ACL deformation. *Proceedings of the 33rd Annual Meeting of the Orthopaedic Research Society*, **12**, 106.

Kannus, P. (1988). Ratio of hamstring to quadriceps femoris muscles' strength in the anterior cruciate ligament insufficient knee. Relationship to long-term recovery. *Physical Therapy*, **68**, 961-965.

Kennedy, J.C., Alexander, I.J., & Hayes, K.C. (1982). Nerve supply of the human knee and its functional importance. *American Journal of Sports Medicine*, **10**, 329-335.

Limbird, T.J., Shiavi, R., Frazer, M., & Borra, H. (1988). EMG profiles of knee joint musculature during walking: Changes induced by anterior cruciate ligament deficiency. *Journal of Orthopaedic Research*, **6**, 630-638.

Markolf, K.L., Mensch, J.S., & Amstutz, H.C. (1976). Stiffness and laxity of the knee. *Journal of Bone and Joint Surgery*, **58A**, 583-594.

Noyes, F.R., Grood, E.S., Suntay, W.J., & Butler, D.L. (1983). The 3D laxity of the anterior cruciate ligament deficient knee as determined by clinical laxity tests. *Iowa Orthopaedic Journal*, **3**, 32-44.

Noyes, F.R., & McGinniss, G.H. (1985). Controversy about treatment of the knee with anterior cruciate laxity. *Clinical Orthopaedics and Related Research*, **198**, 61-76.

Noyes, F.R., Mooar, L.A., Moorman, C.T., & McGinniss, G.H. (1989). Partial tears of the anterior cruciate ligament. *Journal of Bone and Joint Surgery*, **71B**, 825-833.

Pope, D.F., Cole, K.J., & Brand, R.A. (1989). The ACL does not contribute to monosynaptic reflexes of the cat. *Proceedings of the 35th Annual Meeting of the Orthopaedic Research Society*, **14**, 22-23.

Pope, D.F., Cole, K.J., & Brand, R.A. (1990a). Properties of ACL mechanoreceptive afferents. *Journal of Biomechanics*, **23**, 717.

Pope, D.F., Cole, K.J., & Brand, R.A. (1990b). Discharge properties of mechanoreceptive afferents in the anterior cruciate ligament. *Proceedings of the 36th Annual Meeting of the Orthopaedic Research Society*, **15**, 31.

Renstrom, P., Arms, S.W., Stanwyke, T.S., Johnson, R.J., & Pope, M.H. (1986). Strain within the anterior cruciate ligament during hamstring and quadriceps activity. *American Journal of Sports Medicine*, **14**, 83-87.

Rezin, K., Birak, D., Bach, B.R., Jr., & Andriacchi, T.P. (1990). Jogging and cutting in patients following anterior cruciate ligament reconstruction. *Proceedings of the 36th Annual Meeting of the Orthopaedic Research Society*, **15**, 83.

Seto, J.L., Orofino, A.S., Morrissey, M.C., Medeiros, J.M., & Mason, W.J. (1988). Assessment of quadriceps/hamstring strength, knee ligament stability, functional and sports activity levels five years after anterior cruciate ligament reconstruction. *American Journal of Sports Medicine*, **16**, 170-180.

Shiavi, R., Limbird, T., Frazer, M., Stivers, K., Strauss, A., & Abramovitz, J. (1987). Helical motion of the knee: Kinematics of uninjured and injured knees during walking and pivoting. *Journal of Biomechanics*, **20**, 653-665.

Sjolander, P. (1989). *A sensory role for the cruciate ligaments: Regulation of joint stability via reflexes onto the gamma-muscle spindle system*. Unpublished doctoral dissertation, University of Umea, Sweden: Lennart Sjogrens Kvalitetstryckeri.

Solomonow, M., Baratta, R., Zhou, B.H., Shoji, H., Bose, W., Beck, C., & D'Ambrosia, R. (1987). The synergistic action of the anterior cruciate ligament and thigh muscles in maintaining joint stability. *American Journal of Sports Medicine*, **15**, 207-213.

Spencer, J.D., Hayes, K.C., & Alexander, I.J. (1984). Knee joint effusion and quadriceps inhibition in man. *Archives of Physical Medicine and Rehabilitation*, **65**, 171-177.

Stratford, P. (1981). Electromyography of the quadriceps femoris muscles in subjects with normal knees and acutely effused knees. *Physical Therapy*, **62**, 279-283.

Tibone, J.E., Antich, T.J., Fanton, G.S., Moynes, D.R., & Perry, J. (1986). Functional analysis of anterior cruciate ligament instability. *American Journal of Sports Medicine*, **14**, 276-284.

Tracey, D.J. (1980, November). Joint receptors and the control of movement. *Trends in Neuroscience*, pp. 253-255.

Walla, D.J., Albright, J.P., McAuley, E., Martin, R.K., Eldridge, V., & El-Khoury, G. (1985). Hamstring control and the unstable anterior cruciate ligament deficient knee. *American Journal of Sports Medicine*, **13**, 34-39.

Wood, L., Ferrell, W.R., & Baxendale, R.H. (1988). Pressures in normal and acutely distended human knee joints and effects on quadriceps maximal voluntary contractions. *Quarterly Journal of Experimental Physiology*, **73**, 305-314.

Yamaguchi, G.T., & Zajac, F.E. (1989). A planar model of the knee joint to characterize the knee extensor mechanism. *Journal of Biomechanics*, **22**, 1-10.

Yasuda, K., & Sasaki, T. (1987). Exercise after anterior cruciate ligament reconstruction. *Clinical Orthopaedics and Related Research*, **220**, 275-283.

Index

Note. Numbers in italics indicate figures.